Siegfried Schmauder and Immanuel Schäfer (Eds.)
Multiscale Materials Modeling

Also of Interest

Fracture Mechanics in Layered and Graded Solids. Analysis Using Boundary Element Methods
Tian Xiaohong, Quentin Zhong Qi Yue, 2014
ISBN 978-3-11-029787-4, e-ISBN 978-3-11-029797-3

Computer Simulation in Physics and Engineering
Martin Oliver Steinhauser, 2012
ISBN 978-3-11-025590-4, e-ISBN 978-3-11-025606-2

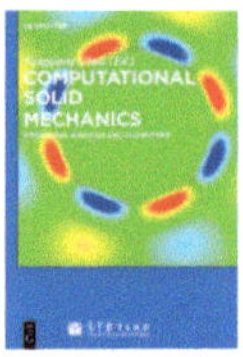

Computational Solid Mechanics. Structural Analysis and Algorithms
Yongqiang Chen (Ed.), 2017
ISBN 978-3-11-028845-2, e-ISBN 978-3-11-028871-1

Science and Engineering of Composite Materials
Suong V. Hoa (Editor-in-Chief)
ISSN 0792-1233, e-ISSN 2191-0359

Journal of the Mechanical Behavior of Materials
Elias C. Aifantis (Editor-in-Chief)
ISSN 0334-8938, e-ISSN 2191-0243

Multiscale Materials Modeling

Approaches to Full Multiscaling

Edited by
Siegfried Schmauder and Immanuel Schäfer

DE GRUYTER

Editors
Prof. Siegfried Schmauder
University of Stuttgart
IMWF
Pfaffenwaldring 32
70569 Stuttgart
Germany
siegfried.schmauder@imwf.uni-stuttgart.de

Immanuel Schäfer
Institute for Materials Testing, Materials Science
and Strength of Materials (IMWF)
University of Stuttgart
Pfaffenwaldring 32
70569 Stuttgart
Germany
Immanuel.Schaefer@imwf.uni-stuttgart.de

ISBN 978-3-11-041236-9
e-ISBN (PDF) 978-3-11-041245-1
e-ISBN (EPUB) 978-3-11-041251-2
Set-ISBN 978-3-11-041246-8

Library of Congress Cataloging-in-Publication Data
A CIP catalog record for this book has been applied for at the Library of Congress.

Bibliographic information published by the Deutsche Nationalbibliothek
The Deutsche Nationalbibliothek lists this publication in the Deutsche Nationalbibliografie; detailed bibliographic data are available on the Internet at http://dnb.dnb.de.

Typesetting: PTP-Berlin, Protago-TEX-Production GmbH, Berlin
Printing and binding: CPI books GmbH, Leck
♾ Printed on acid-free paper
Printed in Germany

www.degruyter.com

List of contributing authors

Peter Binkele
Institute for Materials Testing, Materials Science and Strength of Materials
University of Stuttgart
Pfaffenwaldring 32
70569 Stuttgart
Germany

Erik Bitzek
Department of Materials Science and Engineering
Institute I
Friedrich-Alexander-Universität
Erlangen-Nürnberg (FAU)
Martensstr. 5
91058 Erlangen
Germany

Željko Božić
University of Zagreb
Faculty of Mechanical Engineering And Naval Architecture
I. Lučića
10000 Zagreb
Croatia
zeljko.bozic@fsb.hr

Shu-Wei Chang
Department of Civil Engineering
National Taiwan University
No.188, Xinhai Rd., Da'an Dist.
Taipei City 106
Taiwan

Chuin-Shan (David) Chen
Department of Civil Engineering
National Taiwan University
No.188, Xinhai Rd., Da'an Dist.
Taipei City 106
Taiwan
dchen@ntu.edu.tw

Karthikeyan Chockalingam
Interdisciplinary Centre for Advanced Materials Simulation (ICAMS)
Ruhr-Universität Bochum
44780 Bochum
Germany
karthikeyan.chockalingam@rub.de

Chia-Ching Chou
Department of Civil Engineering
National Taiwan University
No.188, Xinhai Rd., Da'an Dist.
Taipei City 106
Taiwan

Abhik Choudhury
Institute of Applied Materials
Karlsruhe Institute of Technology (KIT)
Haid-und-Neu-Str. 7
76131 Karlsruhe
Germany

T. A. Do
Institute of Technical Biochemistry
Allmandring 31
70569 Stuttgart
Germany

Morad Etier
Institute for Materials Science and Center for Nanointegration Duisburg-Essen (CENIDE)
University of Duisburg-Essen
Universitätsstraße 15
45141 Essen
Germany
morad.etier@uni-due.de

Łukasz Figiel
Centre of Molecular and Macromolecular Studies
Polish Academy of Sciences
Sienkiewicza 112
90-363 Lodz
Poland

and
International Institute for Nanocomposites Manufacturing
WMG
University of Warwick
Coventry CV4 7AL
United Kingdom
l.w.figiel@warwick.ac.uk

Yoshiyuki Furuya
Fatigue Property Group, RCSM
National Institute for Materials Science
1-2-1 Sengen Tsukuba
Ibaraki 305-0047
Japan
FURUYA.Yoshiyuki@nims.go.jp

Alexander Hartmaier
Interdisciplinary Centre for Advanced Materials Simulation (ICAMS)
Ruhr-Universität Bochum
44780 Bochum
Germany
alexander.hartmaier@rub.de

Martin Hummel
Institute for Materials Testing, Materials Science and Strength of Materials (IMWF)
University of Stuttgart
Pfaffenwaldring 32
70569 Stuttgart
Germany

Rebecca Janisch
Interdisciplinary Centre for Advanced Materials Simulation (ICAMS)
Ruhr-Universität Bochum
44780 Bochum
Germany
rebecca.janisch@rub.de

Marc-André Keip
Institute of Applied Mechanics (CE)
Chair I
University of Stuttgart
Pfaffenwaldring 7
70569 Stuttgart
Germany
keip@mechbau.uni-stuttgart.de

Peter Kizler
Staatliche Materialprüfungsanstalt (MPA)
University of Stuttgart
Pfaffenwaldring 32
70569 Stuttgart
Germany

Aenne Koester
Interdisciplinary Centre for Advanced Materials Simulation
Ruhr-Universität Bochum
44801 Bochum
Germany
aenne.koester@rub.de

R. Kuziak
Institute for Ferrous Metallurgy
Gliwice
Poland

Matthias Labusch
Institute of Mechanics
Department of Civil Engineering
Faculty of Engineering
University of Duisburg-Essen
Universitätsstraße 15
45141 Essen
Germany
matthias.labusch@uni-due.de

G. Lasko
Institute for Materials Testing, Materials Science and Strength of Materials (IMWF)
University of Stuttgart
Pfaffenwaldring 32
70569 Stuttgart
Germany

Crystal Liou
Department of Civil Engineering
National Taiwan University
No.188, Xinhai Rd., Da'an Dist.
Taipei City 106
Taiwan

Doru C. Lupascu
Institute for Materials Science and Center for Nanointegration Duisburg-Essen (CENIDE)
University of Duisburg-Essen
Universitätsstraße 15
45141 Essen
Germany
doru.lupascu@uni-due.de

Anxin Ma
Interdisciplinary Centre for Advanced Materials Simulation
Ruhr-Universität Bochum
44801 Bochum
Germany
anxin.ma@rub.de

Łukasz Madej
AGH University of Science and Technology
al. Mickiewicza 30
30-059 Krakow
Poland
lmadej@agh.edu.pl

Ignacio Martin-Bragado
IMDEA Materials Institute
C. Eric Kandel 2
28906, Getafe
Madrid
Spain
ignacio.martin@imdea.org

Marijo Mlikota
Institute for Materials Testing, Materials Science and Strength of Materials (IMWF)
University of Stuttgart
Pfaffenwaldring 32
70569 Stuttgart
Germany

Johannes J. Möller
Department of Materials Science and Engineering
Institute I
Friedrich-Alexander-Universität Erlangen-Nürnberg (FAU)
Martensstr. 5
91058 Erlangen
Germany

David Molnar
Institute for Materials Testing, Materials Science and Strength of Materials
University of Stuttgart
Pfaffenwaldring 32
70569 Stuttgart
Germany

Alejandro Mora
Institute for Materials Testing, Materials Science and Strength of Materials
University of Stuttgart
Pfaffenwaldring 32
70569 Stuttgart
Germany

Rajdip Mukherjee
Institute of Applied Materials
Karlsruhe Institute of Technology (KIT)
Haid-und-Neu-Str. 7
76131 Karlsruhe
Germany
rajdip@gmail.com

Britta Nestler
Institute of Materials and Processes
Karlsruhe University of Applied Science
Moltkestraße 30
76133 Karlsruhe
Germany

Hiroshi Noguchi
Department of Mechanical Engineering
Kyushu University
744 Moto-oka
Nishi-ku
Fukuoka 819-0395
Japan
nogu@mech.kyushu-u.ac.jp

Konrad Perzynski
AGH University of Science and Technology
al. Mickiewicza 30
30-059 Krakow
Poland

J. Pleiss
Institute of Technical Biochemistry
Allmandring 31
70569 Stuttgart
Germany

Aruna Prakash
Department of Materials Science and Engineering
Institute I
Friedrich-Alexander-Universität Erlangen-Nürnberg (FAU)
Martensstr. 5
91058 Erlangen
Germany

Monica Prieto-Depedro
IMDEA Materials Institute
C. Eric Kandel 2
28906, Getafe
Madrid
Spain
monica.prieto@imdea.org

K. Radwanski
Institute for Ferrous Metallurgy
Gliwice
Poland

Manfred Rühle
Max Planck Institute for Metal Research
Heisenberg-Straße 3
70569 Stuttgart
Germany
ruehle@mf.mpg.de

Immanuel Schäfer
Institute for Materials Testing, Materials Science and Strength of Materials (IMWF)
University of Stuttgart
Pfaffenwaldring 32
70569 Stuttgart
Germany
Immanuel.Schaefer@imwf.uni-stuttgart.de

Siegfried Schmauder
Institute for Materials Testing, Materials Science and Strength of Materials (IMWF)
University of Stuttgart
Pfaffenwaldring 32
70569 Stuttgart
Germany
siegfried.schmauder@imwf.uni-stuttgart.de

Jörg Schröder
Institute of Mechanics
Department of Civil Engineering
Faculty of Engineering
University of Duisburg-Essen
Universitätsstraße 15
45141 Essen
Germany
j.schroeder@uni-due.de

Javier Segurado
Department of Materials Science
Polytechnic University of Madrid
E. T. S. de Ingenieros de Caminos
28040, Madrid
Spain
javier.segurado@imdea.org

Michael Selzer
Institute of Applied Materials
Karlsruhe Institute of Technology (KIT)
Haid-und-Neu-Str. 7
76131 Karlsruhe
Germany

Yu-Ching Shih
Department of Civil Engineering
National Taiwan University
No.188, Xinhai Rd., Da'an Dist.
Taipei City 106
Taiwan

Amir Siddiq
Institut für Materialprüfung, Werkstoffkunde und Festigkeitslehre (IMWF)
University of Stuttgart
Pfaffenwaldring 32
70569 Stuttgart
Germany
and
School of Engineering
University of Aberdeen
Fraser Noble, AB24 3UE
United Kingdom
amir.siddiq@abdn.ac.uk

Chandra Veer Singh
Department of Materials Science and Engineering
University of Toronto
184 College St.
Toronto, ON M5S3E4
Canada
chandraveer.singh@utoronto.ca

Mateusz Sitko
AGH University of Science and Technology
al. Mickiewicza 30
30-059 Krakow
Poland

L. Sieradzki
AGH University of Science and Technology
al. Mickiewicza 30
30-059 Krakow
Poland

D. Uhlmann
Staatliche Materialprüfungsanstalt (MPA)
University of Stuttgart
Pfaffenwaldring 32
70569 Stuttgart
Germany

Wolfgang Verestek
Institute for Materials Testing, Materials Science and Strength of Materials (IMWF)
University of Stuttgart
Pfaffenwaldring 32
70569 Stuttgart
Germany
wolfgang.verestek@imwf.uni-stuttgart.de

Derek H. Warner
School of Civil and Environmental Engineering
Cornell University
220 Hollister Hall
Ithaca, NY 14853
USA
dhw52@cornell.edu

Ulrich Weber
Institute for Materials Testing, Materials Science and Strength of Materials (IMWF)
University of Stuttgart
Pfaffenwaldring 32
70569 Stuttgart
Germany

David Weidt
Department of Mechanical, Aeronautical and Biomedical Engineering
University of Limerick
and
Materials & Surface Science Institute
University of Limerick
Ireland

Jing Wiedmaier
Institute for Materials Testing, Materials Science and Strength of Materials (IMWF)
University of Stuttgart
Pfaffenwaldring 32
70569 Stuttgart
Germany
jing.wiedmaier@imwf.uni-stuttgart.de

Preface

The aim of this book is to provide an overview about the growing scientific field of multiscale materials modelling. With increasing computer power and more and more specialized numerical methods an extensive simulation based description of the mechanics of materials can be achieved. For this purpose more than two simulation methods have to be connected for the integral description of material behavior from the nano scale to the meso scale and finally to the macro scale. This book shows the very first available examples from three different topics:

In Part I "**Multi-time-scale and multi-length-scale simulations of precipitation and strengthening effects**" the authors give examples of simulations from the nano to the macro scale. But not only the length scale is important here, also the time scale is naturally involved in this part of the book. From the materials point of view, mostly metals are of interest in this chapter.

The books starts with a paper of Kizler et al. (*"Linking nanoscale and macroscale: Calculation of the change in crack growth resistance of steels with different states of Cu precipitation", Kizler, Uhlmann, Schmauder*). Here the strengthening of steels due to Cu precipitates, is in focus. To analyze this process, a combination of dislocation theory and damage theory is used.

The second example from Molnar et al. (*"Multiscale simulations on the coarsening of Cu-rich precipitates in α-Fe using kinetic Monte Carlo, molecular dynamics and phase-field simulations", Molnar, Mukherjee, Choudhury, Mora, Binkele, Selzer, Nestler, Schmauder*) is based on simulations coupling Monte Carlo, Molecular Dynamics and Phase Field approaches by parameter transfer between the methods for solving the macroscopic precipitation hardening problem on lower length scales.

The article by Verr (*"Multiscale modeling predictions of age hardening curves in AL-Cu alloys", Veer*) is dedicated to precipitation hardening in Al-Cu alloys and investigates this effect using an atomistic-based hierarchical multiscale modeling framework.

The paper by Prieto-Depedro et al. (*"Kinetic Monte Carlo modeling of shear-coupled motion of grain boundaries", Prieto-Depedro, Martin-Bragado, Segurado*) comes next. The authors show how the shear coupled motion of grain boundaries is modelled using molecular dynamics, where relevant results like details about dislocation mechanics are transferred into a Kinetic Monte Carlo (KMC) model that considers grain boundary migration as a result of a sequence of discrete rare events. By using KMC it is possible to impose realistic deformation velocities up to 10 μm/s and overcome the limited time scale of classical molecular dynamics.

Also mangetic-electric composites where analyzed in a multiscale manner. Labusch et al. (*"Product properties of a two-phase magneto-electric composite: Synthesis and numerical modeling", Labusch, Etier, Lupascu, Schröder, Keip*) showed that different states of pre-polarizations of the ferroelectric material (which can be used in applications like in data storage and sensor technology) have a great influence on the overall obtainable magnetic-electric coefficient.

The next paper of Chockalingam et al. (*"Coupled atomistic-continuum study of the effects of C atoms at α-Fe dislocation cores", Chockalingam, Janisch Hartmaier*), covers a topic of carbon and its influences at dislocation cores in α-Fe to determine the Peierls stress, i.e. the critical stress required to move the dislocation at 0 K. For that they included different methods like coupled molecular statics, extended finite element method (XFEM) and atomistically modeled dislocation core.

In Part II **"Multiscale simulations of plastic deformation and fracture"** it is all about plasticity and fracture and the principles behind. In the different simulations and examples the challenges of multiscale simulations with an emphasis on deformation as well as crack nucleation and propagation in different materials such as alumina and iron are shown.

The first example from this middle part of the book from Siddiq et al. (*"Niobium/alumina bicrystal interface fracture: A theoretical interlink between local adhesion capacity and macroscopic fracture energies", Siddiq, Schmauder, Rühle*) establishes a theoretical interlink between local adhesion capacity and macroscopic fracture energies by a multiscale material model which bridges the nano-, meso-, and macro-scales. For this the crystal plasticity theory has been used, combined with a cohesive modelling approach.

The following work of Köster et al. (*"Atomistically informed crystal plasticity model for body-centered cubic iron", Koester, Ma, Hartmaier*) presents a constitutive plasticity model that takes the glide of screw dislocations with non-planar dislocation cores (which dominates the plastic deformation behavior in body-centered cubic iron), the strong strain rate and temperature dependence of the flow stress, the breakdown of Schmid's law and a dependence of dislocation mobility on stress components into account. Validation of the model is done by comparing numerical single crystal tensile tests for a third orientation to the equivalent experimental data from the literature.

Thereafter, the work of Möller et al. (*"FE2AT–finite element informed atomistic simulations", Möller, Prakash, Bitzek*) is about a simple but versatile approach called FE2AT based on finite element calculations to provide appropriate initial and boundary conditions for atomistic simulations. This approach enables to simulate large parts of the elastic loading process, even in the case of complex sample geometries and loading conditions.

The following work of Božić et al. (*"Multiscale fatigue crack growth modelling for welded stiffened panels", Božić, Schmauder, Mlikota, Hummel*) describes the influence of welding residual stresses in stiffened panels on effective stress intensity factor (SIF) values and fatigue crack growth rate. Relevant effects on different length scales such as dislocation appearance and microstructural crack nucleation and propagation are studied using molecular dynamics simulations as well as a Tanaka–Mura approach for the analysis of the problem.

The next contribution ("Molecular dynamics *study on low temperature brittleness in tungsten single crystals", Furuya, Noguchi, Schmauder*) from Furuya et al. is about a numerical model combining molecular dynamics and micromechanics to study the low temperature fracture behavior of tungsten: A pre-crack was introduced on (110) planes and cleavage was observed along the (121) planes which is in good agreement with experiments.

Madej et al. investigate in their work (*"Multi scale cellular automata and finite element based model for cold deformation and annealing of αferritic-pearlitic microstructure.", Madej, Sitko, Perzynski, Sieradzki, Radwanski, Kuziak*) numerical modelling of microstructure evolution during cold rolling and subsequent annealing of a two phase ferritic–pearlitic sample under α/γ phase transformation conditions. By means of the interpolation method based on Smoothed Particle Hydrodynamics (SPH) a link between the finite element and cellular automata models is performed.

The work of Wiedmaier et al. (*"Multiscale Simulation of the Mechanical Behavior of Nanoparticle-Modified Polyamide Composites", Wiedmaier, Verestek, Weber, Schmauder*) describes the multiscale materials behavior of a nano silicate modified polyamide 6 (PA 6). Micromechanical modeling is proposed to predict the mechanical behavior/response of this material combined with a representative volume element to simulate the quasi-static mechanical tensile behavior of the PA 6-composite. The model is modified with the consideration of debonding between the nano silicate platelets and the PA 6 matrix. The interaction between silicate platelets and PA 6 matrix, which is not accessible experimentally, is determined numerically via molecular dynamics (MD) simulations.

These simulation are closing the second chapter of the book and it continues with a chapter with a different focus from the material point of view.

In Part III **"Multiscale simulations of biological and bio-inspired materials, bio-sensors and composites"** the focus changes here from anorganic materials to bio-inspired or bio-connected materials. This chapter shows the importance of a rather new and growing scientific topic, Biomimetics, which also needs to develop scale passing strategies and methods to fully understand the strong or weak interactions of the (partly) biological materials but also covers additional, e.g. functional material behavior on all length scales. The examples presented demonstrate a variety of different results.

Starting with the article of Chen et al. (*"Multiscale modelling of nano-biosensors", Chen, Shih, Chou, Chang, Liou*) with an emphasis on coupling a continuum description with first principles density functional theory calculations or classical molecular dynamics/statics simulations through linking atomic contributions with kinematic constraints imposed by continuum mechanics.

It continues with nanocomposites in the work of Weidt and Figiel (*"Finite strain compressive behaviour of CNT/epoxy nanocomposites: 2D versus 3D RVE-based modelling", Weidt, Figiel*) where the macroscopic finite strain compressive behavior of CNT/

epoxy nanocomposites at quasi-static and high strain-rates was predicted and compared using 2D and 3D RVE approaches.

The book closes with the simulation of a bio-inspired material. Schäfer et al. (*"Peptide–zinc oxide interaction: Finite element simulation using cohesive zone models based on molecular dynamics simulation", Schäfer, Lasko, Do, Pleiss, Weber, Schmauder*) combined molecular dynamics and finite element method simulations to investigate the mechanical properties of a ZnO-peptide material with interface in a multiscale simulation approach.

Contributions of the working group at the Institute for Materials Testing, Materials Science and Strength of Materials (IMWF) in Stuttgart can be found in all three chapters.

With the articles in this book the importance and influence of multiscale materials modeling from atoms to components is shown and we hope that the spark of coupled multiscale materials simulation methods gets lighted in the reader.

Stuttgart, Juli 2016

Siegfried Schmauder
Immanuel Schäfer

Contents

Part III: Multiscale simulations of biological and bio-inspired materials, bio-sensors and composites

Part I: **Multi-time-scale and multi-length-scale simulations of precipitation and strengthening effects**

P. Kizler, D. Uhlmann, and S. Schmauder

1 Linking nanoscale and macroscale

Calculation of the change in crack growth resistance of steels with different states of Cu precipitation using a modification of stress–strain curves owing to dislocation theory

Abstract: In the present study, strengthening of steels due to Cu precipitates in the nanometer scale is investigated. These precipitates lead to a significant increase of the matrix yield stress due to impeded dislocation movement. This effect can be treated quantitatively on the mesomechanical level of dislocation theory. The impact on the macroscopic failure behaviour of ferritic steel is investigated in the framework of damage theory. The presented strategy can be also applied to related materials science problems.

Keywords: crack growth, Cu precipitation, strengthening, nanoscale, macroscale

1.1 Introduction

The linkage of modelling on the nanoscale (nmlengthscale) with such on the macroscale (lengthscale of real specimens) is a current challenge in materials science. Despite cheaper and faster computer resources, it seems unmanageable to inflate atomistic models of crystals to the size of specimens and components. Special problems of crack propagation have been successfully treated by dedicated coupled atomistic continuum methods (e.g. [1, 2]). Nevertheless, a general recipe to connect theoretical calculations on the nanoscale with the macroscale is actually not available. Instead, presently it seems more promising to restrict oneself to selected materials problems, and to connect different length scales by qualified physical laws describing corresponding materials behavior [3, 4]. The terminology of relevant length scales together with referring physical phenomena and methods is illustrated in Fig. 1.1. The topic of the present study is the strengthening of a steel due to copper precipitation. Finite element method (FEM) calculations can be regarded as a 'link-mesoscale-II-to-macroscale module'. On the other hand, a 'link-nanoscale to mesoscale-I module' is able to utilise data from the atomistic scale. In the present study, such 'modules' are linked by characteristic materials parameters, which stem either from direct experimental observations or from micro- and mesoscale calculations using dislocation theory and damage mechanics (hierarchical approach).

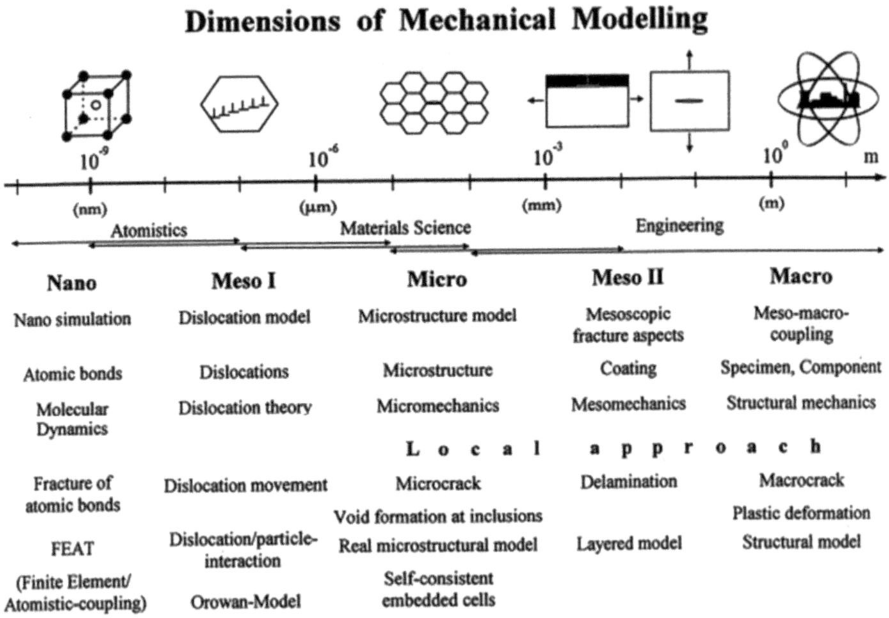

Fig. 1.1: Numerical methods referring to different length scales.

1.2 Nanoscale information from the material

In the present work, the ferritic steel 15 NiCuMoNb 5 with a Cu content of 0.66 % is examined as an example. At $T = 240$ °C and in the circumferential direction in piping, the material possesses a yield stress of about 400 MPa, see Fig. 1.2, and a notch impact energy of 100 J, see Fig. 1.3 [5]. The axial (L) direction with a notch impact energy of 200 J in the upper shelf, see Fig. 1.3, is not discussed further. A very similar melt of 15 NiCuMoNb 5 with a Cu content of 0.64 % was examined in the initial state and in an aged state after 57 000 h of operation at 350 °C, which causes an important increase in strength: At $T = 240$ °C, the yield stress and the ultimate strength of the material in the aged state are increased by 100 MPa and by 69 MPa respectively, due to the formation and growth of copper precipitates. Transmission electron microscopy (TEM) images of these materials are depicted in Figs. 1.4 and 1.5. The precipitated copper particles were evaluated by size and counted by digital image processing and the results are included in the figures. The histogram bars of the figures display the total number of detected particles versus their radius. The data and figures stem from a detailed materials characterization by Schick et al. [6]. Already in the initial state, a considerable fraction of the 0.64 % Cu has precipitated. This fraction of copper particles can be estimated from the distribution of particle sizes, weighted by their volume, and amounts to ~ 40 %.

The matrix strength due to solution hardening [7] is negligibly affected by the small content of dissolved Cu ($\Delta\sigma < 8$ Mpa). Therefore, the initial state can be treated like an aged steel containing 0.26 % Cu, which has completely precipitated.

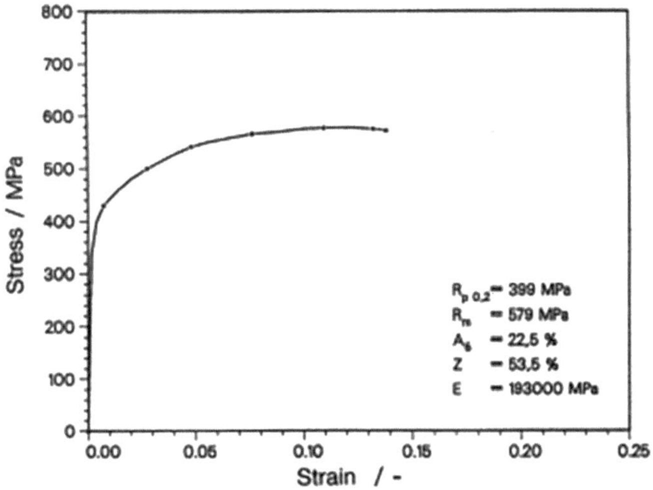

Fig. 1.2: Stress–strain curve of the material 15 NiCuMoNb 5 at $T = 240$ °C, from [5].

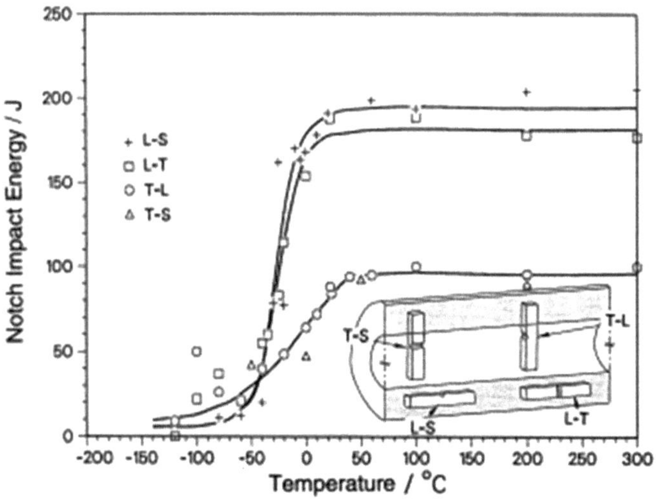

Fig. 1.3: Notch impact energy of the material 15 NiCuMoNb 5 at $T = 240$ °C, from [5].

1.3 Mesoscale theory

Materials can be strengthened by impeding the movement of dislocations with geometrical obstacles such as precipitates. Dispersion strengthening of a material leaves the elastic modulus unaffected but increases the yield stress, and all stress values, beyond the yield stress in the stress–strain relationship $\sigma(\varepsilon)$, by an amount of $\Delta\sigma$ such that $\sigma = \sigma_0 + \Delta\sigma$, where σ_0 is the yield stress of the unstrengthened state. The case of precipitation strengthening of Fe by Cu particles is a particular one. On the one hand, the plastic behaviour of a bcc crystal (e.g. Fe) differs from that of fcc (e.g. Cu) or hcp structures. On the other hand, unlike other obstacles in typical steels such as the carbide particles, the Cu precipitates are softer than the embedding Fe matrix. Thus, a dislocation is not strictly blocked like in the Orowan model but dissociates into sections inside and outside the precipitate, where different energies are encountered and the movement is impeded, but not blocked [8]. This difference causes a restriction for the cutting of precipitates by dislocations. Such resistance, together with the obstacle spacing, forms the basis for the corresponding formula based on a theory of Russell and Brown [8] to describe the increase in yield stress $\Delta\sigma = 2.5\Delta\tau$ (as assumed for Fe [8]) due to cutting of precipitates, which are softer than the embedding matrix:

$$\Delta\tau = \frac{Gb}{D}\left\{1 - \frac{p^2}{E^2}\right\}^{\frac{3}{4}} \tag{1.1}$$

with

$$\frac{P}{E} = \frac{P^{\infty}\log\frac{r_{\mathrm{ppt}}}{r_0}}{E^{\infty}\log\frac{r_{\mathrm{c}}}{r_0}} + \frac{\log\frac{r_{\mathrm{c}}}{r_{\mathrm{ppt}}}}{\log\frac{r_{\mathrm{c}}}{r_0}},$$

where G is the shear modulus of the matrix; b, the burgers vector of the dislocation; D, the distance between the precipitates; E^{∞}, P^{∞}, the energies per unit length of a dislocation in the infinite matrix and particle respectively; r_{ppt}, the precipitate radius; r_{c}, r_0 the outer and inner cut-off radii respectively, used to calculate the energy of the dislocation. The formula points out the impact of the distance D. The dislocations are poorly impeded when they can bow strongly between largely distant obstacles, and thus form sharp angles at the precipitate surfaces to promote their cutting. In the case of Cu in Fe, the numerical values amount to: $P^{\infty}/E^{\infty} \approx 0.6$, $b = 0.248$ nm, $r_0 = 2.5b$, $G = 83$ GPa, and $r_{\mathrm{c}} = 1000r_0$ [8]. In the idealized case with homogeneously distributed particles, D depends on r_{ppt} as $D = 1.77r_{\mathrm{ppt}}f^{-1/2}$ where f is the atomic percentage of Cu. Using this $D(r_{\mathrm{ppt}})$ law, $\Delta\sigma$ can be directly calculated as a function of the particle size. For the present Cu concentrations, these functions peak for r_{ppt} around 1.3 nm, see Fig. 1.7. In real melts like in the present study, neither the sizes nor the distances of precipitates are identical, see Figs. 1.4 and 1.5, and the above equation linking D with r_{ppt} and f does not hold. A detailed treatment of unequal distributions of obstacles like those discussed by Nembach [9] (see also Fig. 1.6) would exceed the scope of the present study. Precipitation means diffusion-driven growth of particles at the expense of emptying the surrounding region from Cu, (see also [10]). Therefore, nucleation of small particles next to bigger ones is improbable. This justifies the assumption that particles grow in mutually exclusive spheres and prefer to claim average distances than distribute randomly. Thus, in a simple approach, average particle sizes and corresponding distances according to the above $D(r_{ppt})$ law are assumed. From the particle size distributions in Figs. 1.4 and 1.5, the effective Cu content, average particle sizes and distances were derived and amount to 0.26 %, 2.6 nm, and 80 nm for the initial state, and 0.66 %, 2.8 nm, and 60 nm for the aged state. Inserting these values into the cited Russell–Brown formula yields $\Delta\sigma = 144$ MPa for the initial and 224 MPa for the aged state (see Fig. 1.7) leaving a total increase in yield stress of 80 MPa. This value has to be reduced by the contribution from solution hardening which vanishes during ageing when the remaining dissolved copper atoms precipitate to form particles. The solution hardening [7] for the initial state amounts to approximately 8 MPa, leaving a total increase in yield stress of 72 MPa for the aged state when all Cu atoms are precipitated.

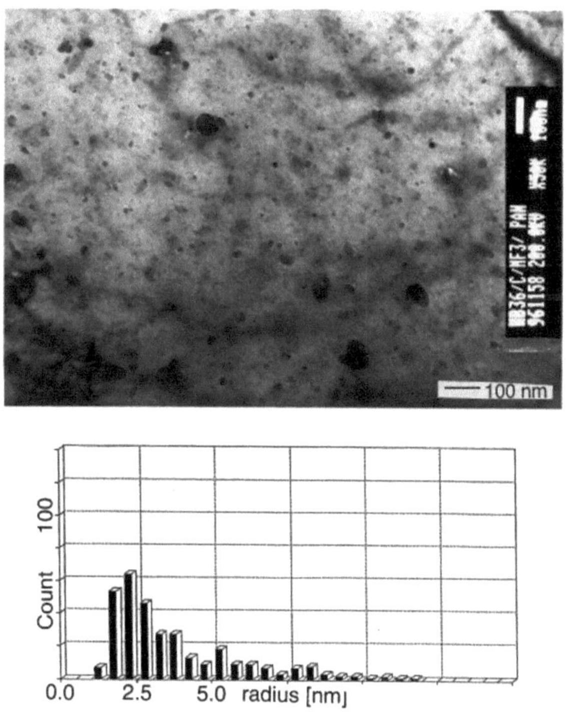

Fig. 1.4: TEM image of 15 NiCuMoNb 5, initial state, with particle size distribution, from [6].

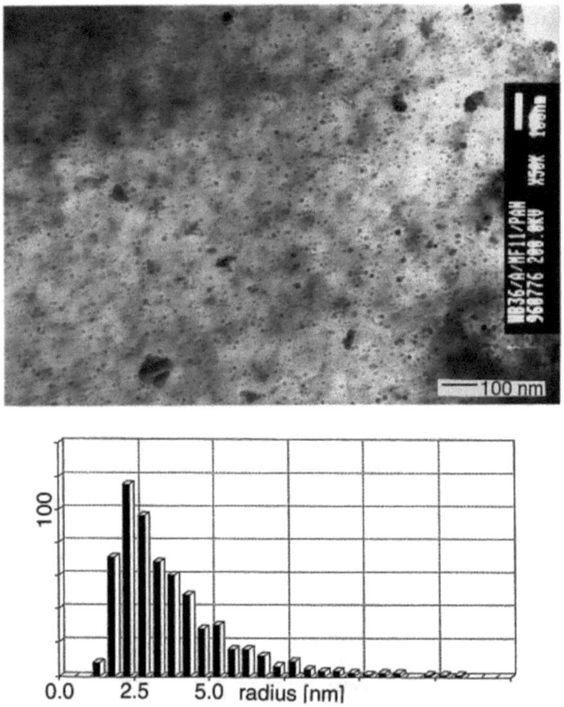

Fig. 1.5: TEM image of 15 NiCuMoNb 5, aged state, with particle size distribution, from [6].

1.4 Micro:macroscale theory

Castem 2000 [11] is a multi-purpose finite element code developed at the CEA. Domains of applications are 2D- and 3D-structural mechanics, fluid mechanics, thermal and magnetic analyses. It can be used not only for fracture mechanical calculations, but also for the simulation of crack initiation and for local approach of ductile crack growth. In the field of structural mechanics, CASTEM 2000 [11] offers special features for elastic-plastic fracture mechanics. It also includes local approach features for modelling the nucleation, growth and coalescence of voids in a metal matrix, which form the underlying mechanism for ductile fracture. Void nucleation occurs essentially at the second phase particles. These second phase particles are MnS with a size of several micrometers. If the load continues to increase, void growth will directly follow from cavity formation. Void coalescence is implicitly included in the formulation of the Rousselier model [12]: as damage increases for increasing values of load, the stresses drop with increasing strains, simulating void coalescence. The parameters f_0, f_c, I_c, D, and σ_k for the Rousselier model are chosen to fit numerically the experimentally determined J_R curve (initial state) for the 20 % side grooved CT20 compound specimen (see Fig. 1.3) – initial void volume fraction $f_0 = 0.011$, critical void volume fraction

Fig. 1.6: Cutting of Cu precipitates in Fe by dislocations, from [9].

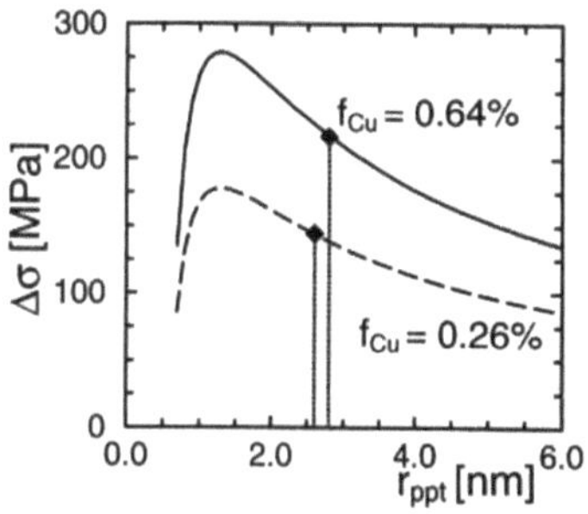

Fig. 1.7: Graph of the Russell–Brown formula: dependence of $\Delta\sigma$ from precipitate radius r_{ppt}, plotted for two different Cu concentrations. Vertical lines: average particle sizes for the two discussed material states as derived from Figs. 1.4 and 1.5.

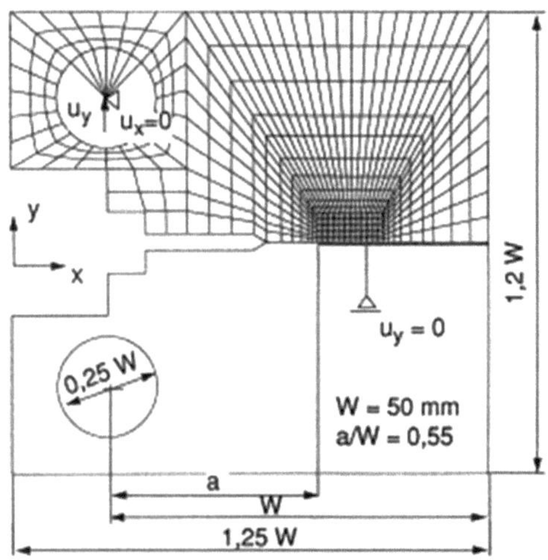

Fig. 1.8: Model of CT specimen showing geometry and finite element discretization.

$f_c = 0.05$, the constants of the Rousselier model $D = 2$, $\sigma_k = 445$ MPa, and the half of the element length near the crack tip of $l_c = 0.1$ mm. Due to symmetry conditions it was sufficient to use a FE mesh with an initial crack length of 27.5 mm for one half of the specimen only, see Fig. 1.8.

Required input parameters for CASTEM 2000 [11] are, besides the FE mesh, the boundary conditions, the load history, and the stress–strain relation of the material for the relevant temperature including the threshold to plastic deformation, characterized by the yield stress. Normally, the required stress–strain relations are derived from smooth tensile specimens and the Rousselier parameters are determined from fits to experimental values obtained from notched tensile specimens and compact tension (CT) specimens. Alternatively, f_0 and I_c can be obtained metallographically from image analyses of material cross sections [13]. The increase in yield stress and ultimate strength due to different material states are traditionally determined from experiments also. For the 2D elastic–plastic displacement controlled plane strain calculations, including geometrical nonlinearities and quadrilateral finite elements with reduced integration, were used.

1.5 Connection of length scales

Figure 1.9 shows experimentally derived true stress–strain relations for the material 15 NiCu-MoNb 5 (WB 36) in the initial and the aged state [5]. They differ by 100 MPa at the onset of the curves and by 70 MPa at higher strains. These experimental data are in good agreement with the value of 72 MPa obtained in Section 1.3 where the increase in yield stress ($\Delta\sigma$) was predicted from mesoscopic dislocation theory making use of nanoscale structural information. This enables one to accurately predict changes in the mechanical behaviour, expressed in the stress–strain curves based on nanoscale data, independent from mechanical experiments. Fig. 1.10 shows a comparison of calculated and experimentally obtained ductile crack growth curves [5]. Such curves visualize the resistance of a CT specimen against crack growth under monotonically increasing tensile load. The calculated curve up to 2 mm for the 'initial' material state agrees well with the displayed experimental behaviour and proofs the successful prediction from the FEM program. Keeping the Rousselier parameters

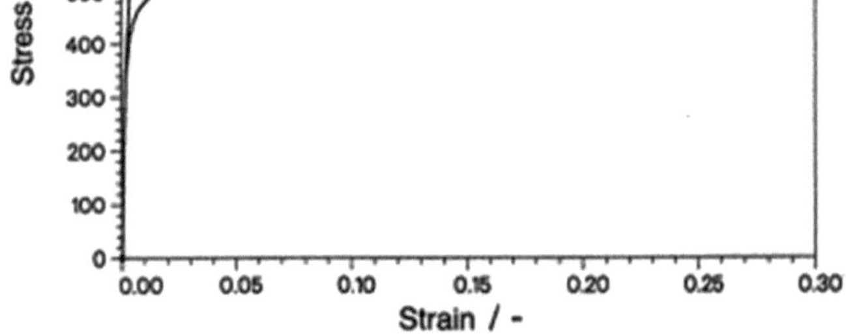

Fig. 1.9: Experimental true stress–strain relations for material 15 NiCuMoNb 5 at $T = 240\,°C$ in the material states 'initial' and 'aged', from [5].

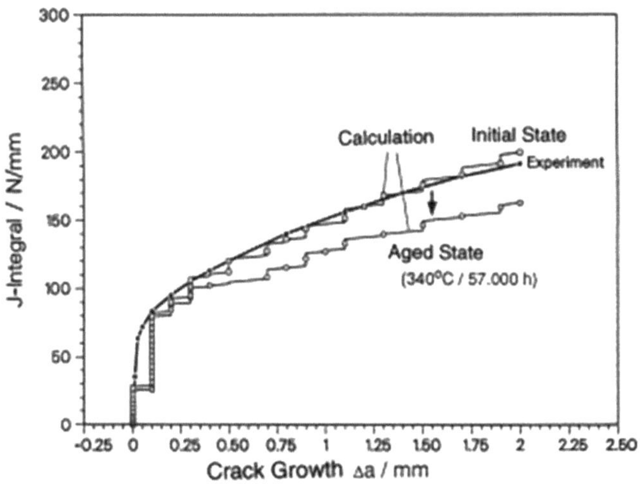

Fig. 1.10: Crack growth curves from experiment at T = 240 °C and FEM calculations, from [5].

constant but changing the stress–strain relationship of the material from 'initial' to 'aged' means to alter the FEM program input data by increasing the yield stress by $\Delta\sigma$, either from experimental results or, as in the present analysis, from theoretical results. Now the consequences for the behaviour of specimens can be studied numerically. For the described CT specimen, the increase in strength by $\Delta\sigma$ leads to a decrease in the corresponding JR-values, see lower curve in Fig. 1.10, which means that with increased yield strength and flow stress a high load induces crack growth rather than plastic deformation. This agrees with experimental results for CT specimens of an almost identical melt where a JR-integral of 150 N : mm corresponded to an increase in crack growth from 1.0 to 1.4 mm for both the 'initial' and the 'aged' states [6], similar to the calculations shown in Fig. 1.10.

1.6 Conclusions

The present study demonstrates how to successfully bridge the gap between nanoscale information and macroscale mechanical behaviour using mesoand micromechanical theoretical analyses. The connecting materials parameter between the theories is $\Delta\sigma$ (Fig. 1.11). The suggested strategy offers several further opportunities for using alternative 'modules' on the different length scales, aiming to import further knowledge from modelling into materials science. Some possibilities shall be mentioned in the following:

1. Nanoscale: Monte Carlo simulations of the formation of obstacles (e.g. [10, 14]) and molecular dynamics simulations of particle:dislocation interactions.

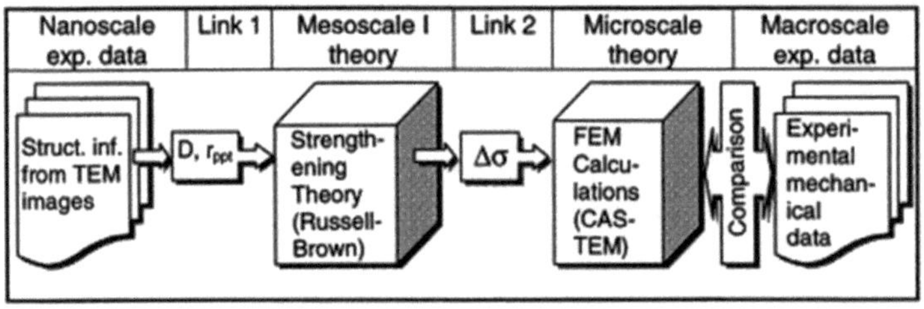

Fig. 1.11: The linkage between the nanoscale and the macroscale.

2. Link 1: Calculated particle sizes and distances, as well as the Orowan angle and in-situ shear modulus from simulations of dislocation pinning as input for the Orowan strengthening formula, as far as the Orowan-model [15, 16] is applicable.
3. Mesoscale: For systems other than Fe(Cu), suitable strengthening models can be applied, for instance those of Orowan (for non-shearable particles) [15], Friedel (for shearable particles) [17] or Nan and Clarke (for large particles, even when fracturing) [18].
4. Link 2a: Two of the Rousselier-parameters (f_0 and l_c) can also be derived from metallographical, instead of purely numerical, investigations [13].
5. Link 2b: A current challenge is to derive absolute yield stress values from modelling on the nano- or mesoscale. Such values might serve as FEM input at this point.
6. Microscale: To simulate ductile crack resistance is an outstanding task for a FEM program, but more common structural mechanical effects such as crack opening displacement (COD) values, or collapse loads can be derived as well. Such results finally correspond to the relevant materials experiments on the macroscale.

Acknowledgment: The support by the German Bundesministerium für Bildung, Wissenschaft, Forschung und Technologie (BMBF) under grant No. 1501029 is gratefully acknowledged.

References

[1] Kohlhoff S, Schmauder S. Atomistic Simulation of Materials: Beyond Pair Potentials. New York: Plenum; 1989. p. 411.
[2] Kohlhoff S, Gumbsch P, Fischmeister H. Crack propagation in b.c.c. crystals studied with a combined finite-element and atomistic model. Phil Mag A. 1991;64:851.
[3] Zohdi T, Oden J, Rodin G. Hierarchical modelling of heterogeneous bodies. Comput Methods Appl Mech Eng. 1996;138:273.
[4] Thirteenth US National Congress of Applied Mechanics (USNCAM). University of Florida; 1998. Abstract Book, ISBN 0-9652609.

[5] Beyer H, Diem H, Iskluth B. Werkstoffuntersuchungen zum Rohrversagensversuch RoRV(B)(T20.2). Prüfungsbericht Nr. 815 508:2 vom 21. Dez. 1988, Staatliche Materialprüfungsanstalt (MPA), Universität Stuttgart; 1988.
[6] Schick M, Wiedemann J, Willer D. Untersuchungen zur sicherheitstechnischen Bewertung von geschweißten Komponenten aus Werkstoff 15 NiCuMoNb 5 im Hinblick auf die Zähigkeitsabnahme unter Betriebsbeanspruchung. BMU-Vorhaben SR2239, Staatliche Materialprüfungsanstalt (MPA), Universität Stuttgart; 1997.
[7] Friedel J. On the entropy of vibration of dislocations. Phil Mag A. 1982;45:271.
[8] Russell KC, Brown LM. A dispersion strengthening model based on different elastic moduli applied to the Fe-Cu system. Acta Met. 1972;20:969–974.
[9] Nembach E. Particle Strengthening of Metals and Alloys. New York: John Wiley & Sons Inc.; 1997.
[10] Soisson F, Barbu A, Martin G. Monte carlo simulations of copper precipitation in dilute iron-copper alloys during thermal ageing and under electron irradiation. Acta Mat. 1996;44(9): 3789–3800.
[11] CASTEM 2000. Code de calcul pour l'analyse de structures par la méthode des éléments finis. Guide d'utilisation. Commissariat à l'Energie Atomique, DRN/DMT/SEMT/ LAMS, F-91191 Gif-sur-Yvette, France; 1996.
[12] Rousselier G. Ductile fracture models and their potential in local approach of fracture. Nucl Eng Design. 1987;105:97–111.
[13] Seidenfuß M. Untersuchungen zur Beschreibung des Versagensverhaltens mit Hilfe von Schädigungsmodellen am Beispiel des Werkstoffs 20 MnMoNi 5 5. Thesis, Staatliche Materialprüfungsanstalt (MPA), Universität Stuttgart; 1992.
[14] BMBF-Vorhaben. Monte Carlo simulation of nucleation and growth of Cu precipitates in steel. BMBF research project, Staatliche Materialprüfungsanstalt (MPA), Universität Stuttgart; 1998. In progress.
[15] Symposium on Internal Stresses in Metals and Alloys. London: The Instititute of Metals; 1948.
[16] Bacon DJ, Kocks U, Scattergood R. Phil Mag. 1973;28:1241.
[17] Friedel J. Dislocations. New York: Pergamon; 1964.
[18] Nan CW, Clarke DR. Modelling the elastic–plastic deformation of Al:Al_2O_3 particulate composites. J Am Ceram Soc. 1997;80:237.

D. Molnar, R. Mukherjee, A. Choudhury, A. Mora, P. Binkele, M. Selzer, B. Nestler, and S. Schmauder

2 Multiscale simulations on the coarsening of Cu-rich precipitates in α-Fe using kinetic Monte Carlo, Molecular Dynamics, and Phase-Field simulations

Abstract: The coarsening kinetics of Cu-rich precipitates in an α-Fe matrix for thermally aged Fe-Cu alloys at temperatures above 700 °C is studied using a kinetic Monte Carlo (KMC) and a Phase-field method (PFM). In this work, KMC adequately captures the early stage of the system evolution which involves nucleation, growth and coarsening, while PFM provides the proper framework to study late stage coarsening at large precipitate volume fraction regimes. Thereby both models complement each other by transferring the results of KMC along with precipitate-matrix interface energies from a broken-bond model to a quantitative PFM based on a grand chemical potential formulation and the CALPHAD database. Furthermore, Molecular Dynamics simulations provide information on the structural coherency of the precipitates and hence justify the sequential parameter transfer. We show that our Phase-field model can be validated quantitatively for the Gibbs–Thomson effect and that it also predicts the coarsening kinetics correctly. It is found that the kinetics closely follow the LSW (Lifshitz, Slyozov, Wagner) law whereas the coarsening rate constant increases with an increase in volume fraction of precipitates.

Keywords: multiscale, precipitation, kinetic Monte Carlo, Molecular Dynamics, Phase-Field methods, particle coarsening, sequential coupling

2.1 Introduction

α-Fe alloyed with Cu among other elements finds application in many areas, e.g. as pipe material in power plants. The alloying with Cu yields an increased flow stress which can be attributed to solid solution strengthening and particle strengthening due to the interaction of dislocations with Cu atoms and Cu precipitates within the material, respectively [1, 2]. The latter strengthening effect depends on the thermal treatment of the material as well as on the service conditions. At elevated temperatures (above 300 °C), Cu-precipitates form within the Fe matrix on a relatively short time scale, yielding first a strengthening of the material. However, as the particles undergo coarsening with time, the failure mechanism due to tensile loading may change which is undesirable for safety reasons. The computational modelling of the precipitate coarsening behaviour requires the understanding of the physical processes on the atomistic scale as well as on intermediate length scales in order to predict material properties on the macroscopic scale.

The classical theory of coarsening of precipitates due to Ostwald ripening was proposed by Lifshitz and Slyozov [3] and Wagner [4] (LSW). Although the theory is valid for an infinitesimally dilute second phase particle, both theoretically [5–9] and experimentally [10–14] it is found that for higher volume fractions of precipitates, the temporal power law is followed with a rate constant larger than predicted by the theory. Furthermore, the size distribution is broader and having a smaller amplitude compared to the distribution predicted by the theory. In this context, our multiscale approach is justified to predict quantitatively the coarsening kinetics for high volume fractions of precipitates along with coalescence events in Fe-Cu systems.

The precipitation of Cu in α-Fe has been observed experimentally [15–17] by means of small angle neutron scattering (SANS) as well as by high resolution tomography and has also been modelled applying a kinetic Monte-Carlo (KMC) approach [18]. On the other hand, Molecular Dynamics (MD) simulations have been performed in order to investigate the interactions of edge dislocations with Cu atoms in a solid solution [19] as well as with Cu precipitates [20, 21] confirming the experimentally observed strengthening in a Cu containing alloy [22].

Both KMC and MD simulations are limited to small sample sizes being in the order of tens of nanometers. In order to reach higher length scales and to simulate late stage coarsening, PFM becomes a necessary tool. A wide range of phenomena described by Phase-field methods can be found in [23–28]. There were several attempts to simulate microstructure evolution using PFM along with atomistic simulations [29, 30], but not for the Fe-Cu precipitation system. Recently, the CALPHAD (*CALculation of PHAse Diagrams*) database has been used for the thermodynamic description for PFM in order to quantitatively predict the microstructure evolution in precipitation systems. Most of these studies for coarsening kinetics are performed in 2D [31–33] and only few in 3D [34–36]. For the Fe-Cu system, limited work has been done using quantitative PFM and these studies mostly involve the spinodal regime [37–39]. In our study, the integration of KMC with PFM makes it possible to study precipitate nucleation as well. In the following section, the sequential multiscale approach is described in detail, followed by the modelling schemes of KMC (Sec. 2.3.1), MD (Section 2.3.2) and PFM (Sec. 2.3.3). In Sec. 2.4, the results of the simulations are discussed. Sec. 2.5 closes the paper with conclusions derived from the simulation results.

2.2 Multiscale Approach

The multiscale approach applied within this study is of a sequential type, i.e., simulation methods are connected via appropriate parameter transfers. In this study, it is chosen to transfer the particle arrangement at late KMC precipitation stages to PFM. By coupling the two methods sequentially, their advantages can be exploited circumventing simultaneously their particular disadvantages. Interface energies derived from a broken-bond model are further input data for PFM. In order to provide structural information of the precipitates, MD relaxation simulations are performed since KMC cannot account for this due to the rigid lattice (see Sec. 2.3.2).

2.3 Simulation Methods and Applied Models

2.3.1 Cu-precipitation – Kinetic Monte-Carlo Simulations

The process of Cu-precipitation in α-Fe is simulated by a KMC method which is based on a thermally activated vacancy diffusion on a rigid bcc crystal lattice Model (RLM) [18]. Although in nature Cu has fcc structure, Cu clusters with sizes smaller than 2 nm are coherently embedded on α-Fe lattice sites [40, 41], justifying the RLM. The KMC simulation used in this study was first proposed by Soisson et al. [18]. A detailed description can be found in [18, 42]. A size of $L = 128$ lattice constants as starting configuration yields $N = 2L^3 = 4194304$ lattice sites and a cubic box with an edge length of 36.7 nm. The box surfaces have normals in {100} directions and periodic boundary conditions are set in all directions. Fe atoms are replaced randomly by Cu atoms to obtain Fe-Cu solid solutions with 1, 2, 5, and 10 at.% Cu, respectively. An empty site represents a single vacancy and the annealing temperature is set to 700 °C.

The chemical binding between atoms is described by first- and second-nearest neighbour pair interactions $\varepsilon^{(i)}_{\text{Fe-Fe}}$, $\varepsilon^{(i)}_{\text{Cu-Cu}}$, and $\varepsilon^{(i)}_{\text{Fe-Cu}}$ with $i \in \{1, 2\}$, where i denotes the i-th nearest neighbour (see Fig. 2.1). The energies $\varepsilon^{(i)}_{\text{Fe-Fe}}$ and $\varepsilon^{(i)}_{\text{Cu-Cu}}$, $i \in \{1, 2\}$ were estimated from the cohesive energies of the pure metals assuming $\varepsilon^{(2)}_{\text{Fe-Fe}} = \varepsilon^{(1)}_{\text{Fe-Fe}}/2$ and

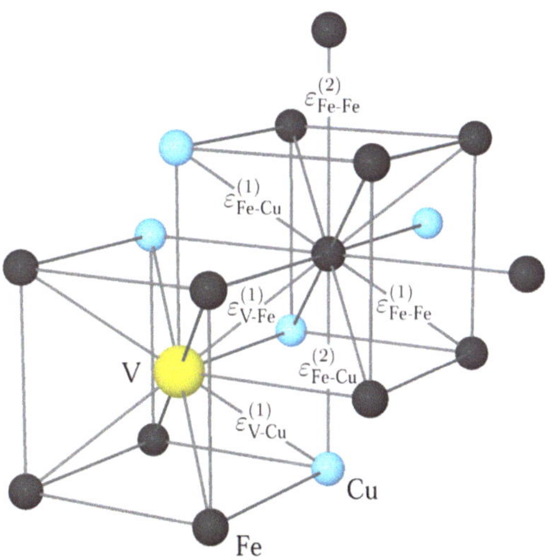

Fig. 2.1: Schematic representation of the bcc lattice and the interaction energies used in the model (Fe = black, Cu = blue, vacancy = yellow).

$\varepsilon^{(2)}_{\text{Cu-Cu}} = \varepsilon^{(1)}_{\text{Cu-Cu}}/2$ (see also [43]). The thermally activated position exchange between the vacancy V and a neighbouring atom A (with A = Fe or Cu) is given by the jump frequencies

$$\Gamma_{A,V} = \nu_A \exp\left(-\frac{\Delta E_{A,V}}{k_B T}\right), \tag{2.1}$$

with T and k_B being temperature and the Boltzmann constant, respectively. ν_A denotes the attempt frequency which is estimated using the diffusion constants of the pure metals. The activation energies for migration which depend on the local configuration are given by

$$\Delta E_{A,V} = E_{\text{SP},A} - \sum_X \varepsilon_{A\text{-}X} - \sum_Y \varepsilon_{V\text{-}Y}, \tag{2.2}$$

where $E_{\text{SP},A}$ is the energy at the saddle point between A and V, $\varepsilon_{A\text{-}X}$ are the interaction energies of the first and second-nearest neighbours of A (X atoms), and $\varepsilon_{V\text{-}Y}$ are the binding energies between the vacancy and its first nearest neighbours (Y atoms) (see Fig. 2.1). For each first nearest neighbour of the vacancy V, the jump frequencies $\Gamma_1, \ldots, \Gamma_8$ are calculated. Applying a rejection-free residence time algorithm [1, 18], one of the eight weighted jump possibilities is selected. This procedure is repeated over 10^{11} times during the simulation of precipitation. The real temporal scale is obtained from

$$t_{\text{real}} = \left(\frac{c_{V,\text{sim}}}{c_{V,\text{real}}}\right) t_{\text{MC}}, \quad \text{with } t_{\text{MC}} = \left(\sum_{j=1}^{8} \Gamma_j\right)^{-1}, \tag{2.3}$$

where $c_{V,\text{sim}}$ and $c_{V,\text{real}}$ denote the vacancy concentrations in the simulation and in the real material, respectively. t_{MC} is the average residence time. The energetic parameters of the KMC simulations are listed in [44] which, in turn, are based on [43, 45–47].

2.3.2 Structural Coherency – Molecular Dynamics Simulations

In order to provide information on the coherency of the embedded Cu precipitates, MD relaxation simulations are carried out using the IMD (ITAP Molecular Dynamics) code [48] which allows for massively parallel computations yielding elastic constants, stress and pressure tensors for relaxed structures. For metals, EAM (Embedded Atom Method) potentials describe the atomic interactions as they include an additional embedding term, besides pair interactions ϕ_{ij}, accounting for the local electron charge density in the lattice, i.e.,

$$V = \frac{1}{2}\sum_{i \neq j} \phi_{ij}(r_{ij}) + \sum_i U_i(n_i) \quad \text{with } n_i = \sum_{i \neq j} \rho_{ij}(r_{ij}), \tag{2.4}$$

where U_i describes the energy of embedding atom i in a density n_i, which is the sum of contributions ρ_j from neighbours j at distances r_{ij}. To describe the Fe-Cu system, the recently published EAM potential by Bonny et al. [49] is applied. Starting from a lattice where all Fe and Cu atoms have the same lattice constant, i.e., the lattice constant of Fe, the structure is relaxed to $T \approx 0\,\mathrm{K}$ (15 000 MD steps with 2 fs per step). During relaxation, the Cu atoms tend to move into an energetically preferred configuration. In order to give the atoms more time to do so, the structure can be heated up to 300 K (50 000 MD steps) and be kept at this temperature for another 50 000 MD steps before relaxation by applying the NpT ensemble (constant number of particles, constant pressure and temperature). In any case, the surrounding α-Fe matrix will affect their relaxation depending on the size of the Cu precipitate. The results will be discussed in Sec. 2.4.2.

2.3.3 Particle Coarsening – Phase-Field Method

2.3.3.1 Model Description

The Phase-field model applied for the investigation of precipitation in the Fe-Cu system is based on the grand-potential functional [50]

$$\Omega(T, \boldsymbol{\mu}, \boldsymbol{\phi}) = \int_{\Omega} \left(\Psi(T, \boldsymbol{\mu}, \boldsymbol{\phi}) + \left(\epsilon \tilde{a}(\boldsymbol{\phi}, \nabla\boldsymbol{\phi}) + \frac{1}{\epsilon} \tilde{w}(\boldsymbol{\phi}) \right) \right) \mathrm{d}\Omega,$$

where Ω is the total grand-potential, Ψ is the grand chemical potential density, T is the temperature, $\boldsymbol{\mu}$ is the chemical potential, $\boldsymbol{\phi} = \phi_\alpha, \phi_\beta, \ldots, N$ is the Phase-field vector consisting of the N Phase-field variables and ϵ is the interface width. The energy densities $\tilde{a}$ and $\tilde{w}$ together represent the interface energy of the system, where the former is the gradient energy and the latter the interface potential. The phase evolution is determined by the phenomenological minimisation of the *grand potential functional.*

The concentration fields are obtained by a mass conservation equation for each concentration field c_i, from the set of $K - 1$ independent concentration variables, K being the number of components in the system. The evolution equation for the N Phase-field variables (ϕ_α, $\alpha = 1, \ldots, N$) can be written as,

$$\begin{aligned} \tau\epsilon \frac{\partial \phi_\alpha}{\partial t} = \epsilon \left(\nabla \cdot \frac{\partial \tilde{a}(\boldsymbol{\phi}, \nabla\boldsymbol{\phi})}{\partial \nabla \phi_\alpha} - \frac{\partial \tilde{a}(\boldsymbol{\phi}, \nabla\boldsymbol{\phi})}{\partial \phi_\alpha} \right) - \frac{1}{\epsilon} \frac{\partial \tilde{w}(\boldsymbol{\phi})}{\partial \phi_\alpha} \\ - \frac{\partial \Psi(T, \boldsymbol{\mu}, \boldsymbol{\phi})}{\partial \phi_\alpha} - \Lambda\,, \end{aligned} \tag{2.5}$$

where Λ is the Lagrange parameter to maintain the constraint $\sum_{\alpha=1}^{N} \phi_\alpha = 1$. The gradient energy density $\tilde{a}(\boldsymbol{\phi}, \nabla\boldsymbol{\phi})$ has the form

$$\tilde{a}(\boldsymbol{\phi}, \nabla\boldsymbol{\phi}) = \sum_{\substack{\alpha,\beta=1 \\ (\alpha<\beta)}}^{N,N} \gamma [a_c(q_{\alpha\beta})]^2 |q_{\alpha\beta}|^2,$$

where $q_{\alpha\beta} = (\phi_\alpha \nabla \phi_\beta - \phi_\beta \nabla \phi_\alpha)$ is a normal vector to the $\alpha\beta$ interface. $a_c\,(q_{\alpha\beta})$ describes the anisotropy of the evolving phase boundary, which for the studies in the present investigation have been assumed to be $a_c(q_{\alpha\beta}) = 1$ for modelling isotropic systems. The obstacle type potential $\tilde{w}(\boldsymbol{\phi})$ which was also previously described in [51, 52] can be written as

$$\tilde{w}(\boldsymbol{\phi}) = \frac{16}{\pi^2} \sum_{\substack{\alpha,\beta=1 \\ (\alpha<\beta)}}^{N,N} \gamma \phi_\alpha \phi_\beta \,,$$

where γ is the interface energy. The parameter τ in Eq. (2.5) is written as

$$\frac{\sum_{\alpha<\beta}^{N,N} \tau_{\alpha\beta} \phi_\alpha \phi_\beta}{\sum_{\alpha<\beta}^{N,N} \phi_\alpha \phi_\beta} \,,$$

where $\tau_{\alpha\beta}$ is the relaxation constant of the $\alpha\beta$ interface. It is chosen in a manner such that interface kinetics vanishes [50]. The evolution equation for the concentration fields can be derived as

$$\frac{\partial c_i}{\partial t} = \nabla \cdot \left(\sum_{j=1}^{K-1} M_{ij}(\boldsymbol{\phi}) \nabla \mu_j \right) . \tag{2.6}$$

Here, $M_{ij}(\boldsymbol{\phi})$ is the mobility of the interface formulated by an interpolation of the individual phase mobilities as

$$M_{ij}(\boldsymbol{\phi}) = \sum_{\alpha=1}^{N-1} M_{ij}^\alpha g_\alpha(\boldsymbol{\phi}) \,.$$

Each of the M_{ij}^α is defined using the expression

$$M_{ij}^\alpha = D_{ij}^\alpha \frac{\partial c_i^\alpha(\boldsymbol{\mu}, T)}{\partial \mu_j} \,.$$

The function $g_\alpha(\boldsymbol{\phi})$ interpolates the mobilities. D_{ij}^α are the interdiffusivities in each phase α. Both the evolution equations (Eq. (2.5) and (2.6)) require the information about the chemical potential $\boldsymbol{\mu}$.

We write the grand potential density Ψ, as an interpolation of the individual grand potential densities Ψ_α, where Ψ_α are functions of the chemical potential $\boldsymbol{\mu}$ and of the temperature T in the system

$$\Psi(T, \boldsymbol{\mu}, \boldsymbol{\phi}) = \sum_{\alpha=1}^{N} \Psi_\alpha(T, \boldsymbol{\mu}) h_\alpha(\boldsymbol{\phi}) \tag{2.7}$$

$$\text{with} \quad \Psi_\alpha(T, \boldsymbol{\mu}) = f_\alpha(\mathbf{c}^\alpha(\boldsymbol{\mu}), T) - \sum_{i=1}^{K-1} \mu_i c_i^\alpha(\boldsymbol{\mu}, T) \,,$$

where $f_\alpha(\mathbf{c}^\alpha(\boldsymbol{\mu}), T)$ is the free energy density of the phase α. The concentration $c_i^\alpha(\boldsymbol{\mu}, T)$ is an inverse of the function $\mu_i^\alpha(\mathbf{c}, T)$ for every phase α and component i. From Eq. (2.7) the following relation can be derived

$$c_i = \sum_{\alpha=1}^{N} c_i^{\alpha}(\boldsymbol{\mu}, T) h_{\alpha}(\boldsymbol{\phi}), \tag{2.8}$$

using $\dfrac{\partial \Psi(T, \boldsymbol{\mu}, \boldsymbol{\phi})}{\partial \mu_i} = -c_i$.

2.3.3.2 Calibration to CALPHAD data

For the Fe-Cu system, we reduce the system to a two phase binary system where the independent concentration is identified by $c = c_{\text{Fe}}$. Similarly, we reduce the chemical potential vector $\boldsymbol{\mu}$ to μ which defines an independent chemical potential with respect to Fe. The Phase-field vector contains only two components $\boldsymbol{\phi} = (\phi_1, \phi_2)$. We start the construction of free energies of the respective phases with the following type of expressions for the free energies

$$f^{\alpha}(T, c) = A^{\alpha}(T)c^2 + B^{\alpha}(T)c + E^{\alpha}(T),$$

where the coefficients $A^{\alpha}(T)$, $B^{\alpha}(T)$, and $E^{\alpha}(T)$ are functions of temperature T. By using a polynomial formulation, we aim to fit a simplified form for the free energies utilising the data obtained from the CALPHAD database. We can determine the terms $A^{\alpha}(T)$ as

$$\left.\frac{\partial^2 f^{\alpha}}{\partial c^2}\right|_{c_{\text{eq}}} \equiv \frac{1}{V_{\text{m}}} \left.\frac{\partial^2 G^{\alpha}}{\partial c^2}\right|_{c_{\text{eq}}},$$

computed at the equilibrium concentration of the phase at the temperature T, where $G^{\alpha}(T, c)$ is the free energy function obtained from CALPHAD. Next we derive the chemical potential

$$\mu_{\text{eq}} = \frac{1}{V_{\text{m}}} \frac{\partial G^{\alpha}}{\partial c}|_{c_{\text{eq}}}$$

from the database and compute $B^{\alpha}(T)$ by equating the first derivative of the constructed free energies to the chemical potential giving

$$B^{\alpha}(T) = \mu_{\text{eq}} - 2A^{\alpha}(T)c_{\text{eq}}.$$

The only term left out is $E^{\alpha}(T)$, which is fitted by equating it to the grand potential which yields

$$E^{\alpha}(T) = \Psi_{\text{eq}} - A^{\alpha}(T)c_{\text{eq}}^2,$$

with $\Psi_{\text{eq}} = \frac{1}{V_{\text{m}}}(G^{\alpha}(T, c_{\text{eq}}) - \mu_{\text{eq}} c)$. With these equations we can adequately fit all the terms in the free energies at the given temperature T. For a non-isothermal description, it is essential to derive the equations in the neighbourhood of the temperature one is simulating and perform a fitting in the temperature space. In most cases, a linear temperature fit suffices. The chemical potential can be derived in terms of the concentration c and volume fractions $\boldsymbol{\phi}_1, \boldsymbol{\phi}_2$ of the phases by using the relation (2.8). The phase-concentrations can be written as functions of the chemical potential as

$$c^\alpha(\mu, T) = \left(\frac{\mu - B^\alpha(T)}{2A^\alpha(T)} \right) .$$

Using the constraint in (2.8), the following relation for μ derives

$$\mu = \frac{c + \sum\limits_{\alpha=1}^{N} \frac{B^\alpha(T) h_\alpha(\boldsymbol{\phi})}{2A^\alpha(T)}}{\sum\limits_{\alpha=1}^{N} \frac{h_\alpha(\boldsymbol{\phi})}{2A^\alpha(T)}} . \tag{2.9}$$

This value of μ is used both in the evolution equations for the Phase-field and in the concentration equations.

2.3.3.3 Mobilities for diffusion

To derive the mobilities, we require the second derivatives of the free energies with respect to the concentrations. This can be realised through the diffusion equation for a binary alloy written as,

$$\frac{\partial c^\alpha}{\partial t} = \nabla \cdot (M^\alpha \nabla \mu) ,$$

where M^α is defined as $D \frac{\partial c^\alpha}{\partial \mu}$, in order to derive Fick's law in the bulk. Taking the partial derivative of $c^\alpha(\mu, T)$ yields

$$\frac{\partial c^\alpha}{\partial \mu} = \frac{1}{2A^\alpha(T)} .$$

2.4 Simulation Results

2.4.1 Kinetic Monte Carlo simulations and Broken-Bond Model

The KMC simulations have been performed as explained in Sec. 2.3.1 for 1, 2, 5, and 10 at.% Cu. At an annealing temperature of 700 °C, $6 \cdot 10^{11}$ KMC steps yield the formation of Cu clusters within the α-Fe matrix.

Fig. 2.2 (a) shows that at the beginning of the ageing process a large number of particles forms in relatively short time due to the high annealing temperature. Hence, these clusters contain only small amounts of Cu atoms, i.e., their mean radius is small. After reaching a maximum, the number of particles decreases again. Nevertheless, the mean radius increases (Fig. 2.2 (b)), a fact that can be attributed to Ostwald ripening, i.e., to the coarsening of particles. Since at the high temperatures considered here, there are almost no clusters for the case of 1 at.% Cu, the corresponding data are not

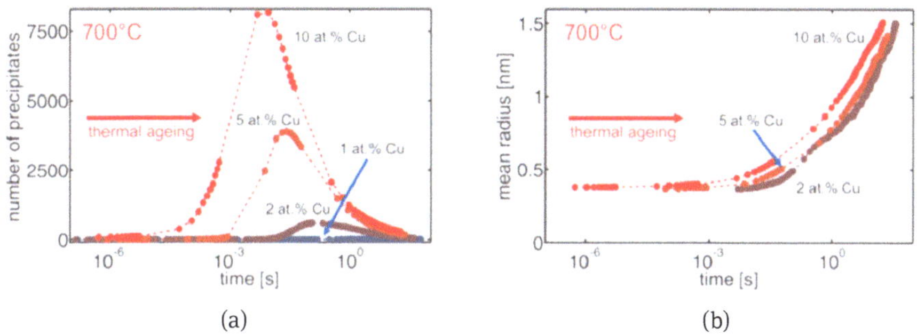

Fig. 2.2: (a) Number of particles and (b) their mean radius, as a function of thermal ageing time. In the case of 1 at.% Cu, almost no particles form, making statistical analysis inaccurate. Therefore, the 1 at.% Cu KMC results are not further considered.

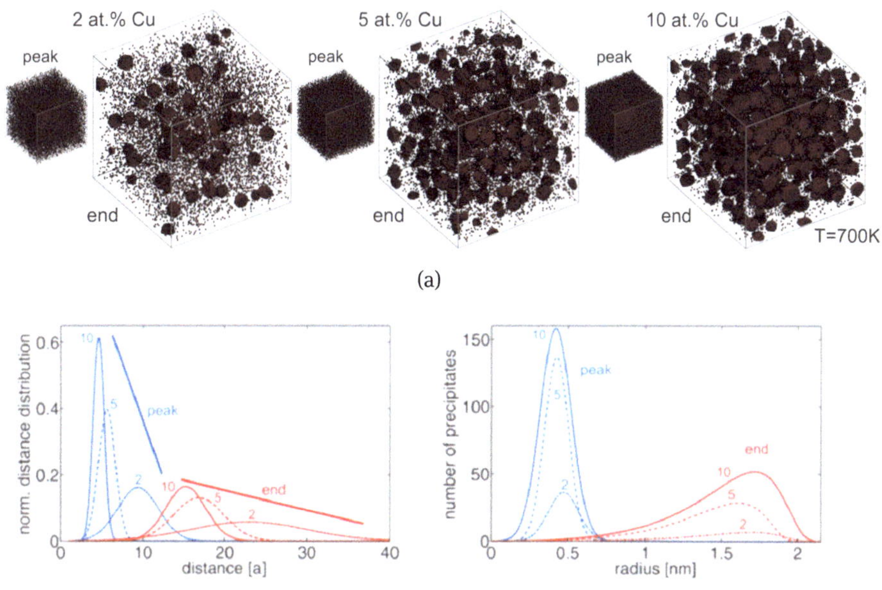

Fig. 2.3: (a) Arrangement of Cu atoms at two distinct times for 2, 5, and 10 at.% Cu. Up to the point where the number of particles reaches the maximum values (see Fig. 2.2), i.e., at the *peak* positions, very small clusters have formed and the resulting images looks nearly like solid solutions. At the *end* of the simulation, several large particles (radii > 1 nm) have formed while the remaining solid solution is in thermal equilibrium. Fitted gaussians represent the distance distributions (b) in lattice constants ($a = 0.2867$ nm) and the *peak* postions of the radius distributions (c) for 2, 5, and 10 at.% Cu whereas LSW distributions are assumed at the *end* positions.

considered in the following. With the KMC method, precipitates with a mean radius of 1.5 nm can be obtained in a reasonable amount of computing time ($\mathcal{O}$(days)). However, reaching mean particle radii of a couple of nanometers would increase the computation time tremendously, justifying the application of PFM in the coarsening regime. Fig. 2.3 (a) shows the atomic configurations at two different ageing times where *peak* denotes the states with the maximum number of particles (peaks in Fig. 2.2 (a)) and *end* denotes the end states of the KMC simulations for 2, 5, and 10 at.% Cu, respectively.

Since at the peak position only small clusters have formed, they are hardly visible among the completely solved Cu atoms. In contrast, at the end of the simulation the precipitates are clearly visible. At this point, the remaining matrix has already reached thermal equilibrium, i.e., almost no further nucleation is expected and Ostwald ripening can be considered as the dominant precipitation process. Transferring information from KMC to PFM can be done in two ways. One possibility is to transfer the whole atomic configuration, i.e., the positions of all Cu atoms, at certain precipitation states to PFM simulations. By this means, no information is lost. The other way is to transfer statistical information on the atomic configuration at certain precipitation states, i.e. fitted distance distributions and radius distributions as shown in Fig. 2.3 (b) and (c). Having a validation after transferring the identical particle arrangement in this study, this kind of information is very useful in order to perform large length scale simulations with PFM.

2.4.1.1 Interface energies – Broken-Bond Model

In order to obtain the interface energy needed in the PFM simulations, the broken-bond model (BBM) is applied [53, 54]. Within this model, the first and second nearest neighbours are considered and all atoms contribute to the interface energy whose coordination number changes when inserting an interface. Deriving the interface energies for interfaces oriented in [100], [110], and [111] directions without relating them to a specific interface area reads

$$E^{\text{int}}_{[100]} = \tfrac{1}{2}(-4\varepsilon^1_{AA} - 2\varepsilon^2_{AA} + 8\varepsilon^1_{AB} + 4\varepsilon^2_{AB} - 4\varepsilon^1_{BB} - 2\varepsilon^2_{BB}) \tag{2.10}$$

$$E^{\text{int}}_{[110]} = \tfrac{1}{2}(-2\varepsilon^1_{AA} - 2\varepsilon^2_{AA} + 4\varepsilon^1_{AB} + 4\varepsilon^2_{AB} - 2\varepsilon^1_{BB} - 2\varepsilon^2_{BB}) \tag{2.11}$$

$$E^{\text{int}}_{[111]} = \tfrac{1}{2}(-5\varepsilon^1_{AA} - 6\varepsilon^2_{AA} + 10\varepsilon^1_{AB} + 12\varepsilon^2_{AB} - 5\varepsilon^1_{BB} - 6\varepsilon^2_{BB}). \tag{2.12}$$

Equations (2.10) and (2.11) can also be found in [55]. Calculating the interface energies using the activation energies from the KMC simulations (see Sec. 2.3.1) yields the interface energies shown in Tab. 2.1.

Thus the preferred interfaces of Cu clusters are {110} interfaces. This is in very good agreement with results that can be found in [18, 42].

Tab. 2.1: Interface energies of Fe-Cu interfaces for different orientations obtained by evaluating Equations (2.10)–(2.12) for A = Fe and B = Cu. By relating the calculated interface energies to the orientation dependent areas per atom, the interface energies per unit areas are obtained.

	Interface energies		
orientation	[100]	[110]	[111]
E^{int} [J]	0.248	0.149	0.396
area per atom	a^2	$\frac{\sqrt{2}}{2}a^2$	$\sqrt{3}a^2$
$\gamma\ [\frac{\text{J}}{\text{m}^2}]$	0.483	0.410	0.447

2.4.2 Molecular Dynamics simulations

From experiments [40, 41] it is known that small Cu clusters (< 2 nm) are coherently embedded into the α-Fe matrix while bigger cluster undergo a structural change, i.e., bcc → 9R → 3R → fcc [40, 41]. In this study, precipitates with radii from 2 to 14 lattice constants, i.e. with radii ranging from 0.57 nm to 4.00 nm are analysed in order to provide qualitative information on the necessity of embedding the structural transition into the PFM simulations. Therefore, each precipitate is sufficiently relaxed according to the modelling scheme described in Sec. 2.3.2. After relaxation, the neighbours of each atom within a certain distance are counted. For the bcc structure, a cut-off distance between the second (0.2855 nm) and third (0.4038 nm) nearest neighbour of the corresponding atom is an appropriate choice, as the first (0.2473 nm) and second nearest neighbours lie very close to each other. In a perfect bcc crystal, the counting would yield 14 nearest neighbours (NN = 14) in total for each atom, i.e., 8 first and 6 second nearest neighbours. Visualising the relaxed structures, atoms with NN = 14 are shown as red points (see Fig. 2.4).

Deviations are shown as spheres. The colouring is explained in Tab. 2.2. The visualisation shows that small precipitates ($r \leq 4$ lattice constants) are coherently embedded within the α-Fe matrix whereas bigger precipitates show severe deviations in their interiors. However, the surfaces remain approximately in bcc structure, i.e., the coordination number of surface atoms does not change. As a conclusion, the transition from bcc to fcc within the clusters does not have to be taken into account in this first coupling of KMC and PFM simulations and the precipitates in the PFM simulation can be assumed as perfectly coherent.

2.4.3 Phase-Field Method Simulations

2.4.3.1 Data Transfer to Phase-field Model

We use a quantitative Phase-field model based on the grand chemical potential formulation (described in Sec. 2.3.3). The model parameters used for our simulations are

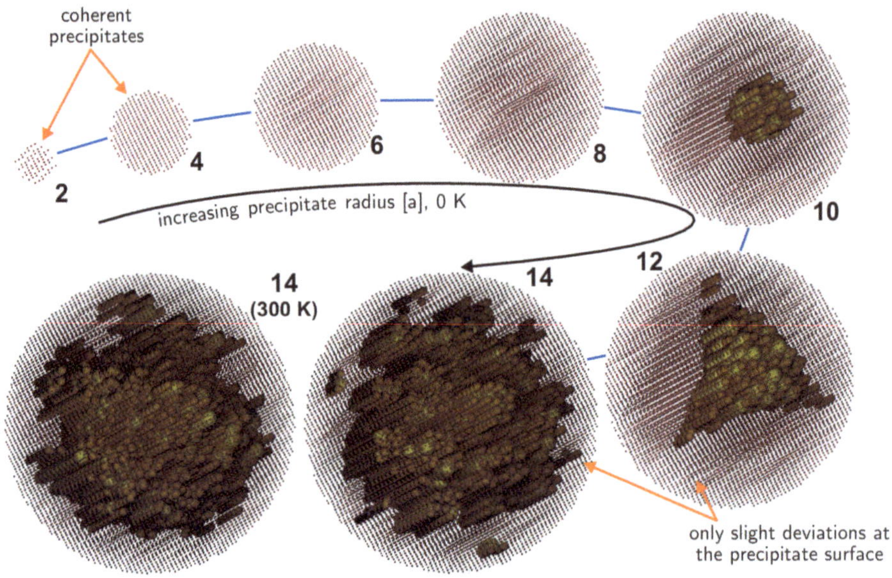

Fig. 2.4: Coherency of the precipitates after relaxation at regions close to the surface. The radii range from 0.57 nm (2 lattice constants) to 4.00 nm (14 lattice constants) where the lattice constant of Fe is taken [49, 56]. Small precipitates are perfectly coherent while deviations develop in bigger spheres. Nevertheless, at the precipitates' surfaces the deviations remain small. This behaviour is also found if the structure is kept at 300 K for some time before relaxation (14, 300 K). Thus, the structural transition does not necessarily have to be incorporated into the PFM simulations in this study.

Tab. 2.2: Atom colouring with respect to the number of atoms within a specified cut-off radius r_C. In bcc crystals, NN = 14 neighbours have to be found. Deviations (NN ≠ 14) indicate defects within the single crystal. An additional identification of atoms with NN = 12 neighbours is due to the fact that the corresponding fcc structure would have 12 neighbours within r_C.

Atom type	Number of neighbours	Colour
Cu	NN = 14	red
Cu	NN < 14 & $n \neq 12$	orange
Cu	NN > 14	purple
Cu	NN = 12	yellow

listed in their dimensional form in Tab. 2.3. As described in Sec. 2.4.1, the size and location of each precipitate cluster obtained in the KMC simulations are transferred as input for the PFM simulations. In PFM simulations, we use a computational box of 128^3 grid points where the grid spacing Δx is equal to the lattice constant resulting in the same system size as used for the KMC simulation described in Sec. 2.3.1. Assuming isothermal aging condition, the free energy for Fe-Cu system at 1100 K (827 °C) is ob-

Tab. 2.3: Parameters used in the PFM simulations. All the paremeters are defined in Sec. 2.3.3.

Parameters	Values	Parameters	Values
A^{α}	$1.71 \cdot 10^{11}\,\mathrm{J\,m^{-3}}$	τ	$6.89 \cdot 10^{18}\,\mathrm{J\,s\,m^{-4}}$
A^{β}	$1.39 \cdot 10^{11}\,\mathrm{J\,m^{-3}}$	D	$10^{-16}\,\mathrm{m^2\,s^{-1}}$
B^{α}	$-3.33 \cdot 10^{11}\,\mathrm{J\,m^{-3}}$	ϵ	$7.0 \cdot 10^{-10}\,\mathrm{m}$
B^{β}	$-6.43 \cdot 10^{9}\,\mathrm{J\,m^{-3}}$	γ	$0.41\,\mathrm{J\,m^{-2}}$
E^{α}	$1.13 \cdot 10^{11}\,\mathrm{J\,m^{-3}}$	Δx	$2.8665 \cdot 10^{-10}\,\mathrm{m}$
E^{β}	$-5.04 \cdot 10^{10}\,\mathrm{J\,m^{-3}}$	Δt	$5 \cdot 10^{-5}\,\mathrm{s}$

tained from the CALPHAD database. We fit this free energy data to the simplified form of the free energy functions using the procedure described in Sec. 2.3.3.2 and calculate the chemical potential μ, which is the driving force for diffusion, using Eq. (2.9).

In our system, the compositions of equilibrium Cu-rich precipitates and Fe-rich matrix are very close to pure Cu and pure Fe, respectively. From the broken-bond model (BBM) calculations (see Tab. 2.1), it is found that the lowest interface energy between a Cu-cluster and the surrounding Fe-matrix, among several possible interfaces, is $0.410\,\mathrm{J\,m^{-2}}$ ([110] interface). We use this interface energy value as a representative and approximated input parameter for the PFM simulations. Internal structural changes within the precipitates as discussed in Sec. 2.4.2 may affect the coarsening kinetics only towards the end of the simulation (mean radius > 3 nm) when distortions get more severe. Hence, in the present PFM simulations we assume the absence of elastic strain fields, which will be incorporated in our future study.

In the Phase-field model, the atomic mobility M can be derived from the interdiffusivity D (described in Sec. 2.3.3.3). D is calculated from the self-diffusivity values of each component using Darken's equation:

$$D = X_{\mathrm{Fe}} D^*_{\mathrm{Cu}} + X_{\mathrm{Cu}} D^*_{\mathrm{Fe}} \approx 10^{-16}\,\mathrm{m^2\,s^{-1}}\,, \tag{2.13}$$

where D^*_{Cu} and D^*_{Fe} are the self-diffusivities, and X_{Cu} and X_{Fe} are the mole fractions for Cu and Fe, respectively. Self-diffusivities of Fe and Cu in α-Fe as a function of temperature are reported in literature [57, 58]. We calculate the value of D at 1100 K from these available experimental data using Eq. (2.13).

2.4.4 Phase-field Results

The equilibrium compositions of precipitate and matrix in the Fe-Cu system at a particular temperature across a flat interface can be calculated using the tie lines from the free energy CALPHAD data. These compositions change from their equilibrium values, across a curved interface due to the Gibbs–Thomson effect. This shift in the surrounding matrix composition provides the driving force for precipitate coarsening. Phase-field models can adequately capture the Gibbs–Thomson effect for the precipitate-matrix microstructure [59, 60]. So, it is an ideal tool for studying coarsening phenomena. In our three dimensional domain, the precipitates are nearly spherical because of the isotropic interface energy. The shift in composition at the precipitate side Δc^{p} is given by

$$\Delta c^{\mathrm{p}} = \frac{2\gamma\kappa}{\xi_{\mathrm{p}}(c^{\mathrm{p}}_{\mathrm{eq}} - c^{\mathrm{m}}_{\mathrm{eq}})}, \tag{2.14}$$

where

$$\xi_{\mathrm{p}} = \left(\frac{\partial^2 f^{\alpha}}{\partial c^2}\right)_{\mathrm{ppt}} = 2A^{\alpha}(T),$$

and the shift in composition at the matrix side Δc^{m} follows as

$$\Delta c^{\mathrm{m}} = (\xi_{\mathrm{p}}/\xi_{\mathrm{m}})\Delta c^{\mathrm{p}}.$$

In the above equations, κ is the interface curvature (inverse of radius), $c^{\mathrm{p}}_{\mathrm{eq}}$ and $c^{\mathrm{m}}_{\mathrm{eq}}$ are the equilibrium compositions across a flat interface at the precipitate and the matrix side, respectively, and γ is the interface energy.

Phase-field simulations are performed with a single spherical precipitate in a three dimensional domain for three different precipitate sizes. For each simulation, we use equilibrium precipitate composition obtained from the free energy function using a common tangent construction, with a very small supersaturation in the matrix. During the simulation, the precipitate and matrix compositions with time evolution reach the composition corresponding to the equilibrium values across a curved interface.

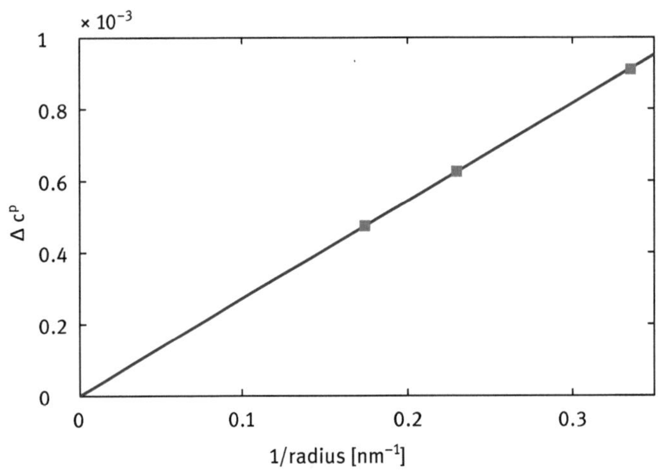

Fig. 2.5: Change in precipitate composition from its equilibrium composition across a flat interface (Δc^{p}) is plotted against inverse of precipitate radius. The straight line for Δc^{p} is given by the Gibbs–Thomson equation (see Eq. (2.14)). The data points represent Δc^{p} measured at the centre of the spherical precipitates obtained from three different PFM simulations.

In Fig. 2.5, we have plotted the shift in composition on the precipitate side from its equilibrium composition, Δc^{p}, as a function of interface curvature. The slope of the line is found to be $2\gamma/(\xi_{\mathrm{p}}(c^{\mathrm{p}}_{\mathrm{eq}} - c^{\mathrm{m}}_{\mathrm{eq}}))$. This test confirms the ability of our model to predict the Gibbs–Thomson effect correctly.

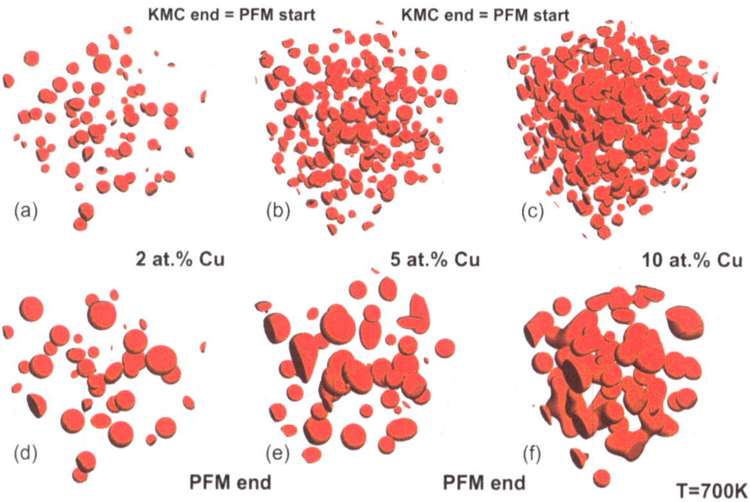

Fig. 2.6: Microstructures obtained at $t = 0$ s (a)–(c) and $t = 175$ s (d)–(f) for 2, 5 and 10 at.% Cu systems (first, second and third columns, respectively).

Fig. 2.6 shows the precipitate-matrix microstructures at times $t = 0$ s and $t = 175$ s for 2, 5, and 10 at.% Cu, respectively. Microstructures at $t = 0$ s correspond to the *end* results of the KMC simulations (see Fig. 2.3). In all systems, the large particles grow at the expense of smaller particles. The mean radius increases with time, while the number of particles decreases with time. During this time interval, after initial transient, the total volume of the particles remains constant for each system, which implies that the matrix supersaturation is zero and the system is in the coarsening regime.

Fig. 2.7 (a)–(c) shows the particle size distributions (PSD) at $t = 0$ s and $t = 175$ s for the three systems. In all the cases, the size distribution shifts to its right and the distribution also broadens with time. An important aspect of LSW theory is dynamic scaling behaviour, which means the whole microstructure is self-similar during steady-state coarsening when scaled by the mean precipitate radius. The PSD predicted by LSW holds only for spherical particles in an infinitesimally small volume fraction of precipitates. As mentioned in Sec. 2.1, several modified theoretical and experimental results showed that, with higher volume fraction of precipitate, the distribution becomes broader having a smaller amplitude. Fig. 2.7 (d) shows the PSDs scaled by the mean radius obtained at different times, for Cu 5 at.% system, follow this trend.

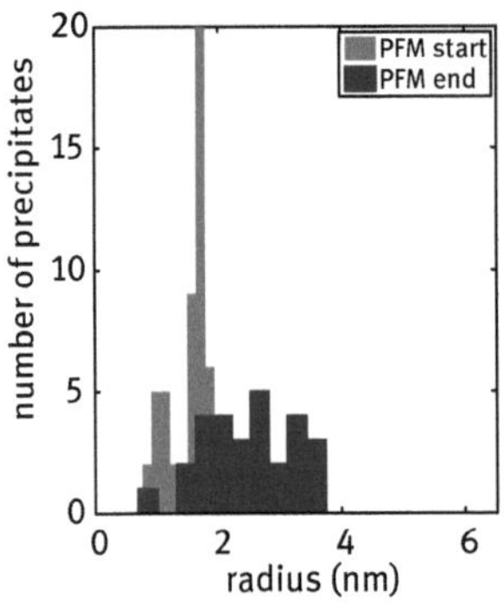

(a) 2 at.% Cu

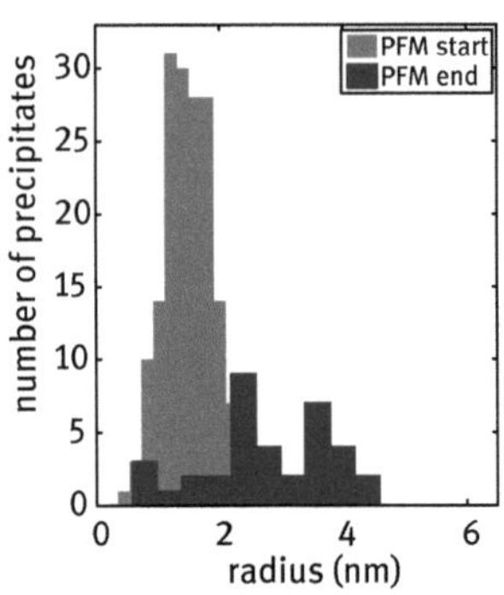

(b) 5 at.% Cu

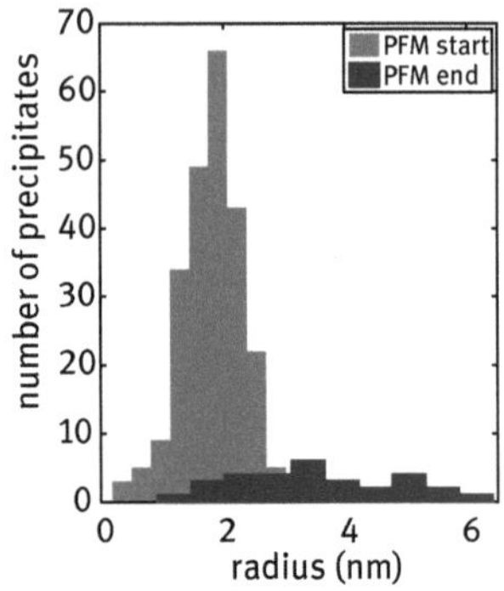

(c) 10 at.% Cu

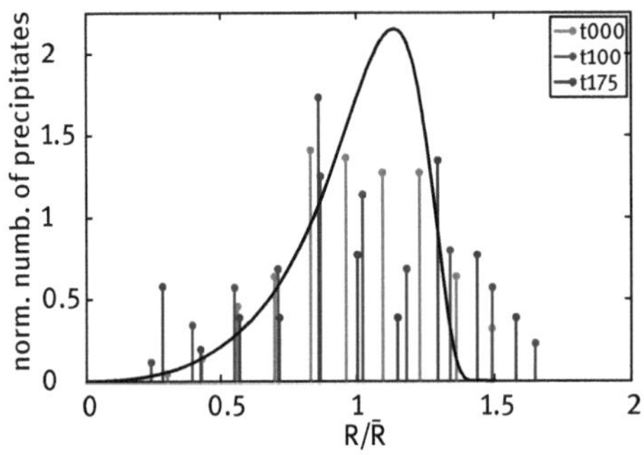

(d) Scaled Distribution for 5 at.% Cu

Fig. 2.7: Particle size distributions (PSD) for (a) 2, (b) 5, and (c) 10 at.% Cu at the beginning of the PFM simulations (start = red) and after 175 s (end = blue). (d) PSDs scaled by mean radius, for Cu 5 at.% system, at three different times (0 s, 100 s, 175 s). The line shows the distribution obtained from the LSW theory.

The mean particle radii $\bar{R}$ are obtained at different stages of coarsening for each system and plotted as a function of ageing time in Fig. 2.8 (a). In Fig. 2.8 (b), the number of precipitates for each of these three systems is plotted as a function of time. For the 2 at.% Cu system, the number of particles in the computational domain reduces from about 55 at $t = 0$ s to 27 at $t = 175$ s. The small number of particles in a domain makes this system statistically unreliable for predicting the overall trend in the temporal evolution of the microstructure. Each denucleation event during evolution causes a gradual decrease in mean radius and sudden increase in the effective supersaturation in the matrix because of the small total number of particles. These events appear as steps in the mean radius vs time plot. For the 10 at.% Cu system, the initial number of particles (at $t = 0$ s) is 260. However, this number decreases rapidly to 126 at $t = 10$ s due to a large number of coalescence events, and then it decreases slowly to 30 at $t = 175$ s. Denucleation events of a group of small particles appear as steps only at a later stage of evolution. For the 5 at.% Cu system, the number of particles

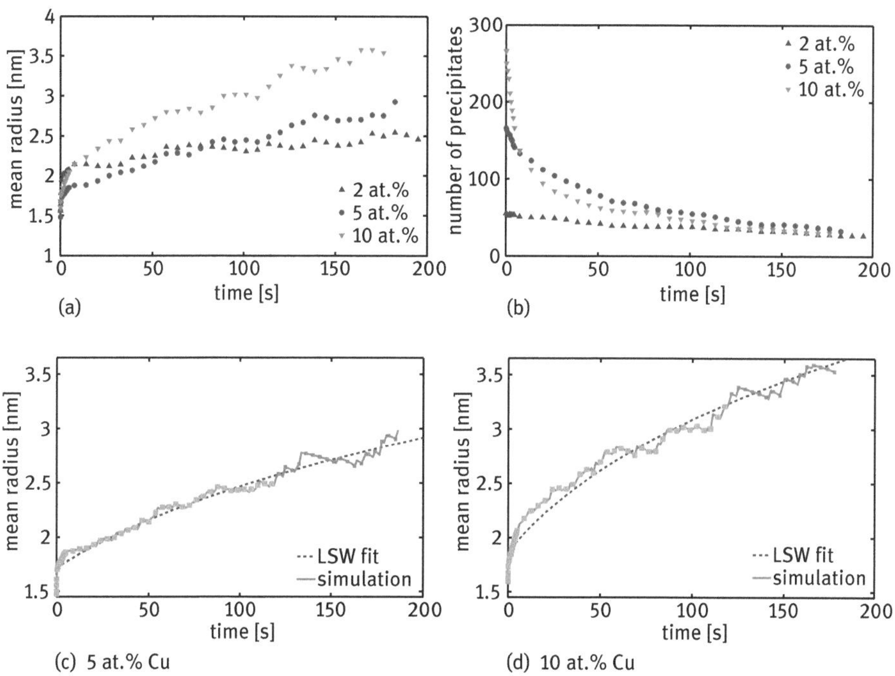

Fig. 2.8: (a) The precipitate mean radius ($\bar{R}$), and (b) the number of particles (n_p) is plotted as a function of time for 2, 5, and 10 at.% Cu. The temporal evolution of the $\bar{R}$ for (c) 5 and (d) 10 at.% Cu, fitted with LSW kinetics at the time interval indicated by the green datapoints. Inside this time interval both systems contain sufficient precipitates for good statistics.

decreases slowly from 166 at $t = 0$ s to 33 at $t = 175$ s. The larger number of particles than in the 2 at.% Cu composition and the less number of coalescence events than in the 10 at.% Cu composition make this system more favourable to study coarsening kinetics following the LSW law.

According to the LSW theory, the cube of mean radius $\bar{R}(t)$ should increase linearly with time t during coarsening regime:

$$\bar{R}(t) = (\bar{R}(0)^3 + Kt)^{1/3}, \tag{2.15}$$

where $\bar{R}(0)$ is the mean radius at $t = 0$ s and K is the coarsening rate constant. In Fig. 2.8 (c) and (d), we fit $\bar{R}(t)$ for the 5 at.% Cu and the 10 at.% Cu systems, respectively, using the above equation in the time interval between $t \approx 0$ s (after the initial transient) and $t = 120$ s when the simulation domain contains reasonable number of particles to describe coarsening kinetics (see Fig. 2.8 (b)). For the 5 at.% Cu system, the mean precipitate radius closely follows the LSW kinetics after an initial transient and the fit is good in this time interval. We found $\bar{R}(0) = 1.72$ nm and $K_{LSW} = 0.1024\,nm^3\,s^{-1}$ for this system. The value of $\bar{R}(0)$ is higher compared to the end results of KMC, because very small precipitates undergo denucleation process during the initial transient.

For the 10 at.% Cu system, the number of coalescence events is very high which leads to an increase in $\bar{R}$. Furthermore, the particles do not remain spherical in shape (see Fig. 2.6 (f)) during time evolution. The fit in this regime is not as good as in the 5 at.% Cu system, because the inherent assumptions of the LSW theory do not hold good for the 10 at.% Cu system. We found $\bar{R}(0) = 1.86$ nm and $K_{\mathrm{LSW}} = 0.2389\,\mathrm{nm}^3\,s^{-1}$ for this system. For our systems, we observe the rate constant K to increase with increasing precipitate volume fraction, which is also reported in the literature.

2.5 Conclusions

In the present survey, we employ kinetic Monte Carlo (KMC) and Phase-field method (PFM) simulations to provide a quantitative description of the different stages of precipitation by suitably using the strengths of each method. While KMC is able to model the process of early nucleation and growth adequately, late stage Ostwald ripening is treated appropriately with PFM. Using this approach we are able to capture not only the coarsening regime with both models, but also the coalescence events at high volume fractions of precipitates which are retrieved in the PFM simulations. It is found that the kinetics closely follows the LSW temporal power law in this regime, but the coarsening rate increases with an increase in volume fraction of precipitates. Noteworthy is that while the analytical theories such as LSW can adequately describe ripening kinetics for specific particle morphologies and configurations, deviations from the assumptions require the employment of simulations. For this reason, computational modelling methods such as KMC and PFM are useful.

Acknowledgment: The authors D. Molnar and S. Schmauder would like to thank the German Research Foundation (DFG) for financial support of the project within the Cluster of Excellence in Simulation Technology (EXC 310/1) at the University of Stuttgart. The authors A. Mora, P. Binkele, S. Schmauder, R. Mukherjee, A. Choudhury, and B. Nestler would like to thank the German Research Foundation (DFG) for financial support of the projects SCHM746/101-1 and NE822/12-1.

References

[1] Schmauder S, Binkele P. Atomistic computer simulation of the formation of Cu-precipitates in steels. Computational Materials Science. 2002;24(1-2):42–53.

[2] Nembach E. Particle Strengthening of Metals and Alloys. 1st ed. John Wiley & Sons, Inc., New York; 1997.

[3] Lifshitz IM, Slyozov VV. The kinetics of precipitation from supersaturated solid solutions. Journal of Physics and Chemistry of Solids. 1961;19(1-2):35–50.
[4] Wagner C. Theory of aging and precipitation through dissolution-Ostwald ripening. Z Elektrochem. 1961;581:65.
[5] Ardell AJ. The effect of volume fraction on particle coarsening: theoretical considerations. Acta Metallurgica. 1972;20(1):61–71.
[6] Brailsford AD, Wynblatt P. The dependence of ostwald ripening kinetics on particle volume fraction. Acta Metallurgica. 1979;27(3):489–497.
[7] Marqusee JA, Ross J. Theory of ostwald ripening: Competitive growth and its dependence on volume fraction. The Journal of chemical physics. 1984;80(1):536–543.
[8] Enomoto Y, Tokuyama M, Kawasaki K. Finite volume fraction effects on ostwald ripening. Acta Metallurgica. 1986;34(11):2119–2128.
[9] Marsh SP, Glicksman ME. Kinetics of phase coarsening in dense systems. Acta Materialia. 1996;44(9):3761–3771.
[10] Kang CH, Yoon DN. Coarsening of cobalt grains dispersed in liquid copper matrix. Metallurgical transactionsA, Physical metallurgy and materials science. 1981;12A(1):65–69.
[11] Bender W, Ratke L. Ostwald ripening of liquid phase sintered cu-co dispersions at high volume fractions. Acta Materialia. 1998;46(4):1125–1133.
[12] Snyder VA, Alkemper J, Voorhees PW. Development of spatial correlations during ostwald ripening: a test of theory. Acta Materialia. 2000;48(10):2689–2701.
[13] Snyder VA, Alkemper J, Voorhees PW. Transient ostwald ripening and the disagreement between steady-state coarsening theory and experiment. Acta Materialia. 2001;49(4):699–709.
[14] Rowenhorst DJ, Kuang JP, Thornton K, Voorhees PW. Three-dimensional analysis of particle coarsening in high volume fraction solid-liquid mixtures. Acta Materialia. 2006;54(8):2027–2039.
[15] Willer D, Zies G, Kuppler D, Föhl J, Katerbau KH. Service-induced changes of the properties of copper-containing ferritic pressure-vessel and piping steels. Final report, MPA Stuttgart; 2001.
[16] Pan F, Ruoff H, Willer D, Katerbau KH. Investigation on the microstructure of a pressure vessel/pipe steel 15 nicumonb 5 (wb36). MPA Report, Stuttgart. 1996.
[17] Schick M, Wiedemann J, Willer D. Untersuchungen zur sicherheitstechnischen Bewertung von geschweissten Komponenten aus Werkstoff 15 NiCuMoNb 5 (WB36) im Hinblick auf die Zähigkeitsabnahme unter Betriebsbeanspruchung. Technischer bericht, MPA Stuttgart; 1997.
[18] Soisson F, Barbu A, Martin G. Monte Carlo simulations of copper precipitation in dilute iron-copper alloys during thermal ageing and under electron irradiation. Acta Materialia. 1996; 44(9):3789–3800.
[19] Schmauder S, Kohler C. Atomistic simulations of solid solution strengthening of alpha-iron. Computational Materials Science. 2011;50(4):1238–124319th International Workshop on Computational Mechanics of Materials (IWCMM 19), Ovidius Univ Constanta, Constanta, ROMANIA, SEP 01-04, 2009.
[20] Kohler C, Kizler P, Schmauder S. Atomistic simulation of precipitation hardening in alpha-iron: influence of precipitate shape and chemical composition. Modelling and Simulation in Materials Science and Engineering. 2005;13(1):35–45.
[21] Bacon DJ, Osetsky YN. Mechanisms of hardening due to copper precipitates in a-iron. Philosophical Magazine. 2009;89(34-36):3333–3349.
[22] Kizler P, Uhlmann D, Schmauder S. Linking nanoscale and macroscale: calculation of the change in crack growth resistance of steels with different states of Cu precipitation using a modification of stress-strain curves owing to dislocation theory. Nuclear Engineering and Design. 2000;196(2):175–183.

[23] Boettinger W, Warren J, Beckermann C, Karma A. Phase-field simulation of solidification. Annual Review of Materials Research. 2002;32:163–194.
[24] Chen LQ. Phase-field models for microstructure evolution. Annual Review of Materials Research. 2002;32:113–140.
[25] Thornton K, Ågren J, Voorhees PW. Modelling the evolution of phase boundaries in solids at the meso- and nano-scales. Acta Materialia. 2003;51(19):5675–5710.
[26] Singer-Loginova I, Singer HM. The phase field technique for modeling multiphase materials. Reports on Progress in Physics. 2008;71(10):106501.
[27] Moelans N, Blanpain B, Wollants P. An introduction to phase-field modeling of microstructure evolution. Calphad: Computer Coupling of Phase Diagrams and Thermochemistry. 2008;32(2): 268–294.
[28] Nestler B, Choudhury A. Phase-field modeling of multi-component systems. Current Opinion in Solid State and Materials Science. 2011;15(3):93–105.
[29] Vaithyanathan V, Wolverton C, Chen LQ. Multiscale modeling of θ' precipitation in Al-Cu binary alloys. Acta Materialia. 2004;52(10):2973–2987.
[30] Kamachali RD, Hua J, Steinbach I, Hartmaier A. Multiscale simulations on the grain growth process in nanostructured materials. International Journal of Materials Research. 2010;101(11): 1332–1338.
[31] Warren JA, Murray BT. Ostwald ripening and coalescence of a binary alloy in two dimensions using a phase-field model. Modelling and Simulation in Materials Science and Engineering. 1996;4(2):215–229.
[32] Ode M, Suzuki T, Kim SG, Kim WT. Ostwald ripening analysis using phase-field model. Materials Transactions. 2001;42(11):2410–2414.
[33] Wen YH, Lill JV, Chen SL, Simmons JP. A ternary phase-field model incorporating commercial calphad software and its application to precipitation in superalloys. Acta Materialia. 2010; 58(3):875–885.
[34] Zhu JZ, Liu ZK, Vaithyanathan V, Chen LQ. Linking phase-field model to calphad: Application to precipitate shape evolution in ni-base alloys. Scripta Materialia. 2002;46(5):401–406.
[35] Zhu JZ, Wang T, Ardell AJ, Zhou SH, Liu ZK, Chen LQ. Three-dimensional phase-field simulations of coarsening kinetics of γ' particles in binary ni-al alloys. Acta Materialia. 2004;52(9):2837–2845.
[36] Kim SG. Large-scale three-dimensional simulation of ostwald ripening. Acta Materialia. 2007; 55(19):6513–6525.
[37] Koyama T, Onodera H. Computer simulation of phase decomposition in Fe-Cu-Mn-Ni quaternary alloy based on the phase-field method. Materials Transactions. 2005;46(6):1187–1192.
[38] Koyama T, Hashimoto K, Onodera H. Phase-field simulation of phase transformation in fe-cu-mn-ni quaternary alloy. Materials Transactions. 2006;47(11):2765–2772.
[39] Koyama T, Onodera H. Phase-field modeling of the microstructure evolutions in fe-cu base alloys. Materials Science Forum. 2007;539-543:2383–2388.
[40] Pizzini S, Roberts K, Phythian W, English C, Greaves G. A fluorescence EXAFS study of the structure of copper-rich precipitates in Fe-Cu and Fe-Cu-Ni alloys. Philosophical Magazine Letters. 1990;61(4):223–229.
[41] Othen P, Jenkins M, Smith G. High-Resolution Electron-Microscopy Studies of the Structure of Cu Precipitates in alpha-Fe. Philosophical Magazine A. 1994;70(1):1–24.
[42] Binkele P. Atomistische Modellierung und Computersimulation der Ostwald-Reifung von Ausscheidungen beim Einsatz von kupferhaltigen Stählen. Phd thesis, Universität Stuttgart, Stuttgart; 2006.
[43] Soisson F, Fu CC. Energetic Landscapes and Diffusion Properties in FeCu Alloys. Solid State Phenomena. 2007;129:31–39.

[44] Molnar D, Binkele P, Hocker S, Schmauder S. Atomistic multiscale simulations on the anisotropic tensile behaviour of copper-alloyed alpha-iron at different states of thermal ageing. Philosophical Magazine. 2012;92(5):586–607.
[45] Vincent E, Becquart CS, Domain C. Solute interaction with point defects in alpha Fe during thermal ageing: A combined ab initio and atomic kinetic Monte Carlo approach. Journal of Nuclear Materials. 2006;351(1-3):88–99.
[46] Vincent E, Becquart CS, Domain C. Atomic kinetic Monte Carlo model based on ab initio data: Simulation of microstructural evolution under irradiation of dilute Fe-CuNiMnSi alloys. Nuclear Instruments & Methods in Physics Research B. 2007;255(1, Sp. Iss. SI):78–84.
[47] Soisson F, Fu CC. Cu-precipitation kinetics in alpha-Fe from atomistic simulations: Vacancy-trapping effects and Cu-cluster mobility. Physical Review B. 2007;76(21).
[48] Stadler J, Mikulla R, Trebin HR. IMD: A software package for molecular dynamics studies on parallel computers. International Journal of Modern Physics C. 1997;8(5):1131–1140.
[49] Bonny G, Pasianot RC, Castin N, Malerba L. Ternary Fe-Cu-Ni many-body potential to model reactor pressure vessel steels: First validation by simulated thermal annealing. Philosophical Magazine. 2009;89(34-36):3531–3546.
[50] Choudhury A, Nestler B. Grand-potential formulation for multicomponent phase transformations combined with thin-interface asymptotics of the double-obstacle potential. Physical Review E – Statistical, Nonlinear, and Soft Matter Physics. 2012;85(2).
[51] Garcke H, Nestler B, Stinner B. A diffuse interface model for alloys with multiple components and phases. SIAM Journal on Applied Mathematics. 2004;64(3):775–799.
[52] Nestler B, Garcke H, Stinner B. Multicomponent alloy solidification: Phase-field modeling and simulations. Physical Review E - Statistical, Nonlinear, and Soft Matter Physics. 2005;71(4).
[53] Janssens K, Raabe D, Kozeschnik E, Miodownik M, Nestler B. Computational materials engineering: an introduction to microstructure evolution. Elsevier Academic Press; 2007.
[54] Sonderegger B, Kozeschnik E. Generalized nearest-neighbor broken-bond analysis of randomly oriented coherent interfaces in multicomponent fcc and bcc structures. Metallurgical and Materials Transactions A. 2009;40:499–51010.1007/s11661-008-9752-6. Available from: http://dx.doi.org/10.1007/s11661-008-9752-6.
[55] Vincent E, Becquart C, Pareige C, Pareige P, Domain C. Precipitation of the FeCu system: A critical review of atomic kinetic Monte Carlo simulations. Journal of Nuclear Materials. 2008;373:387–401. Available from: http://www.sciencedirect.com/science/article/pii/S0022311507008446.
[56] Mendelev MI, Han S, Srolovitz DJ, Ackland GJ, Sun DY, Asta M. Development of new interatomic potentials appropriate for crystalline and liquid iron. Philosophical Magazine. 2003;83(35):3977–3994.
[57] Buffington FS, Hirano K, Cohen M. Self diffusion in iron. Acta Metallurgica. 1961;9(5):434–439.
[58] Anand MS, Agarwala RP. Diffusion of copper in iron. Journal of Applied Physics. 1966;37(11):4248–4251.
[59] Mukherjee R, Abinandanan TA, Gururajan MP. Phase field study of precipitate growth: Effect of misfit strain and interface curvature. Acta Materialia. 2009;57(13):3947–3954.
[60] Mukherjee R, Abinandanan TA, Gururajan MP. Precipitate growth with composition-dependent diffusivity: Comparison between theory and phase field simulations. Scripta Materialia. 2010;62(2):85–88.

C. V. Singh

3 Multiscale modeling predictions of age hardening curves in Al-Cu alloys

Abstract: Precipitation hardening is a common method of manufacturing high strength structural alloys for wide industrial applications. In this chapter, we describe an atomistics based multiscale model to predict macroscopic mechanical strength of precipiatation hardened alloys. Molecular dynamics (MD) is utilized to understand dislocation-precipitate interactions, and to estimate the precipitate strengths. This serves as input to meso-scale dislocation dynamics framework to predict flow strength through the precipitate microstructure at specific aging times. MD simulations also serve as input to a semi-analytical model to compute solute hardening. By combining solute and precipitate hardening contributions together, alloy's yield strength is predicted. A kinetics model based on Avrami theory is developed to obtain chateristics of precipitate microstructure at given aging time. Fitting only the rate constants in the precipitation kinetics model, the macroscopic strength predictions of the hierarchical multiscale model are found to correspond reasonably well with available experimental data.

Keywords: Al alloy, age hardening curve, precipitation hardening, solute hardening, atomistic simulation, molecular dynamics, dislocation-obstacle interaction

3.1 Introduction

Modern high strength aluminum alloys are used for structural components in aerospace, automotive, civil, and sports industries. Precipitation hardening is a common method to obtain high mechanical strength, wherein microscopic precipitates retard the motion of dislocations, thereby retarding the plastic flow and hence increasing the alloy's yield strength. This phenomena was first discovered by Alfred Wilm [1] in the early 20th century. In fact, this phenomenon had previously been inadvertently utilized by the Wright brothers during their historic flight of 1903 [2]. During the last century, phenomenal advancements have taken place to improve our understanding of this behavior [3, 4], and its utilization to manufacture ever strong alloys, with the most recent precipitation hardened Al alloys demonstrating strengths approaching 1 GPa [5, 6]. Initial works involved the development of dislocation theory [7–9] and the observation of nanometer-sized solute clusters by X-ray scattering [10, 11] and transmission electron microscopy (TEM) [12]. Despite such advancements, a physics based model is still lacking that can predict age hardening curves reliably without experimental input, and this chapter presents an atomistics based hierarchical multiscale model to predict age hardening curves and discusses the remaining challenges.

Al-2xxx and Al-6xxx series alloys are prominently utilized in commercial airplanes. Al-2xxx alloys are mainly dilute binary alloys, with up to 5 wt.% Cu in solid solution or precipitate microstructure. In such Al-Cu alloys, the evolution of the microstructure at common aging temperatures (110–150 °C) and Cu concentrations (~ 4 wt.% Cu) consists of [13]: SSSα → GP zones → θ'' (GP2 zones) → θ' → θ ($CuAl_2$), where SSSα represents a supersaturated solid solution. GP zones are Cu disk shaped monolayers on {100} planes, with size 1–10 nm; whereas θ'' precipitates consist of two disk shaped Cu monolayers on {100} planes separated by three planes of Al atoms, with size 10–50 nm. θ' precipitates are Al_2Cu platelets having a tetragonal structure. θ precipitates constitute the equilibrium precipitate phase with Al_2Cu stoichiometry and body centered tetragonal structure. GP zones and θ'' precipitates are known to be fully coherent with the Al matrix, θ' is semi-coherent, and θ is incoherent. Although multilayer GP zones have been observed [14], there is general agreement that they are a single atomic layer thick [15]. The Cu content of GP zones, however, can vary from 40 % to 100 % [14, 16] with an average value of about 80 % [16].

The interaction of dislocations with precipitates can be quite complex, depending upon dislocation character, slip plane, precipitate size, shape, orientation and density. These interactions could be elastic in nature, due to lattice misfit or mistfit of elastic moduli between the precipitate and the matrix [17–20]. Inelastic chemical interactions, such as discrepancy in dislocation core energy in precipatate versus matrix, can also play an important role [21, 22]. Experimental efforts have also shown the complexity of these interactions, but a clear consensus is not available [23–27]. In this regard, atomistic simulations have started providing valuable insights. Notable examples of atomistic simulations concern dislocation interactions with material defects such as grain boundaries [28–30], with voids and precipitates in irradiated α-iron [31–33] and other alloys [34–37]. A comprehensive review of dislocation–obstacle interactions at the atomic level can be found in [38]. Without going into details, an overarching finding of these reports is that dislocation-defect interactions are indeed very complex, and thus require detailed atomistic understanding.

Current approaches to design and development of age hardened alloys are empirical in nature, i.e. based on extensive trial and error experimental testing. Existing models are based on continuum theories that do not include complex atomistic details of dislocation-precipitate interactions [39–41]. Here we present a physics based multiscale model that can predict age hardening curves of Al-Cu alloys without much experimental input. We first present atomistic computation of precipitate and solid-solution strengths. Subsequently, dislocation dynamics model to predict macroscopic precipitate strength from atomistic results is detailed. A kinetics model to obtain precipitate microstructural growth is described next. Combining, age hardening predictions are made in the athermal limit, and the effect of aging environment is understood. Finally, age hardening is predicted at room temperature by accounting for thermal activation of dislocation-defect interactions, and comparisons are made with experiments.

This book chapter is taken from a previously published article [42], with kind permissions from Springer.

3.2 Atomistic modeling of precipitation hardening

Dislocation–precipitate interactions and the resultant critical resolved shear stress (τ_c) are inherently linked to microstructural parameters, that include: (a) orientation of dislocation–precipitate interaction, (b) precipitate size, (c) the location at which the slip plane of the dislocation intersects the precipitate (offset), (d) the influence of neighboring precipitates, (e) temperature/strain rate, (f) dislocation character, and (g) previous dislocation–precipitate interactions. In the following, we present results of molecular dynamics (MD) simulations to understand the role of above factors on τ_c for nano-sized precipitates (GP zones and θ'') in binary Al-Cu alloys. It is noted that the larger θ' and θ precipitates are not currently accessible to atomistic simulation. Nonetheless, they are likely impenetrable to dislocations, thus their interaction with dislocations can be accurately studied by simple continuum models.

3.2.1 Methodology

MD simulations were carried out using the freely available open source LAMMPS code [43]. The code was modified to use a recently developed Al-Cu empirical potential developed by Apostol and Mishin [44]. The potential is an angular dependent extension of the embedded atom method (EAM) [45, 46], where the potential energy of the system is written as

$$E_{\text{tot}} = \frac{1}{2}\sum_{i,j(j\neq i)} \Phi_{ij}(r_{ij}) + \sum_i F_i(\bar{\rho}_i) + \frac{1}{2}\sum_{i,\alpha}(\mu_i^\alpha)^2 + \frac{1}{2}\sum_{i,\alpha,\beta}(\lambda_i^{\alpha\beta})^2 - \frac{1}{6}\sum_i \nu_i^2 \,. \tag{3.1}$$

where i and j enumerate atoms and the superscripts $\alpha, \beta = 1, 2, 3$ refer to the Cartesian directions. The first two terms, taken together, represent regular EAM contributions, where Φ_{ij} is the pair-interaction potential, and F_i is the embedding energy of atom i with $\bar{\rho}_i$ loosely representing the electron density. The three remaining terms represent the angular dependence through dipole and quadrupole distortions. The details of the potential can be found in Apostol and Mishin [44] and Mishin et al. [46]. Our implementation was validated by comparing relaxed cohesive energies, vacancy formation energies, surfaces energies, heats of solution, and formation energy for the θ'' (Al_2Cu) phase with independent values provided by Apostol and Mishin [44].

The simulation cell (Fig. 3.1) consisted of an FCC Al lattice, bounded by $(\bar{1}10)$, (111), and $(11\bar{2})$ faces in X, Y, and Z directions, respectively. Periodic boundary conditions were applied in the X and Z directions, and the Y surfaces were used to apply the load. Starting from a perfect lattice, a precipitate was created by simply changing

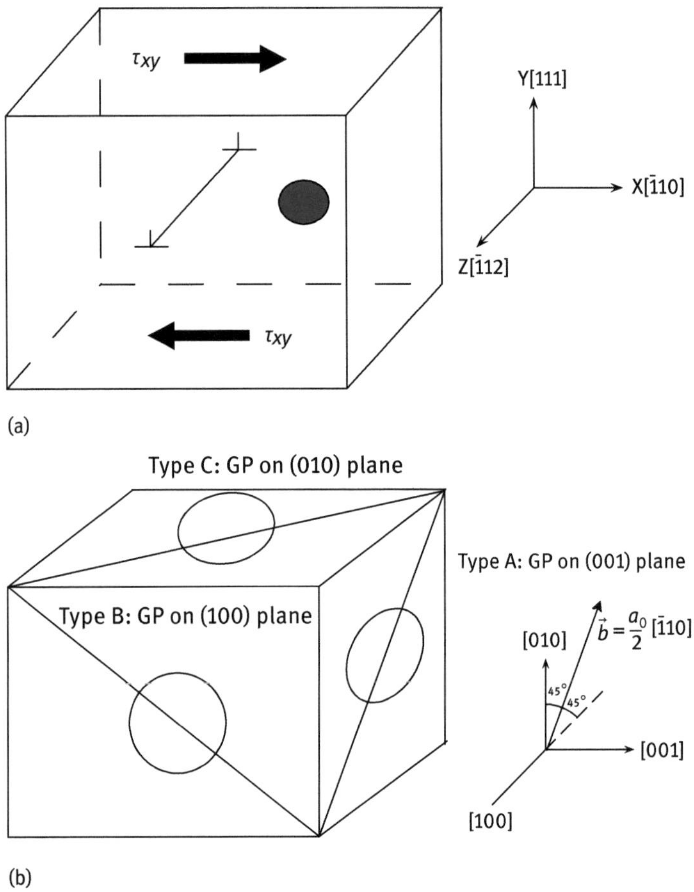

Fig. 3.1: (a) Simulation cell with an edge dislocation and GP zone, (b) 3 possible GP zone orientations with respect to the Burgers vector of the dislocation. Type A is referred to as a 0° interaction while types B and C are considered equivalent and refered to as 60° interactions.

selected atom types to Cu. The continuum displacement field of an edge dislocation was used to create a dislocation in the center of the simulation cell ($X = Y = 0$) with a line direction parallel to the Z-axis and the burgers vector $\vec{b} = 1/2[\bar{1}10]$.

Because GP zones and θ'' precipitates only form on {100}-type planes, there are only three physically relevant orientations by which they interact with edge dislocations; two of which are crystallographically equivalent. The Burgers vector of the edge dislocation will either be at an angle of 60° from the precipitate plane, or it will be parallel to it (Fig. 3.1). Thus we will refer to dislocation–precipitate interactions as either being 60° or 0° in nature. As a simplification, we assume that GP zones are monolayers of Cu-atoms with a 100 % Cu content, while θ'' precipitates are two Cu layers separated by three Al layers [14, 16].

The minimum cell size was approximately $34 \times 42 \times 16\,\text{nm}^3$ and contained approximately 1.4 million atoms. For the simulations with large GP zones and θ'' precipitates a larger cell size was used (with largest cell size being approximately $68 \times 44 \times 64\,\text{nm}^3$ consisting of 11 million atoms). In all cases, box size studies were performed to verify that simulation cell boundary forces were negligible in the X and Y directions. The effect of cell size in the Z direction will be discussed later, however we note that the cell sizes in the Z direction that we explored are consistent with experimental observations of GP zone and θ'' spacing [27, 47].

Prior to loading, NPT dynamics were performed for 50 ps to relax the system and alleviate out-of-balance forces and net stresses. The system was then loaded in shear by subjecting the atoms near the top and bottom Y surfaces to a constant traction in the X-direction given by

$$f_{\text{top}} = \frac{\tau_{xy} A_{xz}}{N}, \quad f_{\text{bottom}} = -\frac{\tau_{xy} A_{xz}}{N} \tag{3.2}$$

for top and bottom Y-surfaces, respectively, where τ_{xy} is applied shear stress, A_{xz} is the surface area, and N is the number of atoms on the respective surfaces. The applied shear stress was increased quasi-statically until the dislocation breaks free from the precipitate. At each load step of 5 MPa, nonlinear conjugate gradient was performed to minimize the out-of-balance forces to less than 10^{-8} eV/Å.

3.2.2 GP zone strengthening

Traditional wisdom suggests that dislocations overcome precipitates either by cutting or Orowan looping [48, 49]. Our atomistic simulations of dislocation-GP zone interactions display both of these mechanisms and variations of the two, dependent upon the specific parameters of the interaction. Fig. 3.2 shows two cases of edge dislocation interacting with GP zones of 4.4 nm diameter at 0 K where the slip plane of the dislocation intersects the center-point of the GP zone, i.e., offset = 0. Movies of these two interactions are provided in the online supplementary material. In order to visualize the interaction, we only plot the atoms not in perfect FCC stacking, i.e., having a large "centro-symmetry parameter" [50], using the AtomEye visualization software [51]. For the 60° interaction (Fig. 3.2 (a)) the leading partial dislocation cuts the GP zone while the trailing partial loops around it. The Orowan looping of the trailing partial is the critical event controlling τ_c. For 60° interactions with smaller GP zones (diameter < 3 nm) both partial dislocations overcome the GP zone via precipitate cutting. On the other hand, in case of the 0° interaction (Fig. 3.2 (b)), the edge dislocation overcomes the precipitate by full Orowan looping. This mechanism has been observed across all GP zone sizes that have been investigated for the 0° interaction. As reported previously [36], the offset between the dislocation glide plane and GP zone center can

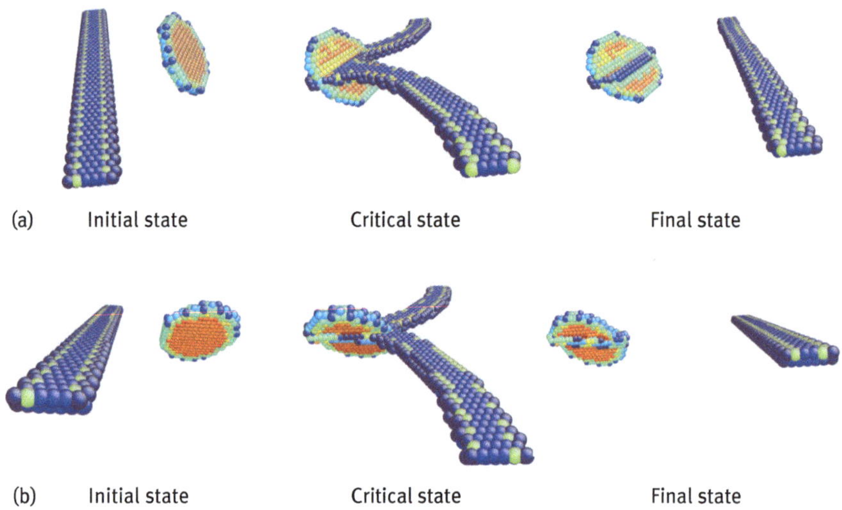

Fig. 3.2: Dominant mechanisms by which an edge dislocation overcomes GP zones: (a) 60° interaction involving leading partial cutting and trailing partial looping, (b) 0° interaction involving full dislocation looping. Movies of the dislocation-GP zone interaction are provided in the supplementary online material.

significantly influence the interaction mechanism, e.g., diffusionless climb and defect nucleation at the dislocation-GP zone contact point have also been observed at high offsets.

To make the atomistic simulation-based precipitation hardening predictions performed here computationally feasible, we have limited the modeling to edge dislocation interaction. The role of dislocation character has been examined in detail in our earlier report [37]. For screw dislocation-GP zone interactions, the 60° orientation yields two prevalent mechanisms depending upon offset. The first involves full Orowan looping, similar to the 0° edge dislocation interaction described above. The second entails the screw dislocation cross-slipping onto a plane that allows it to glide past the GP zone without intersecting it. Cross-slip was observed at both 0 K and finite temperatures, indicating that it can be an important mechanism even in the absence of thermal activation. For the 0° interaction involving a screw dislocation, cross-slip was also observed. However, the plane onto which the dislocation cross-slips is also impeded by the GP zone. In this case, the cross-slip mechanism leads to an enhancement of precipitate strengthening [37].

A key parameter that effects τ_c is the spacing between neighboring precipitates along the dislocation line. The most common approach for incorporating the effect of spacing involves two assumptions: (1) the precipitate spacing, L, is much larger than both the precipitate diameter, D, and the dislocation core width; and (2) the applied force acting to move the dislocation forward at the precipitate results solely from its line tension and configuration. These assumptions provide a one-to-one relationship between the precipitate's ability to inhibit dislocation glide (precipitate strength) and the configuration of the dislocation just before it overcomes the precipitate, described by its critical bowing angle, ϕ_c.

Using a simple line tension model and ignoring dislocation self interactions, the critical stress required for a dislocation to shear a precipitate can be written as

$$\tau_c = \frac{2T}{bL}\cos(\phi_c/2) \tag{3.3}$$

with T being the line tension of the dislocation. Note that this relation suggests τ_c scales as $1/L$. In the limit of an impenetrable precipitate, the dislocation will bow into a configuration with neighboring segments on each side of the precipitate perpendicular to the dislocations original glide direction, i.e., $\phi_c = 0$. In this situation, the dislocation segments will collapse via the Orowan looping mechanism [52], with a critical stress of

$$\tau_c^{\text{Orowan}} = \frac{2T}{bL} \tag{3.4}$$

From continuum elasticity theory, $T \approx Gb^2/2$, where G is the shear modulus of the matrix material. This leads to $\tau_c^{\text{Orowan}} = \frac{Gb}{L}$ and $\tau_c = \frac{Gb}{L}\cos(\phi_c/2) = \tau_c^{\text{Orowan}}\cos(\phi_c/2)$ for Orowan looping and shearing, respectively. A simple modification is often used to extend these relations into a domain where D is not negligible with respect to L [25]

$$\tau_c = \frac{Gb}{L - D_p}\cos(\phi_c/2), \tag{3.5}$$

with $L - D_p$ representing the distance which is available between neighboring precipitate edges for the dislocation to bow-out and D_p being the projected cross-sectional diameter of the precipitate along the dislocation line. Because the above expressions ignore dislocation self-interactions, they overestimate precipitate strengthening. Specifically, the attraction between the two segments of dislocation on each side of the precipitate will promote a more acute bowing angle as well as Orowan looping before $\phi = 0°$. A semi-empirical expression incorporating self-interaction effects has been given by Scattergood and Bacon [53] by fitting to numerical results,

$$\tau_c^* = \frac{Gb}{2\pi AL}\left[\ln\left(D^{-1} + L^{-1}\right)^{-1} + B\right], \tag{3.6}$$

where B is a fitting parameter, and $A = 1$ if the dislocation is initially pure edge and $(1 - \nu)$ if pure screw, where ν is the Poisson's ratio.

Fig. 3.3 reports τ_c values measured from atomistic simulations having various cell dimensions in Z, equivalent to different GP zone spacings, L. The simulation results correspond to the 60° edge dislocation-GP zone interaction with a 4.4 nm diameter GP zone at 0 K. We find that the data for $L \gg D$ can be well described with both the simple $\tau_c \propto 1/L$ relation and the Scattergood & Bacon relation, Eq. (3.6). It is interesting to note that for $L/D < 2$, $1/L$ rule does better than the Scattergood & Bacon relation;

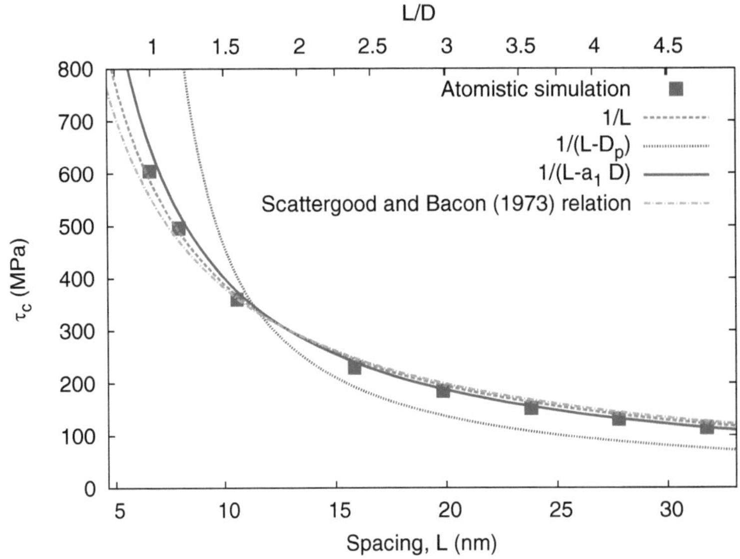

Fig. 3.3: Resolved shear stress required for an edge dislocation to overcome a periodic array of equally spaced 4.4 nm GP zones at 0 K (60° interaction) as a function of GP zone spacing. Solid lines show several popular analytic models which were fit to the atomistic data.

and $\tau_c \propto 1/(L - D_p)$ relation does not seem to work at all (here $D_p = D\cos(30°)$). For the former relation, ϕ_c was considered a fitting parameter with $G = 31.6$ GPa and $b = 2.86$ Å while for the latter B was fit to 2.89. For the $\tau_c \propto 1/L$ relation, ϕ_c was fit as 132°, which compares well to the maximum bowing angle observed in the simulations (Fig. 3.2 (a)). In the spirit of Eq. (3.5), we find that a particularly good fit is given by the relation $\tau_c \propto 1/(L - a_1 D)$ with $a_1 = 0.19$ and $\phi_c = 134°$. Based on these observations, particularly the utility of the simple $\tau_c \propto 1/L$ model down to small precipitate spacing relative to precipitate diameters, $L/D \approx 1$, $\tau_c \propto 1/L$ is used in the subsequent analysis.

τ_c is also controlled by the precipitate size. Fig. 3.4 reports τ_c from simulations involving edge dislocations interacting with GP zones at 0 K with a simulation cell depth (spacing between GP zones centers) of $L = 15.9$ nm. In all cases, the slip plane of the dislocation intersects the center of the GP zone. We note that while the values of τ_c reported in Fig. 3.4 correspond to $L = 15.9$ nm, larger simulation cells were used for the largest GP zone sizes ($D > 6$ nm) to ensure sufficient spacing between periodic GP zone images. In these cases τ_c values were then scaled to $L = 15.9$ nm using $\tau_c \propto 1/L$. For 60° interactions, the simulations suggest that strength increases linearly with GP zone diameter for small to medium sized GP zones, but asymptotically saturates as the diameter becomes large ($D > 8$ nm). Considering that τ_c is controlled by Orowan looping of the trailing partial for the 60° interaction when the GP zone diameter is greater than 3 nm, we attribute the size dependence to the attractive interaction between the neighboring segments of the bowed-out dislocation. Following the $(1/r)$

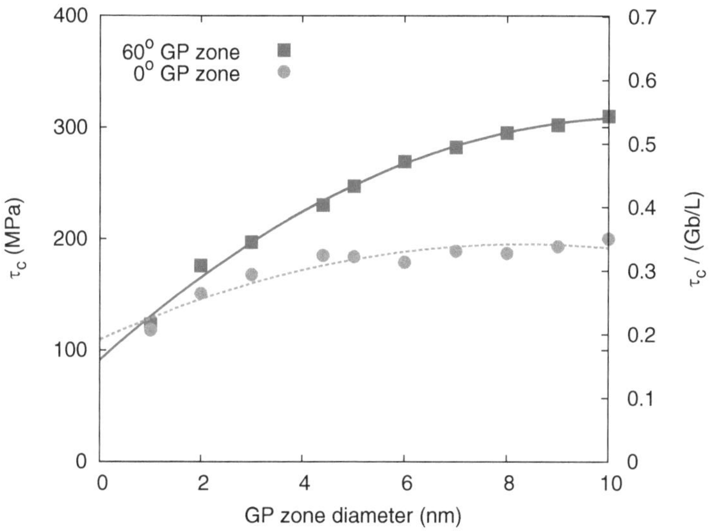

Fig. 3.4: Resolved shear stress required for an edge dislocation to overcome a periodic array of equally spaced GP zones at 0 K as a function of GP zone diameter. The atomistic simulation data, corresponding to $L = 15.9$ nm, is described by phenomenological fits, i.e. $\tau_c = -1.92D^2 + 40.85D + 91.27$ for the 60° interaction and $\tau_c = -1.23D^2 + 20.55D + 109.42$ for the 0° interaction.

stress field associated with a dislocation, the attractive interaction between the neighboring bowed-out dislocation segments is a decaying function of the distance between the segments, which scales with the diameter of the GP zone. We note that traditional chemical hardening arguments for dislocation-precipitate interactions would suggest a linear size dependence for this orientation, inconsistent with the behavior reported here. Also, the chemical hardening models are not consistent with the cutting mechanism that we observe, i.e., a zipping motion of the dislocation across the face of the GP zone, which would not yield a linear size dependence. τ_c for 0° interactions is generally smaller and exhibits a smaller dependence on GP zone diameter. This is consistent with the interaction geometry and mechanism of Orowan looping since the precipitate diameter does not influence the dislocation in its bowed-out state. The maximum values of τ_c from our simulations are $0.35\tau_c^{\mathrm{Orowan}}$ for the 0° interaction and $0.55\tau_c^{\mathrm{Orowan}}$ for the 60° interaction, where $\tau_c^{\mathrm{Orowan}} = 570$ MPa for this box size.

The location at which a dislocation intersects the GP zone relative to its center (offset) has been observed to substantially influence τ_c for a particular interaction [36]. Building from our previous work and ignoring the special cases of defect generation and diffusionless climb, we include the effect of offset into our dislocation line tension model as $\tau_c(O) = \tau_{c0} + O\delta\tau_c$. τ_{c0} represents τ_c when the dislocation intersects the GP zone through its center. O represents the offset, i.e., the location of the active slip plane relative to the center of the GP zone, normalized with respect to the $\{111\}$ interplanar spacing distance. Based on our previous work [36], the offset effect is taken as

$\delta\tau_c = 0.012\tau_c^{\text{Orowan}}$ and $0.015\tau_c^{\text{Orowan}}$ for 60° and 0° dislocation-GP zone interactions, respectively. Finally, we point out that successive dislocation-GP zone interactions are ignored in this work. We direct the interested reader to [36] where the complexity of successive interactions is documented.

3.2.3 θ'' strengthening

As mentioned previously, θ'' precipitates are modeled as two layers of Cu atoms separated by three atomic layers of Al. Fig. 3.5 depicts the mechanism by which they are overcome by edge dislocations at 0 K for all diameters studied here, 8–30 nm. For the 60° interaction, the Cu atoms are displaced by a partial burgers vector after the dislocation overcomes the precipitate. In this regard the interaction is similar to that observed for the 60° edge dislocation interactions with large GP zones (Fig. 3.2 (a)). However, in the case of the GP zone, the leading partial clearly cuts the GP zone before the critical state whereas the leading partial does not cut the θ'' precipitate before the critical state. Thus, the 60° θ'' mechanism could entail full Orowan looping at the critical state with the leading partial component of the full dislocation loop collapsing through the θ'' precipitate after it is formed. Or alternatively, leading partial cutting may occur. The critical state of the 60° θ'' dislocation interaction involves a completely bowed-out dislocation where the screw segments of the bowed dislocation have cross-slipped. The cross-slip of the screw segments of a bowed edge dislocation at a precipitate was first envisioned by Hirsch [54] and has been more recently observed in atomistic simulations of impenetrable precipitates in Cu [55]. Here, the cross-slipped segments of the dislocation are observed to cross-slip back to their original plane as the dislocation overcomes the precipitate, making the later stages of the mechanism observed here distinctly different than the Hirsch mechanism [54, 56]. It is also worth noting that the cross-slip observed in our simulations takes place at 0 K, in the absence of thermal activation. The 0° interaction exhibits full Orowan looping for all sizes studied (8–16 nm), Fig. 3.5 (b), analogous to the 0° edge dislocation-GP zone interaction. Interestingly, these results viewed as a whole suggest that for plate shaped precipitates the competition between Orowan looping and cutting mechanisms may be more influenced by orientation than the precipitate size, at least in the athermal limit.

τ_c as a function of θ'' precipitate diameter is shown in Fig. 3.6 for both the 60° and 0° interaction orientations. θ'' diameters spanning 8–30 nm for the 60° interaction and 8–16 nm for the 0° interaction were investigated. The minimum simulation cell size for these studies was approximately $34 \times 42 \times 32$ nm^3 and it was increased for larger precipitates. In Fig. 3.6, the simulation results are normalized to $L = 31.8$ using $\tau_c \propto 1/L$. In all cases the θ'' precipitate provides significantly greater strengthening than GP zones, with maximum values of $\tau_c/\tau_c^{\text{Orowan}} = 1.754$ and $\tau_c/\tau_c^{\text{Orowan}} = 0.495$ for the 60° and 0° orientations respectively. The most notable feature of the results is

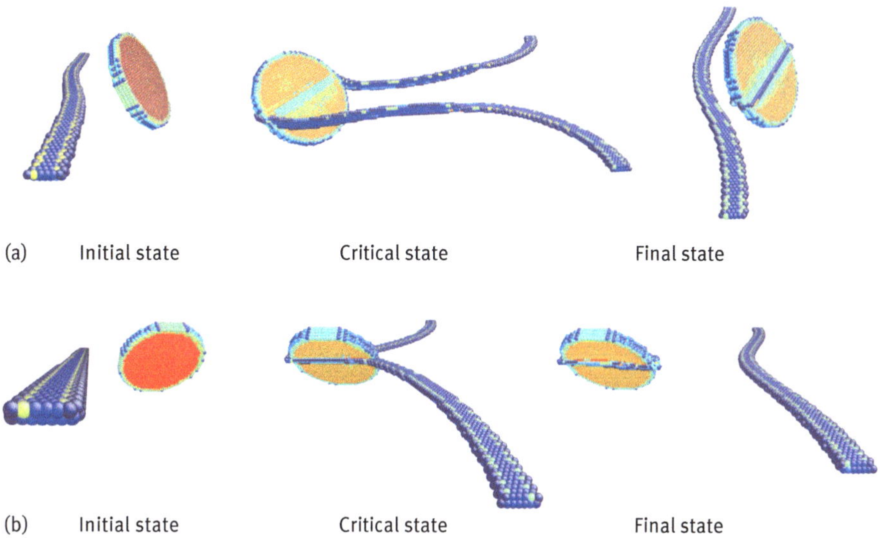

Fig. 3.5: Dominant mechanisms by which an edge dislocation overcomes θ''-precipitates: (a) 60° interaction involving leading partial cutting and trailing partial looping, note the cross-slipped segments of the dislocation in the critical state, (b) 0° interaction involving full dislocation looping.

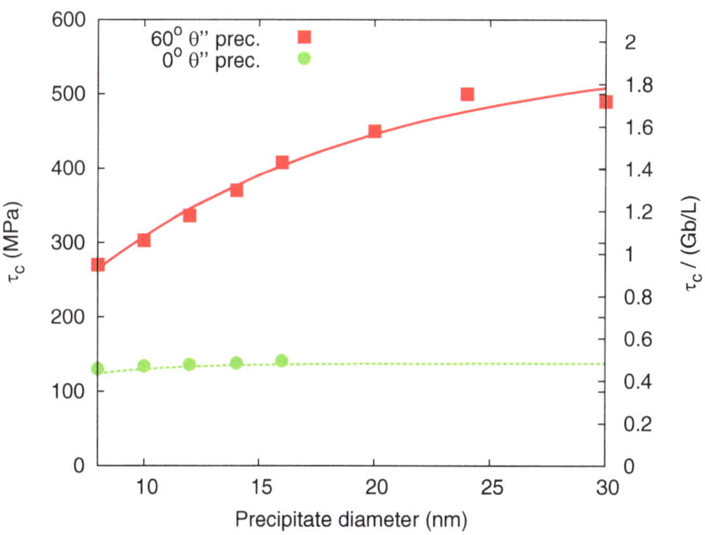

Fig. 3.6: Resolved shear stress required for an edge dislocation to overcome a periodic array of equally spaced θ''-precipitates at 0 K as a function of their diameter. The atomistic simulation data, corresponding to $L = 31.8$ nm, is described by phenomenological fits, i.e. $\tau_c = 558.36(1 - \exp(-0.08D))$ for the 60° interaction and $\tau_c = 137.40(1 - \exp(-0.29D))$ for the 0° interaction.

the super-Orowan strengthening associated with the 60° interaction. This can be attributed to the controlling mechanism of this interaction, i.e. the cross-slip of the screw dislocation segments back onto their original slip plane. While the size of simulation cell prohibited an extensive finite temperature investigation of this occurrence, experience with cross-slip strengthening of screw dislocation-GP zone interactions [37] suggests that thermal activation can significantly reduce the stress at which the second controlling cross-slip event occurs. Thus, the high values of τ_c of the 60° θ'' interaction cited here may not be representative of behavior at typical experimental time scales and temperatures.

Orientation has a significant effect on τ_c for the dislocation-θ'' precipitate interactions. τ_c for the 0° interactions is relatively insensitive to precipitate size while τ_c for the 60° precipitate interactions approximately doubles as diameter increases from 8 to 30 nm. While the size independence of the 0° interaction can easily be attributed to the controlling mechanism, the behavior of the 60° interaction is not as easily understood. For θ'' precipitates, the effect of offset on τ_c is unknown. Thus, we assume the effect is similar to what we have observed for the GP zone interactions and incorporate its effect in an identical manner.

3.3 Atomistic modeling of solute hardening

To model solute hardening in Al-Cu we have utilized a semi-analytic method recently developed by Leyson et al. [57]. The method stems from the framework developed by Labusch [58, 59] for understanding solute hardening from the perspective of dislocation glide being inhibited by the heterogeneous energy landscape that a dislocation encounters in a field of randomly positioned solute atoms. Relying on input from atomistic simulation, both elastic and chemical contributions to solute hardening are included with no adjustable parameters. Accordingly, the shear stress needed to move a typical segment of dislocation forward through an array of randomly distributed solute atoms, in the athermal limit, is given by

$$\tau_{ys} = \frac{\pi}{2} \frac{\Delta E_b}{b \xi_c w_c} \tag{3.7}$$

where ΔE_b represents the energy barrier to the dislocation glide posed by solute atoms, and ξ_c and w_c signify the dislocation segment length, and the incremental glide distance of a dislocation segment in a field of solute atoms, respectively. w_c is taken to be the splitting distance of the dislocation core, which was found in Leyson et al. [57] to correspond to the most energetically favorable choice. ΔE_b is in turn is a function of the interaction energy between a dislocation and Cu solute atoms, ΔE_p. Noting that both ξ_c and ΔE_b are functions of c_s, Eq. (3.7) can be rewritten [57] to show that $\tau_{ys} \propto c_s^{2/3}$, consistent with the experimental data.

As with the precipitation hardening simulations, the newly developed Al-Cu angular dependent EAM interatomic potential [44] was used to compute interaction energy between dislocation and Cu solute at different lattice sites. A simulation cell of $11 \times 11 \times 10\,\mathrm{nm}^3$ with 69,320 atoms and an edge dislocation at its center was utilized. Following [60], the atoms in several X–Y planes around $Z = 0$ were held fixed to prevent the dislocation from freely moving towards (or away from) the Cu atom. The interaction energy, ΔE_p, as a function of Cu solute position is shown in Fig. 3.7. In the figure, unfilled circles represent the lattice sites at which the interaction energies were calculated. Away from the dislocation core, the interaction energies correspond well with the continuum linear elasticity solution, which is based on the interaction energy of the hydrostatic stress fields. Near the dislocation core, where the interaction energies are the strongest, they were found to range between −0.15 and 0.25 eV. Using Eq. (3.7), $\tau_{\mathrm{ys}}/c_\mathrm{s}^{2/3}$ is found to be 1138 MPa. A comparison with DFT reported value and experimental data is coherent with traditional understanding that the solute hardening is significantly dependent on the solute misfit volume and the heat of solution.

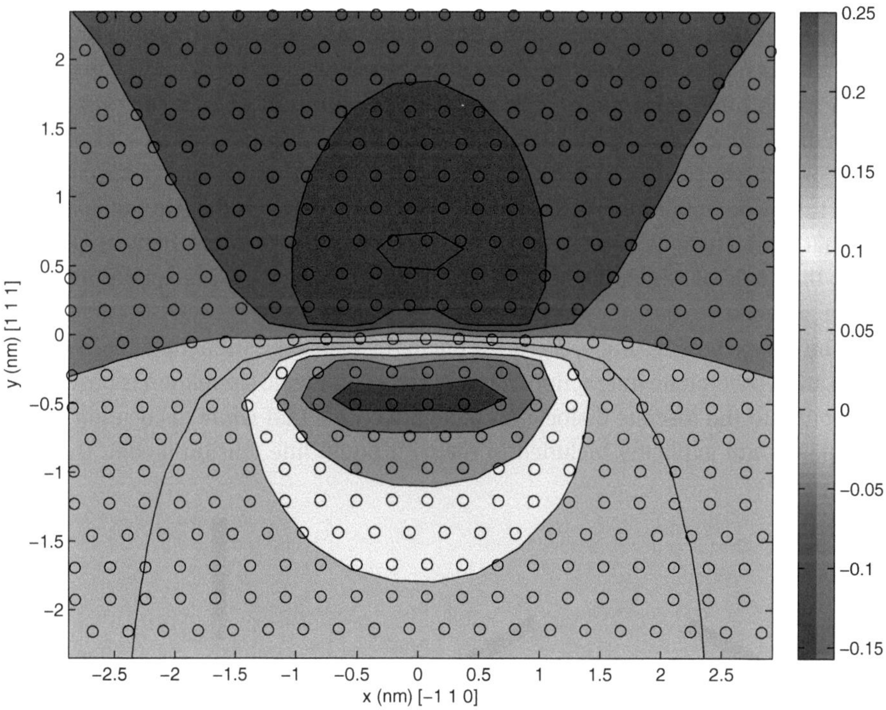

Fig. 3.7: The interaction energy between dilute Cu solute atoms and an edge dislocation lying on the *xy*-plane and centered about (0, 0). Contour colors are in eV and were calculated via atomistic simulation at the lattice sites (unfilled circles).

3.4 Dislocation dynamics model for macroscopic precipitate strength predictions

In order to estimate the CRSS required for a dislocation to propagate through a field of precipitates with different types, sizes, orientations, offsets, and random positions, τ_{yp}, we use a discrete dislocation model governed by continuum line tension mechanics [61]. The model takes as input, τ_{c_i}'s from the atomistic simulations, where i denotes a different precipitate distributed in the alloy microstructure. The developed is then used to generate a set of τ_{yp} values corresponding to different time points during age hardening. To generate continuous aging curves, the values generated at discrete time points from the discrete dislocation model are interpolated using a standard analytic model with a single fitting parameter.

The continuum line tension–discrete dislocation modeling performed here is based on the algorithm originally devised by [62] and most recently employed for point obstacles by [63–65]. Briefly, the algorithm consists of representing a dislocation line in a field of precipitates by a series of circular segments connecting the precipitates that the dislocation is in contact with. The segments have a circular radius of curvature, $R = \frac{Gb}{2\tau_{\text{app}}}$, that is a function of the applied resolved shear stress, τ_{app} (Fig. 3.8). As τ_{app} is increased the dislocation segments bow further and can come in contact with other precipitates, creating two segments out of one. The propagation of the dislocation line is controlled by its detachment from precipitates. This occurs when one of two conditions are met: (1) a precipitate is connected to a segment that has a radius of curvature less than half the distance between the two precipitates to which it is attached, i.e., a segment of the dislocation becomes mechanical unstable, or (2) the angle at which dislocation segment(s) connect to a precipitate reaches a critical value.

The critical bowing angle, ϕ_{c_i}, is the means by which the strength, τ_{c_i}, of a specific dislocation-precipitate interaction, i, measured in atomistic simulation is provided as an input into the discrete dislocation model. We note that while dislocation self interaction is not explicitly included in the continuum line tension model, the effect

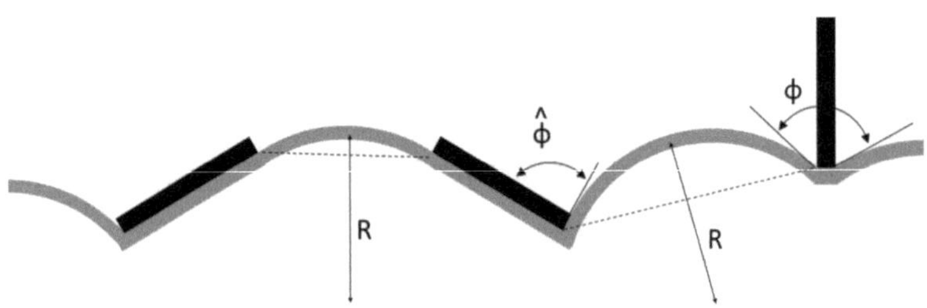

Fig. 3.8: Schematic illustration of key components of dislocation line tension modeling. Here a dislocation is pinned by 3 plate precipitates. Between pinning points, it bows out by a radius *R*. The two measures of angle that dictate precipitate strength in the model are shown.

is implicitly included in the value of τ_{c_i} extracted from the atomistic modeling. For the model to meaningfully simulate plate precipitates, two definitions of angle and critical angle are required (Fig. 3.8). The first ($\hat{\phi}$) corresponds to the case when the dislocation is only in contact with one edge of the plate, which is equivalent to the point obstacle case. The second (ϕ) corresponds to the case when the dislocation is interacting with both edges of the precipitate. In both cases, once the bowing angle has reached the critical angle at an edge of the precipitate, the precipitate is considered inactive. The critical bowing angle for a specific interaction is related to τ_{c_i} via $\phi_{c_i} = 2\cos^{-1}(\tau_{c_i}\hat{L}_i/Gb)$ and $\hat{\phi}_{c_i} = \cos^{-1}(\tau_{c_i}\hat{L}_i/Gb)$ for the two cases, respectively. $\hat{L}_i$ represents the precipitate spacing in the atomistic simulation cell in which τ_{c_i} is measured. An initial validation of the implementation was performed on a system composed of periodic arrays of precipitates consisting of identical interactions, i.e. $\tau_{yp} = \tau_{c_i}$.

The super-Orowan strengths associated with 60° dislocation–θ'' precipitate interactions pose a specific challenge as they correspond to $\tau_{c_i}\hat{L}_i/Gb > 1$ leading to imaginary critical angles. To accomodate this, we use an artificially high value of $Gb = 22.6$ N/m in the discrete dislocation simulations. This allows a meaningful critical bowing angle to be defined for the 60° dislocation–θ'' precipitate interactions, while only mildly influencing the simulation results (< 5 % error).

Consistent with previous discrete dislocation studies of dislocation-point obstacle interactions [63], τ_{yp} is found to depend on the simulation cell size. Considering that the propagation of the dislocation line is controlled by the weakest pinning point along its length, τ_{yp} decreases with dislocation length and increases with glide distance (noting that τ_{yp} is the CRSS at which the dislocation traverses the entire simulation cell). In the simulations performed here, the cell sizes are chosen such that the measured value of τ_{yp} is within 5% of the value that would be obtained if the simulation cell dimensions matched the grain size, based on our own experience and Nogaret and Rodney [63].

The simulation cell of the discrete dislocation model was populated with precipitates, both GP zones and θ'' precipitates, following the sizes and densities suggested by the precipitate growth kinetics model that will be described in the next section. The orientation and offset of the precipitates were distributed uniformly in the simulation cell. The precipitates were positioned randomly under the constraint that they neither touch each other nor have centers within $0.3(w_i + w_j)$, where w_i and w_j are the projected diameters of precipitates i and j, respectively. This approach is qualitatively consistent with the physics of the precipitation process [66] and simplifies the dislocation dynamics algorithm.

As mentioned at the beginning of this section, the discrete dislocation model is employed to evaluate τ_{yp} at a set of time points during the aging process corresponding to a single set of aging parameters, i.e., Cu concentration and aging temperature. The points are then fit to an aging curve following a standard analytic expression with a single fitting parameter. This approach not only serves to limit the computational expense of the discrete dislocation model (the algorithm has not been optimized), but more importantly, it provides a simple and effective means to estimate age hardening from the strength contributions of individual components. Specifically, at a given aging time we consider the total resolved shear strength of a field of precipitates τ_{yp} to be

$$\tau_{\mathrm{yp}} = \eta \left[\sum_{i=1}^{N} \left(\tau_{\mathrm{c}_i} \frac{\hat{L}_i}{L_i} \right)^{\alpha} \right]^{1/\alpha} \tag{3.8}$$

where the index i cycles through all precipitate types and orientations present in the material. $\hat{L}_i$ represents the precipitate spacing used in the atomistic simulation to calculate τ_{c_i}, while L_i represents the spacing between precipitates with characteristics i in the microstructure. η is a fitting parameter that accounts for the collective effect of offset, random precipitate positions, and the interactions between the various precipitates that are not included in the summation. Due to the dense microstructure of precipitates studied here, we choose $\alpha = 2$ consistent with the recent work of [65].

Once τ_{yp} is obtained for a given distribution of precipitates (whether using the analytic or the discrete dislocation model), the combined strengthening due to both precipitates and solutes can then be written as

$$\tau_{\mathrm{y}}^{\alpha} = \tau_{\mathrm{ys}}^{\alpha} + \tau_{\mathrm{yp}}^{\alpha} \tag{3.9}$$

Consistent with the Dong et al. (2010) analysis and the relatively high density of precipitates studied here, we again use $\alpha = 2$ to combine solute strengthening with precipitate strengthening.

The total strength of an Al-Cu alloy results from dislocation motion not only being inhibited by Cu solute atoms and precipitates, but also grain boundaries, other dislocations, and other crystallographic and chemical obstacles. The strengthening contribution from these miscellaneous obstacles, $\sigma_{\mathrm{y0}} = 20\,\mathrm{MPa}$, is considered to be constant in this work, i.e., independent of aging time, temperature, and Cu concentration. Thus, we consider the uniaxial yield strength of a polycrystalline Al-Cu alloy as

$$\sigma_{\mathrm{y}} = \sigma_{\mathrm{y0}} + T\,\tau_{\mathrm{y}}\,, \tag{3.10}$$

where T is the Taylor factor which for FCC materials is $T = 3.06$ [67–69]. It is notable that this value of Taylor factor is usually considered as the upper limit and may yield somewhat higher hardness predictions.

In the most heavily referenced experimental studies on Al-Cu age hardening [47], which we will compare to in this work, strengthening is expressed in terms of Viker's hardness, which is related to uniaxial yield strength by

$$H_{\mathrm{v}} = c_{\mathrm{vp}}\sigma_{\mathrm{y}} \tag{3.11}$$

where c_{vp} is a constant determined by geometrical factors during experimentation and usually ranges between 2 and 4. Here we use $c_{\mathrm{vp}} = 3.0$.

3.5 Modeling of precipitate kinetics

The kinetics of the precipitation process in Al-Cu alloys is complicated by the existence of several precipitate phases, i.e., GP zones, θ'', θ', and θ precipitates, as outlined in the introduction. Following classical theory of phase transformation kinetics [70, 71], the precipitate concentration, $c_p(t)$, at time t relative to the equilibrium concentration of precipitates, c_{pe}, is taken to follow the Johnson, Mehl, and Avrami (JMA) equation,

$$f_p(t) = \frac{c_p(t)}{c_{pe}} = 1 - \exp\left(-k_1 t^n\right), \tag{3.12}$$

where k_1 is the rate constant at the prescribed aging temperature, and n is the transformation exponent. The equilibrium precipitate concentration is expressed as $c_{pe} = c - c_{se}$, where c is the total Cu concentration in the alloy and c_{se} is the equilibrium solute concentration at a given temperature. c_{se} is obtained by linearly interpolating the solute equilibrium concentrations of 0.1 wt.% Cu at 100 °C and 0.2 wt.% Cu at 200 °C, as reported in Hornbogen [72] and Massalski [73]. Recent first-principles investigations [74] support the applicability of Eq. (3.12) for Al-Cu alloys with $n = 1.5$, which is followed here. Ignoring the over-aged precipitates, the JMA equation is also used to describe the concentration of θ'' precipitates, $c_{\theta''}$, relative to the total precipitate concentration

$$f_{\theta''}(t) = \frac{c_{\theta''}(t)}{c_p(t)} = 1 - \exp\left(-k_2 t^n\right) \tag{3.13}$$

with k_2 being a second rate constant related to the rate at which Cu atoms belonging to GP zones become associated with θ'' precipitates. When $k_1 \gg k_2$, k_1 controls the rate of formation of GP zones and k_2 controls the rate of formation of θ'' precipitates. When $k_2 \gg k_1$, GP zones do not form and k_1 controls the rate of formation of θ'' precipitates. The model's ability to capture these key phenomena make it sufficient for the objectives of this paper. The total number of GP zones is assumed to be fixed, allowing the GP zone diameter at time t to be expressed as

$$D_{GP}(t) = (D_{GP})_{max} f_{GP}{}^{1/2}(t) \tag{3.14}$$

where $f_{GP}(t) = 1 - f_{\theta''}(t)$; and $(D_{GP})_{max}$ corresponds to the maximum GP zone diameter which we take as 10 nm [16]. Following experimental observation [16], θ'' precipitates are assumed to have a constant diameter for $f_p < 0.8$. When $f_p \geq 0.8$, Ostwald ripening of θ'' precipitates is included with their diameter evolving according to the Wagner rule [75]

$$D_{\theta''}^2 - (D_{\theta''})_{min}^2 = k_3(t - t_0), \tag{3.15}$$

where $(D_{\theta''})_{min} = 8$ nm and t_0 corresponds to the time at which $f_p = 0.8$. k_3 represents the rate of the ripening process and is considered the third fitting parameter of the kinetics model. With k_3 being a fitting parameter, the exact choice of t_0 is not influential so long as it corresponds to a high f_p value. For plate precipitates with uniformly distributed offsets, the average distance, L_i, between precipitate centers of a particular type and orientation, i, is [76]

$$L_i = 0.931(0.306\pi D_i h_i / f_i)^{1/2} \tag{3.16}$$

where h_i, D_i, and f_i represent the thickness, diameter, and relative concentration of the precipitates with characteristics, i, respectively.

3.6 Age hardening predictions of Al-4 wt.% Cu aged at 110 °C

The rate constants k_1, k_2, and k_3 are chosen so that the distinguishing features of the age hardening curve for Al-4 wt.% Cu aged at $T_{age} = 110\,°C$ match experiments. Specifically, k_1 is chosen so that the GP zone concentration starts saturating at t = 3.5 days; k_2, which controls the transformation from GP zones to θ'' precipitates, is chosen so that the strength plateau ends at t = 20 days; and k_3, which controls the rate of coarsening of θ'' precipitates, is chosen so that the peak hardness occurs at t = 230 days. Accordingly, $k_1 = 2.0{\cdot}10^{-8}\,s^{-1}$, $k_2 = 1.9{\cdot}10^{-11}\,s^{-1}$ and $k_3 = 2.4{\cdot}10^{-3}\,Å^2\,s^{-1}$ for Al-4 wt.% Cu aged at $T_{age} = 110\,°C$. The evolution of the concentrations of solute, GP zones, and θ'' precipitates are given in Fig. 3.9 (a); and the evolution of the precipitate diameters and effective spacings are giving in Fig. 3.9 (b) and (c), respectively. The relative concentrations of each constituent correspond well with the diffraction observations [47] and are in accordance with traditional understanding. With the kinetics rate constants known, L_i can be computed for the various precipitate types for use in the analytic model and the microstructure can be constructed for the discrete dislocation model as described in Sec. 3.4. Two images of representative microstructures used in the line tension simulations are shown in Fig. 3.10. The offsets of the precipitates in the slip plane were uniformly distributed.

Ten discrete dislocation line tension simulations were performed at various aging times for microstructures corresponding to Al-4 wt.% Cu aged at $T_{age} = 110\,°C$. The minimum size of the dislocation dynamics model was taken as $0.6 \times 0.6\,\mu m^2$. The solute hardening contribution is added to the discrete dislocation results, Eq. (3.8); and the resultant τ_y's are shown in Fig. 3.11. At aging times of 400 days and 1000 days, two realizations were performed to highlight the variation due to the randomness in precipitate microstructure. Fitting the free parameter in the analytic model to the data points generated with the discrete dislocation model, i.e. setting $\eta = 0.64$ in Eq. (3.8), reveals the shape of the well known Al-Cu aging curve demonstrating both a plateau and peak in strength with aging. For comparison the analytic model predictions with

▸ **Fig.3.9:** Precipitation kinetics of age hardening for Al-4 wt.% Cu at $T_{age} = 110\,°C$: (a) evolution of precipitate and solute concentrations, (b) evolution of average precipitate diameter, and (c) evolution of average center-to-center precipitate spacing.

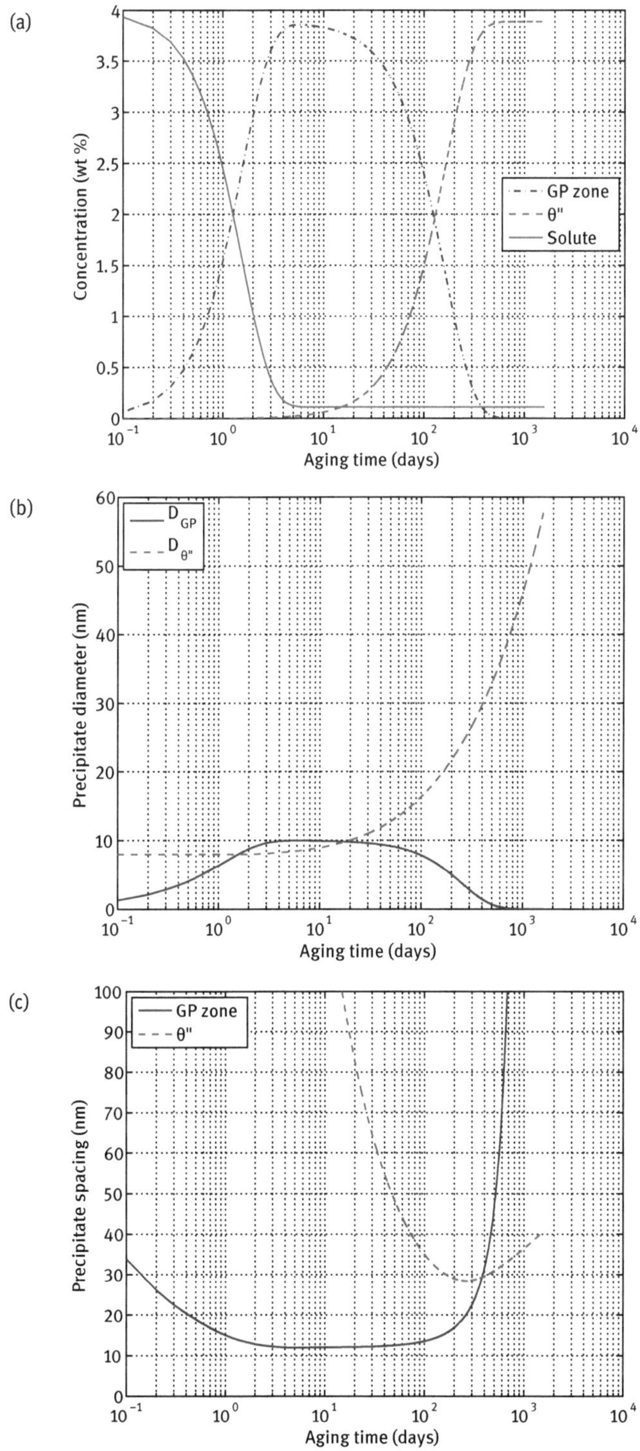
(a)
Concentration (wt %)
Aging time (days)
GP zone
θ"
Solute
(b)
Precipitate diameter (nm)
Aging time (days)
D GP
D θ"
(c)
Precipitate spacing (nm)
Aging time (days)
GP zone
θ"

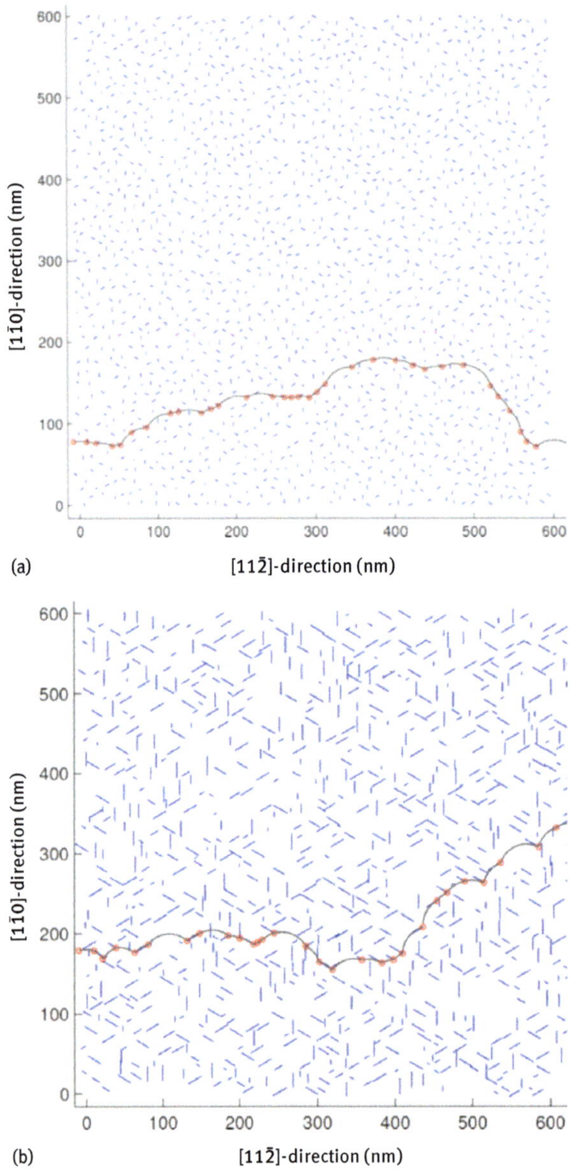

Fig. 3.10: Snapshots of dislocation line tension model for Al-4 wt.% Cu alloy aged at T_{age} = 110 °C: (a) underhaged, 1 day, (b) at optimum hardness, 230 days. The blue line segments represent sections of precipitates which intersect the slip plane. The black line represents the dislocation and the red circles represent its contact points with the precipitates.

$\eta = 0.60$ and $\eta = 0.70$ are also plotted. It is interesting to point out that a constant value of η provides a relatively accurate approximation across the entire range of microstructures that exist during aging. This observation provides justification for our use of the same value of η at different aging temperatures and Cu concentrations.

In addition to enabling quick predictions of age hardening by avoiding discrete dislocation simulations, the analytic model allows the contribution of individual constituents to be assessed. Fig. 3.11 shows that the initial increase in strength with aging is associated with an increase in GP zone strengthening consistent with experiment [47]. The increase in strength as Cu atoms transfer from solute to GP zones is a result of $\frac{\partial \tau_{y\mathrm{GP}}}{\partial c_{\mathrm{GP}}} > \frac{\partial \tau_{\mathrm{ys}}}{\partial c_{\mathrm{s}}}$, where c_{GP} represents the concentration of Cu belonging to GP zones and $\tau_{y\mathrm{GP}}$ represents the total strengthening due to GP zones. The strength of the Al-Cu alloy will increase as Cu atoms transfer from solute to GP zones if $\frac{\partial \tau_{y\mathrm{GP}}}{\partial c_{\mathrm{GP}}} > \frac{\partial \tau_{\mathrm{ys}}}{\partial c_{\mathrm{s}}}$ with $c = c_{\mathrm{GP}} + c_{\mathrm{S}}$ where c_{GP} represents the concentration of Cu belonging to GP zones and $\tau_{y\mathrm{GP}}$ represents the strengthening due to GP zones. While $\frac{\partial \tau_{\mathrm{ys}}}{\partial c_{\mathrm{s}}}$ can be explicitly ex-

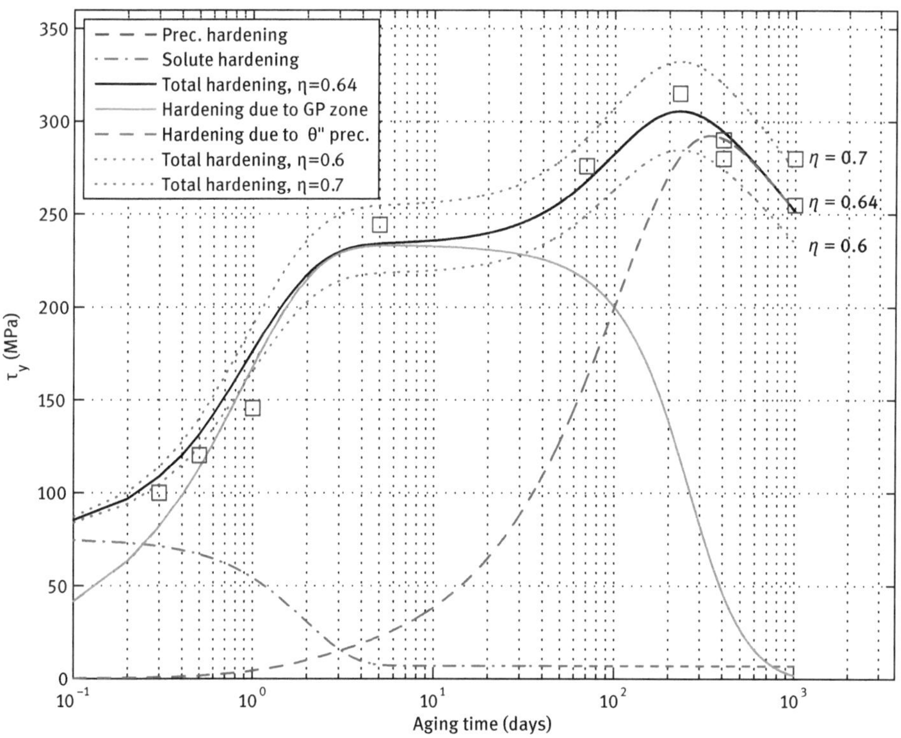

Fig. 3.11: 0 K resolved shear strength vs. aging time predictions for Al-4wt.%Cu aged at $T_{\mathrm{age}} = 110\,°\mathrm{C}$. The data points were obtained using the dislocation line tension simulations and the curves represent the fitted simplified continuum model described in the text. The solute and precipitate hardening curves correspond to $\eta = 0.64$ in the simplified continuum model.

pressed in terms of c_s, i.e. $\tau_{ys} \propto c_s^{2/3}$ (see Sec. 3.3 and [57]), $\frac{\partial \tau_{yGP}}{\partial c_{GP}}$ must be approximated due to its non-analytic dependence on GP zone diameter. For the Al-4 wt.% Cu alloy aged at 110 °C, $\tau_{yGP}(c_{GP})$ can be described reasonably well with a simple power law

$$\tau_{yGP} = \tau_{yGP}^0 c_{GP}^{3/7} \tag{3.17}$$

with the fitting constant $\tau_{yGP}^0 = 1400$ MPa. Considering this relation and $\tau_{ys}/c_s^{2/3} = 1138$ MPa (from Sec. 3.3), the inequality, $\frac{\partial \tau_{yGP}}{\partial c_{GP}} > \frac{\partial \tau_{ys}}{\partial c_s}$, implies that strength will increase as Cu atoms transfer from solute to GP zones if $c_{GP} < 0.663 c_s^{7/12}$.

The second increase in strength, beginning at ~30 days of aging time, is due to both an increase in the concentration and size of θ''-precipitates, consistent with experiment [47]. In general, the GP zone to θ''-precipitate transformation would be expected to bring a strength increase when $\tau_{c_{\theta''_i}} > \tau_{c_{GP_i}} \sqrt{2D_{\theta''}/D_{GP}}$, considering Eqs. (3.3) and (3.16), where $\tau_{c_{GP_i}}$ and $\tau_{c_{\theta''_i}}$ represent values of τ_{c_i} from the atomistic simulations detailed in Sec. 3.2 with a consistent simulation cell size. While the diameters of the two precipitate types evolve during the transformation, at its peak, ~9 nm GP zones transform to ~15 nm θ''-precipitates. This sufficiently satisfies the inequality given above.

The increasing size of the θ''-precipitates is governed by the coarsening law, Eq. (3.14). As the precipitates coarsen, they individually provide greater resistance to dislocation glide (Figs. 3.4 and 3.6); however, the individual strengthening is counteracted by an increase in precipitate spacing (Eq. (3.16)) which reduces their collective effect (Fig. 3.3). In general for plate precipitates, coarsening leads to strengthening when $\frac{\partial \ln \tau_c}{\partial \ln D} > \frac{1}{2}$. Here, the results of Fig. 3.6 combined with Eq. (3.8) predict that coarsening will strengthen when $D_{\theta''_i} < 15$ nm. Thus, the strengthening observed in Fig. 3.11 after 100 days of aging can be solely attributed to the transformation of Cu from GP zones to θ''-precipitates, as coarsening is detrimental to strength when $D_{\theta''_i} > 15$ nm.

After the majority of precipitates have transformed into θ''-precipitates at ~300 days, the strength of Al-4 wt.% Cu decreases with time as coarsening is the only active precipitate evolution mechanism. At ~1000 days of aging at 110 °C, θ''-precipitates would grow to the size where θ'-precipitate formation would be expected. Considering the dislocation interaction mechanisms observed for θ''-precipitates, it is reasonable to assume that θ'-precipitates would have similar strengths. Thus, we hypothesize that the formation of θ'-precipitates (not considered in the model) in the later stages of aging would expedite the rate of softening relative to the current model.

3.7 Effect of Cu concentration and aging temperature

The aging temperature and the Cu concentration are two key parameters that are often adjusted to tailor alloy performance. Both affect strength via precipitation kinetics and equilibrium concentrations. In this section, these effects are examined within the modeling framework developed in the previous sections. First the effects of Cu concentration are discussed, then aging temperature. The rate, k_1, at which Cu atoms transform from solute to GP zones in a solutionized and then quenched Al-Cu alloy is a complex function of the Cu concentration. k_1 depends on the homogeneous nucleation rate of GP zones and the GP zone growth rate. The factors affecting k_1 are discussed in [41, 71, 77] with a first order estimate derived as

$$k_1 = k_{1_0}(c - c_{se})^{2.5} \exp\left(-\frac{E_{eff}}{k_B T_{age}}\right) \tag{3.18}$$

where k_{1_0} is a constant, c is the initial concentration of Cu in solute in the quenched alloy, c_{se} is the equilibrium solute concentration, k_B is the Boltzmann constant, and E_{eff} is an effective activation energy barrier for the process, which is a function of the activation energy barrier for diffusion and the activation energy barrier for GP zone nucleation. Here, we assume that k_2 takes a similar form to k_1. The rate constant for coarsening, k_3, is expected to take a different form. Following Wagner [75] (see page 21 of [78]),

$$k_3 = \frac{64 K V_a}{81 k_B T_{age}} c_{se} , \tag{3.19}$$

where V_a is the atomic volume of Cu. The constant K includes the diffusion coefficient of Cu in bulk Al and the precipitate-matrix interface energy density. It can be written in the form

$$K = K_0 \exp\left(-\frac{E^w_{eff}}{k_B T_{age}}\right)$$

with K_0 being a constant, and E^w_{eff} being the corresponding effective activation energy barrier.

Fig. 3.12 depicts the age hardening curves from the model with $\eta = 0.64$. The variation of the curves with Cu content (2.5–4.5 wt.%) displays the same qualitative features as observed in experiment [47]. The aging time to the first strength plateau and the time to the peak hardness increase with decreasing Cu content. The strength level along the entire aging curve, and the values associated with the plateau and the peak decrease with decreasing Cu content. Quantitatively, the change in strength with concentration is reasonably consistent with experiment. However, at short aging times (~ a few hours) the ratio between the 2.5 and 4.5 wt.% strengths is considerably different than experiment. While this difference could be due to not considering thermally activated plasticity in the model, it might also be indicative of a varying value of σ_{y0} with solute concentration and inaccuracies in the solute hardening model.

A more pronounced disagreement between the model predictions shown in Fig. 3.12 and the experimental data [47] is associated with the width of the strength plateau, i.e. the duration of aging time for which the strength is approximately constant. The experimental data displays a decreasing strength plateau width with decreasing Cu concentration on the standard semi-log aging plot. The decrease in the strength plateau width is driven by slower GP zone precipitation at decreased

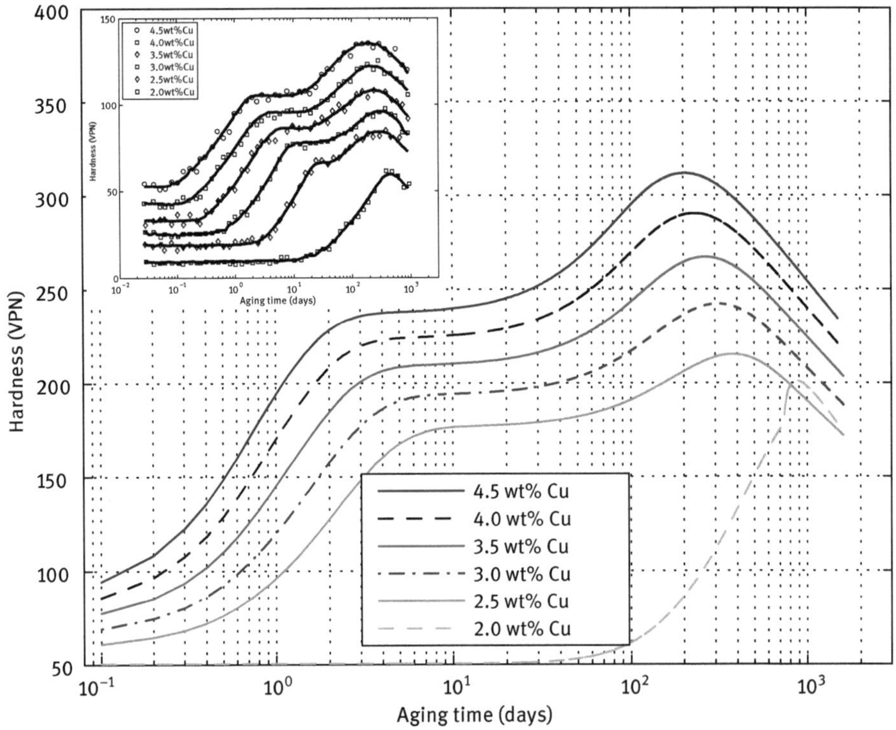

Fig. 3.12: 0 K hardness vs. aging time predictions for Al-Cu aged at T_{age} = 110 °C with various Cu concentrations. Inset shows experimental data at 300 K from [47].

Cu concentrations (noting that the aging time to the end of the strength plateau is governed by θ''-precipitate formation which is more mildly affected by Cu concentration both in our model, Fig. 3.12, and experiments [47]). At concentrations below 2.5 wt.% Cu, a strength plateau no longer exists in the experiments. This occurrence is in accordance with the GP zone solvus concentration at 110 °C [73]. The linear scaling of the GP zone formation rate kinetics model used here does not predict the rate going to zero as the solvus concentration is approached. This represents a shortcoming of the current kinetics modeling formulation. In future work it would make sense to formulate the kinetics in a manner that is consistent with the thermodynamic stability of the precipitate phases. One way to avoid this heuristic approach would be to perform a first principles based prediction of the formation of the precipitate phases at given temperature and solute concentration, similar to the study reported by Wang et al. [74], which is out of scope of the present work. With the current model, the creation of a qualitatively accurate aging curve at concentrations below 2.5 wt.% Cu can only be created by explicitly prohibiting GP zone formation, as was done in Fig. 3.12.

The effect of aging temperature on the age hardening curve has also been examined by modeling the aging process at 130 and 165 °C (Fig. 3.13). The temperature scaling of the kinetics rate constants associated with GP zone nucleation and growth, θ''-precipitate nucleation and growth, and θ'' coarsening was obtained from Eqs. (3.18) and (3.19) with estimated values of the effective activation energies. As a whole, the aging process in Al-Cu alloys is typically associated with an activation energy ranging from 0.78 to 1.35 eV via differential scanning calorimetry (DSC) experiments [41, 79, 80]. However, more recent DSC work by Bassani et al. [81] has been able to specifically associate calorimetric data with GP zone and θ'' formation. Here we will use their values of 0.57 eV for GP zone formation and 0.86 eV for θ''-precipitate formation. In the absence of data, we use 0.86 eV for coarsening as well.

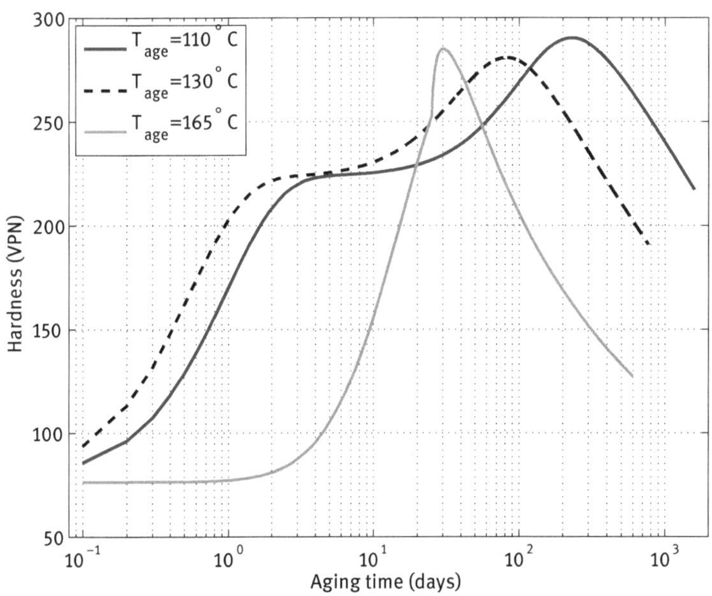

Fig. 3.13: 0 K hardness vs. aging time predictions for Al-4 wt.% Cu aged at different temperatures.

For the 4 wt.% Cu alloy, increasing the aging temperature from 110 °C to 130 °C accelerates the aging process by accelerating the formation, growth, and coarsening of precipitates as seen in Fig. 3.13. The acceleration of the entire aging curve over this temperature range is consistent with experimental observations [47]. In the range of parameters investigated here, the peak strength is a function of the correspondence between the θ''-precipitate formation rate and the coarsening rate, as well as the amount of Cu available for precipitation. All three are affected by aging temperature. Nonetheless, only slight changes in the peak strength are observed in the model between 110 °C and 130 °C aging temperatures, consistent with experimental observations.

For 4 wt.% Cu, the GP zone solvus temperature is approximately $T_{solvus} = 150\,°C$ [73]. Therefore, aging above 150 °C will not produce GP zones, leading to aging curves that do not display strength plateaus [47]. The current model as formulated does not capture this feature as the kinetics modeling predicts GP zone formation at all realistic aging temperatures. In future efforts, this inconsistency could be addressed by using a kinetics modeling that accounts for the thermodynamic stability of precipitate phases, as performed in [74] using first principles calculations.

Noting this deficiency in the model, we produced an aging curve at 165 °C (Fig. 3.13) by explicitly prohibiting the growth of GP zones. The qualitative features of the curve agree with experiment [47], with the exception of the kink that exists at 14 days aging time. The kink results from the sudden start of coarsening kinetics in our model when 80 % of the Cu atoms have transformed from solute to precipitates. Obviously the abrupt start of coarsening in the model is unrealistic. Further, the model predicts the formation of unrealistically large θ''-precipitates, (diameters larger than 100 nm). In reality these large θ''-precipitates would transform into tetragonal θ'-precipitates [47]. Experimental observations suggest that optimum hardness occurs for 4 wt.% Cu aged at 165 °C during the transition from θ'' to θ'-precipitate formation [47].

3.8 Role of thermal activation and direct comparison to experiment

The age hardening predictions presented in the previous sections significantly overpredicted alloy hardness because they ignored the role of thermal activation in dislocation-precipitate and dislocation-solute interactions. Specifically, the predictions were formulated directly from atomistic simulations of dislocation-precipitate interactions conducted at zero temperature and an analytic solute hardening model that did not include temperature and strain rate effects. The aim of this section is to interpret the athermal predictions relative to typical experimental conditions, i.e. $T \approx 300\,K$ and $\dot{\gamma} \approx 10^{-3}\,s^{-1}$ [27].

While thermal activation is thought to play a significant role in dislocation-solute interactions, its significance in dislocation-precipitate interactions is unclear. A collection of experiments suggest that precipitate strengthening is relatively independent of thermal activation [22, 25, 82], while others show evidence to the contrary [27, 83]. We note that the study by [27] is of particular significance in that it examined underaged Al-Cu alloys with large populations of GP zones and θ'' precipitates. Atomistic simulations have not necessarily clarified the issue. Bacon and co-workers' simulations [31, 38] suggest that precipitate strengths are significantly affected by thermal activation with 300 K strengths differing by a factor of 1.5–2.0 compared to 0 K, while [84] report that dislocation-nanovoid interactions are largely temperature independent (noting that there are many similarities between dislocation-void and dislocation precipitate interactions).

Here, thermal activation is investigated using the same underlying framework and assumptions for both solute and precipitation hardening. The applied load τ is related to plastic strain rate $\dot{\gamma}$ in the presence of thermal activation [57, 83] by

$$\frac{\tau}{\tau_{\text{inst}}} = 1 - \left(\frac{k_B T}{\Delta E_0} \ln \frac{\dot{\gamma_0}}{\dot{\gamma}}\right)^{2/3} . \tag{3.20}$$

where τ_{inst} is the load at which the thermally activated events occur instantaneously and ΔE_0 is the activation energy barrier. The parameter $\dot{\gamma_0}$, which represents the plastic strain rate that would occur if the controlling thermally activated events were to happen instantaneously is take here as $\dot{\gamma_0} = 10^4$, consistent with Leyson et al. [57] Using $T = 300\,\text{K}$, $\Delta E_0 = 2.78\,\text{eV}$, $\dot{\gamma} = 10^{-3}\,\text{s}^{-1}$, and $\dot{\gamma_0} = 10^4\,\text{s}^{-1}$, a thermal activation correction factor of $\tau/\tau_{\text{inst}} = 0.72$ is obtained for solute hardening.

For precipitate hardening, the unloaded energy barrier is not readily extractable from the 0 K simulations preformed in Sec. 3.2. It must either be measured directly using a chain of states method [85] or inferred from a set of direct molecular dynamics simulations [86, 87]. In this work we have taken the latter approach. The investigation focused specifically on a 60° interaction between an edge dislocation and a 4.4 nm diameter GP zone, as studied in Sec. 3.2.2. The molecular dynamics simulation cell size in the Z-directions was $L \approx 7.8\,\text{nm}$. At each fixed load multiple molecular dynamics simulations were performed at $T = 300\,\text{K}$ with different initial random velocity seeds. The simulation time before the dislocation overcame the GP zone is shown in Fig. 3.14. At a fixed load, the mean time, $\bar{t}$, for the dislocation to overcome the GP zone is the inverse of the rate, $\tilde{\nu}$, associated with the thermally activated event. First, the rate at which the controlling thermally activated events occur is assumed to follow an Arrhenius relation

$$\tilde{\nu} = \tilde{\nu}_0 \exp\left(\frac{-\Delta E}{k_B T}\right) . \tag{3.21}$$

The exponential prefactor, $\tilde{\nu}_0$, is assumed to be a constant and the energy barrier, ΔE, is assumed to only depend upon the ratio of the applied load, τ, over the load at which the thermally activated events occur instantaneously[1], τ_{inst}. Assuming that the shape of the energy barrier, with respect to the reaction coordinate, is sinusoidal in form, the functional form of ΔE can be approximated to the first order as

$$\Delta E = \Delta E_0 \left(1 - \frac{\tau}{\tau_{\text{inst}}}\right)^{3/2} . \tag{3.22}$$

Finally, the rate at which the controlling thermally activated events occur, ν, is assumed proportional to the plastic strain rate, $\dot{\gamma}$.

1 We refer the interested reader to Nguyen et al. [85] for insight into the origin and significance of these assumptions.

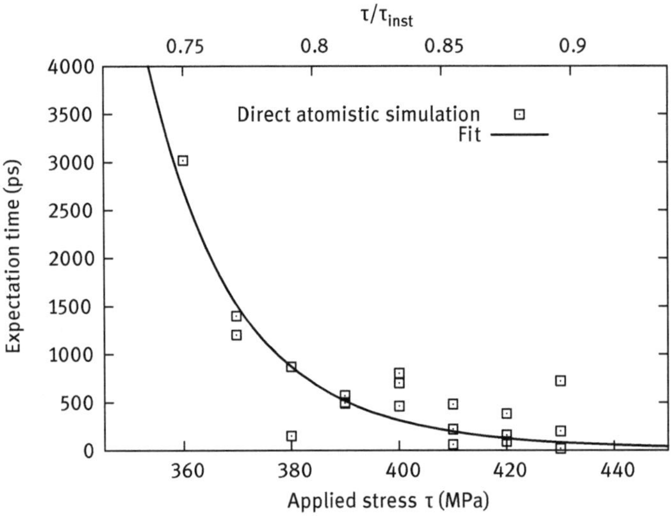

Fig. 3.14: Expectation time for an edge dislocation to overcome a 4.4 nm GP zone (60° interaction and $L = 7.94$ nm) as a function of fixed applied shear stress at 300 K. The instantaneous applied shear stress, τ_{inst}, is 460 MPa. The fit corresponds to Eq. (3.22).

Assembling the above assumptions and Eqs. (3.21) and (3.22), we will get back an expression identical to Eq. (3.20) for relating the applied load τ to the plastic strain rate $\dot{\gamma}$ in the presence of thermal activation [57, 83]. Accordingly, the data in Fig. 3.14 can be described by Eqs. (3.21) and (3.22). The best fit is obtained with $\Delta E_0 = 0.98$ eV and $\tilde{\nu}_0 = 1/\bar{t}_0 = 4.25 \cdot 10^{10}\,\text{s}^{-1}$. With ΔE_0 in hand, Eq. (3.20) can be used to predict a thermal activation correction factor for a specific temperature and strain rate as was done for solute hardening. Using $T = 300$ K, $\Delta E_0 = 0.98$ eV, $\dot{\gamma} = 10^{-3}\,\text{s}^{-1}$, and $\dot{\gamma_0} = 10^{4}\,\text{s}^{-1}$, a factor of $\tau/\tau_{inst} = 0.44$ was obtained. Interestingly, this suggests that thermal activation plays a more significant role in this dislocation-GP zone interaction than in solute hardening.

The thermal activation correction factor computed above is assumed to apply to all dislocation-precipitate interactions considered in the age hardening model. This assumption was required in light of our finite computational and human resources, and is not necessarily consistent with previous experimental [27] and atomistic simulation [37] results which indicate that the role of thermal activation is highly specific to the particular features of individual dislocation-precipitate interactions. An additional assumption implicit to the generation of finite temperature aging curves is that the effect of thermal activation on individual dislocation-precipitate interactions is proportional to the effect of thermal activation on an entire glide plane with a collection of dislocations and precipitates. We point the interested reader to Xu and Picu [64] for further discussion of this point.

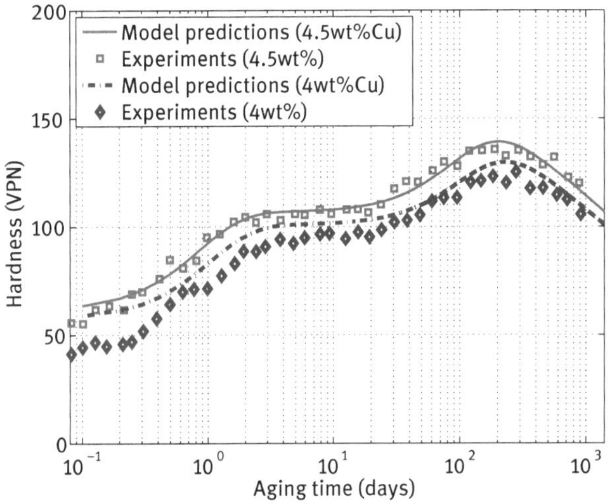

Fig. 3.15: 300 K hardness vs. aging time predictions for Al-Cu aged at T_{age} = 110 °C. Experimental data is from Silcock et al. [47].

Utilizing the thermal activation correction factor for both solute and precipitate hardening, age hardening predictions for both 4 and 4.5 wt.% Cu aged at T_{age} = 110 °C and deformed at T = 300 K and a strain rate of 10^{-3} s^{-1} have been produced. The curves are directly compared to the experimental data from [47] in Fig. 3.15. Overall the model predictions are found to agree very well with experiments considering the number and complexity of ingredients involved in the model and the lack of mechanics fitting parameters. The agreement of the strength plateau and peak strength is particularly noteworthy in light of the significant assumptions noted in the previous paragraph. The most significant discrepancy between the model predictions and the experimental curves involves the effect of Cu concentration on strength at the earliest aging times. As with the athermal predictions in Sec. 3.3, the solute hardening model underpredicts this effect.

3.9 Summary and conclusion

Significant improvements in computational power and methodologies [57, 88] in the last decade have led to unparalled advancements in the quest to quantitatively connect mechanical behavior to underlying atomic scale processes. Nonetheless, a direct prediction of age hardness solely from atomistic principles has remained a formidable challenge. As a step towards this goal, this chapter described a hierarchical multiscale approach in which atomistic modeling was used to provide inputs on mechanical strength contributions from solid solution and precipitate microstructure, and did not involve any free parameters. To capture the complex microstructural evolution of an

Al-Cu alloy (Cu solute → GP zones → θ''-precipitates) well established kinetics models were utilized. The kinetics modeling involved three fitting parameters which governed the rate at which the precipitates formed, grew, and coarsened. The fitting parameters were chosen so that key features of the aging curve aligned in time with experimental data for 4 wt.% Cu aged at 110 °C. Overall, the strength values of the predicted aging curve corresponded remarkably well with experimental data. Considering the scale of the model and lack of mechanical fitting parameters, its general agreement with experiment is remarkable. With that said, the primary value of this effort is not that the model reproduced experimental data, but what was learned in the quest to do so.

At the lowest scale of the model, the atomistic simulations of dislocation–GP zone and dislocation–θ''-precipitate interactions revealed significant complexity beyond textbook understanding. Full dislocation looping, full dislocation cutting, and leading partial cutting with trailing partial looping were observed depending upon the details of the interaction. Contrary to traditional wisdom, dislocation looping occurred in some interactions involving very small precipitates. However, the lack of thermal activation and the validity of the interatomic potential motivate caution when connecting this observation to a real Al-Cu alloy deformed at room temperature. Precipitate cutting occurred by an unzipping mechanism, which is not considered in many traditional models for precipitate strengthening. On the whole, the variety and complexity of the dislocation-precipitate interactions complicate continuum precipitate strengthening predictions and motivate atomistic analysis.

Precipitate strengthening was found to scale quite well with the simple $1/L$ relation even when L was comparable to the precipitate diameter. This suggested that more sophisticated relations between L, D, and strength may not always be warranted. In general, precipitate strength was found to be higher for 60° dislocation-precipitate interactions than for 0° interactions. The maximum strength of dislocation-GP zone interactions was found to be 0.35 and 0.55 of the theoretical Orowan strength depending upon the orientation of the interaction. The maximum strength of dislocation-θ''-precipitate interactions was found to be 0.50 and 1.75 of the Orowan strength depending upon the orientation of the interaction. The super-Orowan strengths of the dislocation-θ''-precipitate interactions result from the cross-slip of the bowed-out dislocation segments. Based on observations of cross-slip strengthening in screw dislocation precipitate interactions in our previous work [37], we are uncertain of the existence of this mechanism and the subsequent super-Orowan strengths under ordinary experimental conditions. In this regard it is important to note that it is the cross-slip strengthening that makes larger (< 15 nm) θ''-precipitates significantly stronger than GP zones, and thus leads to the peak in the age hardening curve. If the Orowan strength is taken as the maximum dislocation-precipitate interaction strength in the model, the strength increase due to the transition from GP zones to θ''-precipitates would be small. This would produce an aging curve qualitatively different than observed in experiment, without the plateau-then-peak shape. Based on our experience, two further studies would go a long way towards illuminating this matter. First, a thorough analysis of the role of thermal activation is needed for the dislocation–θ''-precipitate interactions with super-Orowan strengths. Second, both screw and edge dislocations must be considered.

For solute hardening, directly simulating the motion of a dislocation through a field of solute atoms using molecular dynamics simulations resulted in an unrealistically high athermal strength, mainly due to a very high Al-Cu bond strength in the interatomic potential. Therefore, a newly developed semi-analytic solute hardening model [57] that uses atomistically computed Cu solute–edge dislocation interaction energies as input was utilized. While these predictions were still significantly larger than corresponding DFT values in the literature [57], they were usable in the age hardening model. Broadly, the solute hardening section highlights the importance of both matrix-solute bond strengths and dilute heat of solution energies in interatomic potentials when solute hardening predictions are desired.

While the formulation of the solute hardening model directly provides the shear strength for a dislocation to propagate through a field of solute atoms, the shear stress required for a dislocation to propagate through a field of precipitates requires additional consideration beyond the atomistic studies of a dislocation interacting with an equally-spaced periodic array of identical precipitates. Although a considerable literature exists on this topic, the majority of the work has focused on point precipitates rather than plates. This motivated the construction of a continuum dislocation line model. A particular challenge in employing the model involved the incorporation of the super-Orowan strengths of the dislocation θ''-precipitate interactions. This was overcome by using an artificially high value of Gb in the model. Parametric studies suggest that this induced an error of less than 5 % in the predictions. In addition to the continuum dislocation line model providing a means for connecting the atomistic simulations of dislocation-precipitate interactions with macroscopic strengths, it illuminated the utility of simple analytic approaches at doing the same. Specifically, we found that a quadratic law addition of individual precipitate mean strengths weighted by a knockdown factor can describe the strength to propagate an edge dislocation on a slip plane populated by a set of randomly positioned plate precipitates with different offsets and strengths. A factor of 0.64 was found to work reasonably well for the span of the Al-Cu aging curve examined here.

To quantitatively compare the model to experiments, thermally activated plasticity must be considered. This task is straight forward with the semi-analytic solute hardening model as it directly provides the energy barrier associated with dislocation motion through a field of solutes. However, for dislocation-precipitate interactions, additional atomistic simulations are required at finite temperature. Focusing on a specific edge dislocation-GP zone interaction, we found that thermal activation can play a large role in this interaction with the strength under typical experimental conditions, i.e. 300 K and $\dot{\gamma} = 10^{-3}\,s^{-1}$, being 44 % of the athermal strength. The generality of this result to other dislocation-precipitate interactions is unclear, providing motivation for further investigation.

Considering the development of the model in total and the comparison of its predictions to experimental age hardening curves, two substantial outstanding issues exist. First, the initial alloy strengths predicted by the model (due primarily to solute hardening) do not scale as strongly with Cu concentration as does experimental data. Second, the plateau-then-peak behavior of the predicted aging curve is the result of a cross-slip strengthening mechanism that may not occur under ordinary experimental conditions. Specifically, for the plateau-then-peak to exist in the aging curve, the θ''-precipitate strength must be significantly greater than the GP zone strength. One way for the plateau-then-peak behavior to exist, without the cross-slip super-Orowan strengthening mechanism of θ''-precipitate interactions, would be for the actual GP zone strengths to be lower than our atomistic simulations predict. This is plausible considering that the Al-Cu interactions are likely over predicted with the interatomic potential that we have used here, as indicated from the solute hardening model. Thus, GP cutting may be more prevalent than predicted, making GP zone strengths significantly lower than the Orowan strength of θ''-precipitates.

Overall, this work highlights that mechanism-based plasticity modeling is still faced with considerable challenges of both breadth and depth. The complexity of dislocation-precipitate interactions call for many more simulations to be performed before the interactions can be fully characterized. More broadly, the characterization of the vast space of dislocation-grain boundary interactions will likely prove even more challenging. At the same time, more accurate (and most likely computationally demanding) interatomic interaction models are needed to obtain results that better connect to relevant engineering alloys. Furthermore, the effects of thermal activation must be considered, something which often requires a considerable amount of additional computational cost and analysis beyond what is required for athermal simulations. With that said, computational methods and resources continue to improve making the above challenges appear evermore surmountable in the future.

Acknowledgment: The author gratefully acknowledges Prof. Derek Warner at Cornell University for advice and guidance, and Ed Glaessgen and Steve Smith at NASA for financial support (Grant No. NNX08BA39A).

References

[1] Wilm A. Physical-metallurgical experiments on aluminium alloys containing magnesium. Metallurgie. 1911;8:225–227.
[2] Gayle FW, Goodway M. Precipitation hardening in the first aerospace aluminum alloy: The wright flyer crankcase. Science. 1994;266(5187):1015–1017.
[3] Hornbogen E. Hundred years of precipitation hardening. Journal of Light Metals. 2001;1(2): 127–132.
[4] Polmear I. Light alloys: from traditional alloys to nanocrystals. Butterworth-Heinemann; 2006.

[5] Liddicoat PV, Liao XZ, Zhao Y, Zhu Y, Murashkin MY, Lavernia EJ, et al. Nanostructural hierarchy increases the strength of aluminium alloys. Nature Communications. 2010;1(6):1–7.
[6] Valiev RZ, Enikeev NA, Murashkin, Kazykhanov VU, Sauvage X. On the origin of the extremely high strength of ultrafine-grained al alloys produced by severe plastic deformation. Scripta Materialia. 2010;63(9):949–952.
[7] Orowan E. Zur Kristallplastizität. III. Über den Mechanismus des Gleitvorganges. Z Phys. 1934; 89:634–659.
[8] Polanyi M. Über eine Art Gitterstörung, die einen Kristall plastisch machen könnte. Z Phys. 1934;89:660–664.
[9] Taylor GI. The mechanism of plastic deformation of crystals. Part I. Theoretical. Proc Royal Soc London A. 1934;145(855):362–387.
[10] Guinier A. La diffraction des rayons x aux tres petits angles; application a l'etude de phenomenes ultramicroscopiques. Ann Phys (Paris). 1939;12:161–237.
[11] Preston GD. LXXIV. The diffraction of X-rays by an agehardening alloy of aluminium and copper. The structure of an intermediate phase. The London, Edinburgh, and Dublin Philosophical Magazine and Journal of Science. 1938;26(178):855–871.
[12] Polmear IJ, Hardy HK. Some metallographic observations on aged aluminium-copper alloys. J Inst Metals. 1952–1953;81:427–432.
[13] Guinier A. Heterogeneities in Solid Solutions, Solid State Physics, vol. 9. Elsevier; 1959. pp. 293–398.
[14] Hono K, Satoh T, Hirano KI. Evidence of multi-layer GP zones in Al-1·7Cu alloy. Philosophical Magazine A. 1986;53:495–504.
[15] Ringer SP, Hono K. Microstructural evolution and age hardening in aluminium alloys: Atom probe field-ion microscopy and transmission electron microscopy studies. Materials Characterization. 2000;44(1-2):101–131.
[16] Karlik M, Bigot A, Jouffrey B, Auger P, Belliot S. HREM, FIM and tomographic atom probe investigation of Guinier-Preston zones in an Al-1.54Cu alloy. Ultramicroscopy. 2004;98(2–4): 219–230.
[17] Report of a conference on the strength of solids. The Physical Society, London; 1948.
[18] Gerold V, Haberkorn H. On the critical resolved shear stress of solid solutions containing coherent precipitates. Physica Status Solidi (b). 1966;16(2):675–684.
[19] Knowles G, Kelly PM. Effect of second phase particles on the mechanical properties of steel. In: BSC/ISI Conference, Scarborough, The Iron and Steel Institute, London. p. 9.
[20] Nembach E. Precipitation hardening caused by a difference in shear modulus between particle and matrix. Physica Status Solidi (a). 1983;78:571–581.
[21] Hirsch PB, Kelly A. Stacking-fault strengthening. Philosophical Magazine. 1965;12(119):881–900.
[22] Kelly A, Nicholson RB. Precipitation hardening. Progress In Materials Science. 1963;10:151–391.
[23] Kelly A, Fine ME. The strength of an alloy containing zones. Acta Metallurgica. 1957;5:365–367.
[24] Harkness SD, Hren JJ. An investigation of strengthening by spherical coherent GP zones. Metall Trans. 1970;1:43–49.
[25] Brown LM, Ham RK. Strengthening methods in crystals. Applied Science Publishers, London; 1971. p. 9.
[26] Eto T. Hardening due to oriented GP zones in Al-Cu single crystals. Scripta Metallurgica. 1980; 14(1):133–135.
[27] Muraishi S, Niwa N, Maekawa A, Kumai S, Sato A. Strengthening of al-cu single crystals by stress-oriented guinier-preston zones. Philosophical Magazine A. 2002;82:2755–2771.

[28] Van Swygenhoven H, Derlet PM. Grain-boundary sliding in nanocrystalline fcc metals. Physical Review B (Condensed Matter and Materials Physics). 2001;64(22):224105–9.
[29] Rao SI, Parthasarathy TA, Dimiduk DM, Hazzledine PM. Discrete dislocation simulations of precipitation hardening in superalloys. Philosophical Magazine. 2004;84:3195–3215.
[30] Dewald MP, Curtin WA. Multiscale modelling of dislocation/grain-boundary interactions: I. edge dislocations impinging on $\Sigma 11$ (113) tilt boundary in Al. Modelling and Simulation in Materials Science and Engineering. 2007;15(1):S193–S215.
[31] Osetsky YN, Bacon DJ. Void and precipitate strengthening in α-iron: what can we learn from atomic-level modelling? Journal of Nuclear Materials. 2003;323(2-3):268–280.
[32] Kohler C, Kizler P, Schmauder S. Atomistic simulation of precipitation hardening in α-iron: influence of precipitate shape and chemical composition. Model Simul Mater Sci Eng. 2005;13: 35–45.
[33] Shim JH, Cho YW, Kwon SC, Kim WW, Wirth BD. Screw dislocation assisted martensitic transformation of a bcc Cu precipitate in bcc Fe. Applied Physics Letters. 2007;90:021906.
[34] Terentyev D, Bonny G, Malerba L. Strengthening due to coherent Cr precipitates in Fe-Cr alloys: Atomistic simulations and theoretical models. Acta Materialia. 2008;56(13):3229–3235.
[35] Takahashi A, Ghoniem NM. A computational method for dislocation-precipitate interaction. J Mech Phys Solids. 2008;56(4):1534–1553.
[36] Singh CV, Warner DH. Mechanisms of guinier-preston zone hardening in athermal limit. Acta Materialia. 2010;58(17):5797–5805.
[37] Singh CV, Mateos A, Warner DH. Atomistic simulations of dislocation-precipitate interactions emphasize importance of cross-slip. Scripta Materialia. 2011;64:398–401.
[38] Bacon DJ, Osetsky YN, Rodney D. Dislocation–obstacle interactions at the atomic level, vol. 15, ch. 88; 2009. pp. 1–90.
[39] Shercliff H, Ashby M. A process model for age hardening of aluminium alloys I. the model. Acta Metallurgica et Materialia. 1990;38(10):1789–1802.
[40] Schmauder S, Binkele P. Atomistic computer simulation of the formation of Cu-precipitates in steels. Computational Materials Science. 2002;24:42–53.
[41] Starink MJ, Gao N, Davin L, Yan J, Cerezo A. Room temperature precipitation in quenched Al-Cu-Mg alloys: A model for the reaction kinetics and yield strength development. Philosophical Magazine. 2005;85(13):1395–1417.
[42] Singh CV, Warner DH. An atomistic-based hierarchical multiscale examination of age hardening in an al-cu alloy. Metallurgical and Materials Transactions A. 2013;44(6):2625–2644.
[43] Plimpton SJ. Fast parallel algorithms for short-range molecular dynamics. J Comp Phys. 1995; 117:1–19.
[44] Apostol F, Mishin Y. Interatomic potential for the Al-Cu system. Phys Rev B. 2011;83:054116.
[45] Daw MS, Baskes MI. Embedded-atom method: Derivation and application to impurities, surfaces, and other defects in metals. Physical Review B. 1984;29(12):6443–6453.
[46] Mishin Y, Mehl M, Papaconstantopoulos D. Phase stability in the Fe-Ni system: Investigation by first-principles calculations and atomistic simulations. Acta Materialia. 2005;53:4029–4041.
[47] Silcock JM, Heal TJ, Hardy HK. Structural ageing characteristics of binary al-cu alloys. J Inst Metals. 1953-54;82:239–248.
[48] Dieter GE. Mechanical Metallurgy. 3rd ed. Materials Science and Engineering, New York: McGraw-Hill; 1986.
[49] Hull D, Bacon D. Introduction to dislocations. 4th ed. Butterworth-Heinemann; 2001.
[50] Kelchner C, Plimpton S, Hamilton J. Dislocation nucleation and defect structure during surface indentation. Physical Review B. 1998;58:11085–11088.
[51] Li J. Atomeye: An efficient atomistic configuration viewer. Modelling and Simulation in Materials Science and Engineering. 2003;11(2):173–177.

[52] Symposium on Internal Stresses in Metals and Alloys; 1948.
[53] Bacon DJ, Kocks UF, Scattergood RO. The effect of dislocation self-interaction on the Orowan stress. Philosophical Magazine. 1973;28(6):1241–1263.
[54] Hirsch PB. The interpretation of the slip pattern in terms of dislocation movements. J Inst Metals. 1957;86:13–14.
[55] Hatano T. Dynamics of a dislocation bypassing an impenetrable precipitate: The Hirsch mechanism revisited. Physical Review B. 2006;74(2):020102+. Available from: http://dx.doi.org/10.1103/PhysRevB.74.020102.
[56] Humphreys FJ, Hirsch PB. The deformation of single crystals of copper and copper-zinc alloys containing alumina particles. II. Microstructure and dislocation-particle interactions. Proceedings of the Royal Society of London Series A, Mathematical and Physical Sciences. 1970;318(1532):73–92.
[57] Leyson GP, Curtin WA, Hector LG, Woodward CF. Quantitative prediction of solute strengthening in aluminium alloys. Nat Mater. 2010;9(9):750–755.
[58] Labusch R. A statistical theory of solid solution hardening. physica status solidi (b). 1970; 41(2):659–669.
[59] Labusch R. Statistical theories of solid-solution hardening. Acta Metallurgica. 1972;20(7):917–927.
[60] Olmsted D, Hectorjr L, Curtin W. Molecular dynamics study of solute strengthening in Al/Mg alloys. Journal of the Mechanics and Physics of Solids. 2006;54(8):1763–1788.
[61] Friedel J. Les Dislocations. Paris, France: Gauthier-Villars; 1956.
[62] Foreman AJE, Makin MJ. Dislocation movement through random array of obstacles. Canad J Phys. 1967;45:511–517.
[63] Nogaret T, Rodney D. Finite-size effects in dislocation glide through random arrays of obstacles: Line tension simulations. Phys Rev B. 2006;74(13):134110+.
[64] Xu Z, Picu RC. Thermally activated motion of dislocations in fields of obstacles: The effect of obstacle distribution. Phys Rev B. 2007;76(9):094112+.
[65] Dong Y, Nogaret T, Curtin WA. Scaling of dislocation strengthening by multiple obstacle types. Metallurgical and Materials Transactions A. 2010;41(8):1954–1960.
[66] Binkele P, Schmauder S. An atomistic Monte Carlo simulation of precipitation in a binary system. Z Metallkd. 2003;94(8):858–863.
[67] Taylor GI. Plastic strain in metals. Journal of the Institute of Metals. 1938;62:307–324.
[68] Taylor GI. Strains in crystalline aggregates. In: Proceedings of the Colloquium on Deformation and Flow of Solids (Madrid, 1955). Berlin: Springer, pp. 3–12.
[69] Stoller RE, Zinkle SJ. On the relationship between uniaxial yield strength and resolved shear stress in polycrystalline materials. Journal of Nuclear Materials. 2000;283–287(Part 1):349–352.
[70] Avrami M. Kinetics of phase change. i general theory. The Journal of Chemical Physics. 1939; 7(12):1103–1112.
[71] Christian J. The theory of phase transformations in metals and alloys. 3rd ed. Pergmon Press, Oxford; 2002.
[72] Hornbogen E. The electrom microscope investigation of precipitation in aluminum-copper solid solutions ii. Aluminum. 1967;43:115–121.
[73] Massalski T. The al–cu (aluminum-copper) system. Journal of Phase Equilibria. 1980;1(1):27–33.
[74] Wang J, Wolverton C, Muller S, Liu Z, Chen L. First-principles growth kinetics and morphological evolution of Cu nanoscale particles in Al. Acta Materialia. 2005;53(9):2759–2764.
[75] Wagner C. Theory of precipitate change by redissolution. Z Elektrochem. 1961;65:581–591.

[76] Nie JF, Muddle BC, Polmear IJ. The effect of precipitate shape and orientation on dispersion strengthening in high strength aluminium alloys. Materials Science Forum. 1996;217-222: 1257–1262.
[77] Starink M. Analysis of aluminium based alloys by calorimetry: quantitative analysis of reactions and reaction kinetics. Int Mater Rev. 2004;49:191–226.
[78] Martin JW. Precipitation Hardening. 2nd ed. Butterworth-Heinemann; 1998.
[79] Badini C, Marino F, Verne E. Calorimetric study on precipitation path in 2024 alloy and its SiC composite. Materials Science and Engineering A. 1995;191(1-2):185–191.
[80] Smith G. Precipitation kinetics in solutionized aluminum alloy 2124: Determination by scanning and isothermal calorimetry. Thermochimica Acta. 1998;317(1):7–23.
[81] Bassani P, Gariboldi E, Vimercati G. Calorimetric analyses on aged Al–4.4Cu–0.5Mg–0.9Si–0.8Mn alloy (AA2014 grade). Journal of Thermal Analysis and Calorimetry. 2007;87(1):247–253.
[82] Ardell A. Precipitation hardening. Metallurgical and Materials Transactions A. 1985;16(12): 2131–2165.
[83] Argon A. Strengthening Mechanisms in Crystal Plasticity (Oxford Series on Materials Modelling). Oxford University Press, USA; 2007.
[84] Hatano T, Matsui H. Molecular dynamics investigation of dislocation pinning by a nanovoid in copper. Physical Review B. 2005;72(9):094105+.
[85] Nguyen L, Baker K, Warner D. Atomistic predictions of dislocation nucleation with transition state theory. Phys Rev B. 2011;84:024118.
[86] Warner DH, Curtin WA, Qu S. Rate dependence of crack-tip processes predicts twinning trends in f.c.c. metals. Nature Materials. 2007;6(11):876–881.
[87] Warner D, Curtin W. Origins and implications of temperature-dependent activation energy barriers for dislocation nucleation in face-centered cubic metals. Acta Materialia. 2009;57(14): 4267–4277.
[88] Trinkle DR, Woodward C. The chemistry of deformation: how solutes soften pure metals. Science. 2005;310(5754):1665–7.

M. Prieto-Depedro, I. Martin-Bragado, and J. Segurado*

4 Kinetic Monte Carlo modeling of shear-coupled motion of grain boundaries

Abstract: The shear coupled motion of grain boundaries (GBs) is modeled by using two different atomistic simulation techniques: molecular dynamics (MD) and kinetic Monte Carlo (KMC). MD simulations are conducted to identify the elementary mechanisms that take place during the coupled motion of GBs. This process is described on the one hand, in terms of the geometrical approach of the dislocation content in the boundary; and on the other hand, by the thermodynamics of the dislocation passage, shown as a thermal activated process. Relevant MD output is extended into a KMC model that considers the GB migration as a result of a sequence of discrete rare events. The independent motion of each structural unit forming the boundary conforms a single event, having a rate per unit of time to move to the next stable position computed according to the transition state theory. The limited time scale of classical MD is overcome by KMC, that allows to impose realistic deformation velocities up to 10 μm/s.

Keywords: upscale modeling, molecular dynamics, kinetic Monte Carlo, grain boundary motion, dislocation, nickel

4.1 Introduction

Grain boundaries play an important role in determining the mechanical properties of nanocrystalline (nc-) materials [1], e.g. enhancement of the initial resistance to yielding in nc-aggregates by the presence of GBs [2]. Several investigations have been carried out in order to understand the GB structure [3], thermodynamics [4] and related deformation mechanisms [5, 6].

In particular, atomistic simulations have provided a further insight into the microscopic mechanisms and fracture nucleation at interfaces under applied mechanical loads. Shear behaviours of GBs have been studied in detail, and grain boundary (GB) motion coupled to shear deformation has been evidenced to be an important mode of plastic deformation, as a dominant behaviour or competing with other GB mechanisms in a wide range of temperatures, such as GB sliding [7–9] or dislocation mediated slips [10–13].

Basically this mechanism is described through the coupling of the GB to mechanical loads which in turn induces a shear deformation in the crystal region swept by the motion. Stress-induced GB migration was first observed in symmetrical low-angle GBs (LAGBs) in Zn bicrystals [14]. Read and Shockley [15] determined the mechanism of this motion to be due to the collective glide of the array of parallel edge dislocations forming the LAGB. This mechanism was limited to boundaries with low tilt angle where the core of individual dislocations could be resolved. However, this model has been extended to high-angle GBs (HAGB) in the case of metals. Goryaka et al. [16] confirmed experimentally the coupled motion of GBs for both low- and high-angle symmetrical tilt GBs with a misorientation angle within 0–90° in Al bicrystals. Also, computer simulations evidence that the coupling between normal boundary motion and the rigid translation of the adjacent grains can occur for HAGBs [17–19]. The coupling effect was proposed to be a generic property of many atomically ordered GBs in the unified approach reported by Cahn and Taylor [20] and confirmed over a large catalogue of 388 GBs [21].

To date, GB shear-coupled motion has been subjected to several studies to explore the impact of this mechanism in the mechanical properties of nc- materials, as well as its relation to other deformation processes [22–25]. The major challenge relies on the multiscale modeling, as discussed by Berbenni et al. [26] where a complete micromechanical model based on molecular dynamics (MD) simulations is proposed. Taupin et al. [27] describe the shear coupled boundary migration in a continuous manner by using an elasto-plastic theory of disclination and dislocation fields, and results are found to be in good agreement with atomistic simulations and experiments.

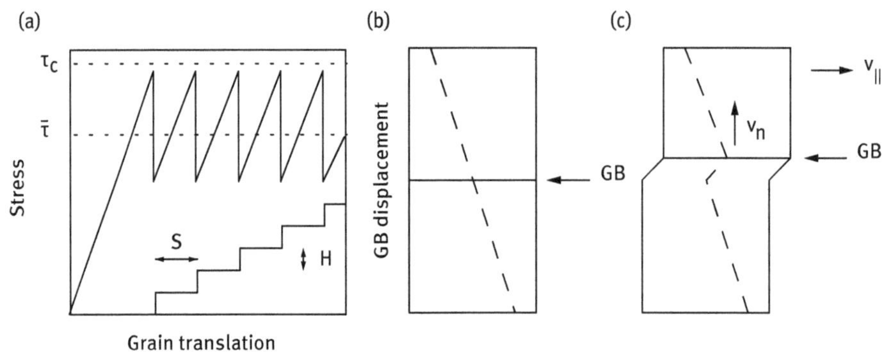

Fig. 4.1: Stick-slip behaviour. (a) In a perfect coupled motion, the normal motion of GB by increments of H is accompanied by rigid translation of the grains by increments of S. After each jump of the GB to the next stable position the stress drops. (b) Initial configuration of the system and (c) sheared lattice.

In this paper, we continue to investigate the dynamics of this phenomenon within a multiscale framework for its comprehensive understanding. Two different atomistic simulation techniques are used throughout this work. On the one hand, MD simulations are used to identify the elementary mechanisms of GB migration, and on the other hand, we have extended the outcome of MD simulations to a kinetic Monte Carlo (KMC) model. KMC models the shear coupled motion of GBs based on the thermodynamics of the dislocation motion, in terms of individual resolved cores forming the GB. The main goal is to be able to reproduce the MD simulation results at a reduced computational cost by means of KMC. We apply this multiscale approach in bicrystals containing one individual [001] symmetrical tilt GB in Ni in a range of misorientation angles between 0 and 90°.

The structure of the present paper is the following. In Sec. 4.2 a complete description of the dynamics of coupling and the geometric characterization of GBs are introduced. MD simulations are carried out in different temperature regimes and reported in Sec. 4.3. To reach our final goal of extending the MD output, the main concern before moving to KMC is to validate such output. Due to the lack of experimental data, in Sec. 4.3.2 and 4.3.3 simulated values are compared in detail to theoretical models previously illustrated. The minimum energy path of the shear-coupled GB motion is computed using the nudged elastic band method in Sec. 4.3.4, which is the most relevant input in the KMC method. Within the multiscale approach, the latter is proposed in Sec 4.4. Numerical examples of KMC are presented in Sec. 4.4.2, where results are contrasted to previously shown MD results.

4.2 Dynamics of shear-coupled motion of grain boundaries and coupling modes

The simple model proposed by Ivanov and Mishin [28] assumes the coupled GB motion as a motion through a periodic landscape. Following this approach, the system has a set of equivalent stable states which are separated by an energy barrier. At 0 K, in the absence of any external force applied, the GB is trapped in a particular equilibrium position. By applying a shear stress through imposing a parallel velocity ($v_{\|}$), the GB is elastically deformed and when a critical value τ_c^0 is reached, the relevant barrier vanishes to zero and the GB makes a transition by increments of H. This normal motion of the GB occurs simultaneously with the rigid-body translation of the adjacent grains (S), producing a permanent shear deformation of the lattice and the subsequent drop of the stress. The ratio $\beta = S/H$ is called coupling factor, discussed later. The first stress relaxation is followed by a new elastic loading step until the critical stress is reached again, and the GB makes another step, see Fig. 4.1, identified as a stick-slip behaviour.

At finite temperature, the GB could move before the critical stress is reached. Thermal fluctuations assist the GB to overcome the energy barrier, so critical stress τ_c is then expected to decrease with temperature. This decreasing is intensified as temperature increases, and at a fixed $v_{\|}$ the peak stress is expected to be linear in $T^{2/3}$ as [28]:

$$\tau_c = \tau_c^0 - BT^{2/3}, \tag{4.1}$$

being B a constant dependent on the attempt frequency and other factors. As temperature continues to increase, spontaneous movements of the GB may occur, and the stick-slip behaviour can be destroyed or replaced by other dominant mechanisms, such as sliding events.

At a fixed temperature, critical stress is expected to increase with the imposed grain translation velocity since the time to overcome the barrier is reduced. This variation is given by [29, 30]:

$$v_{\parallel} = \frac{\tau_c^0}{\eta}\left(1 - \frac{\tau_c}{\tau_c^0}\right)^{1/2} \exp\left[-\frac{(1-\tau_c/\tau_c^0)^{3/2} E_0}{k_B T}\right], \tag{4.2}$$

where η represents an effective friction coefficient, k_B is the Boltzman constant and E_0 is the energy barrier in the absence of temperature. This variation was tested by using accelerated MD methods that allow to cover a wide range of velocities that are not accessible by classical MD [31].

Regarding the GB crystallography, GBs consists of identical structural units. When resolving individual cores is possible, GBs are identified as an array of parallel edge dislocations. As a result, GB shear coupled motion is understood as a result of the passage of the dislocation content inducing itself plastic deformation of the latice [15].

To define the dislocation content of a GB the Frank-Bilby equation (FBE) is solved [32, 33], and the coupling factor is determined to be related to GB interfacial Burgers vector and slip plane. The crystal symmetry leads to the multiplicity of the dislocation content, and so to different coupling factors. By exploring the two possible solutions for the [001] tilt GBs considered in the present work, two possible Burgers vectors are identified, and as a result two different coupling modes are predicted. When the misorientation angle (θ) tends to 0, the set of dislocations corresponds to Burgers vector $\boldsymbol{b} = [100]$, being the Frank-Bilby dislocation slip planes (100). This model is extended to GB with θ approaching 90°, with Burgers vector $\boldsymbol{b} = -\frac{1}{2}[110]$ and glide along (110) planes. Assuming the dislocations move in increments of $\boldsymbol{b}$, the two branches of the misorientation dependence of β are obtained [17]. There is one branch called ⟨100⟩, for $\theta \to 0°$, given by:

$$\beta_{\langle 100\rangle} = 2\tan\left(\frac{\theta}{2}\right). \tag{4.3}$$

The other branch corresponds to $\theta \to 90°$, as:

$$\beta_{\langle 110\rangle} = -2\tan\left(\frac{\pi}{4} - \frac{\theta}{2}\right). \tag{4.4}$$

In between, boundaries can move according to either of the two possible modes. These modes compete with each other, existing a transition between them at a critical θ depending on the temperature. The expansion of the ⟨110⟩ mode as temperature decreases was observed in Cu bicrystals [17], even it may be the only active mode at 0 K for the whole range of θ. To activate the ⟨110⟩ mode, the mirror symmetry due to equivalent row translations by lattice vectors $\frac{1}{2}[001]$ and $\frac{1}{2}[00\bar{1}]$ has to be broken, and this is only feasible at finite temperatures when the GB develops ledges and steps assisting in breaking such symmetry [26]. Boundaries tend to move in the most energetically favored mode, and as a consequence those boundaries corresponding to ⟨110⟩ mode move instead according to ⟨110⟩ at low temperatures. At this point, it is obvious that the switch between modes is related with a change in the dislocation content. Note that the two coupling modes have different signs, and then GBs are expected to move in opposite directions depending on the active mode.

4.3 Molecular Dynamics

4.3.1 Computational procedure

Atomistic simulations are performed using an embedded-atom method potential fit to a large database including both experimental and first-principles data for Ni [34]. A series of [001] symmetrical tilt GBs with different misorientation angle are created, see Tab. 4.1. Two separated crystals with the desired crystallographic orientation are joined along a plane normal to x_2-direction, see Fig. 4.2. The simulation block contains two grains forming a bicrystal configuration with only one planar GB in the middle. Periodic boundary conditions are imposed in the parallel directions to the GB plane, x_1 and x_3. In the normal direction, x_2, dynamic atoms are sandwiched between two slabs whose thickness is about twice the cutoff radius of the interatomic potential. Atoms at the slabs are held fixed in their perfect lattice positions and do not participate in computing data from the simulations. Each symmetrical tilt GB is identified by the indices (hkl) of the GB plane and by the reciprocal value of the total number of coincidence sites, Σ [35]. Throughout the text, the notation will be Σ(hkl).

MD simulations are carried out using LAMMPS [36] simulator. As a first step, the ground-state structure of each GB is determined by static minimization of the potential energy of the whole system. The canonical NVT ensemble (number of particles,

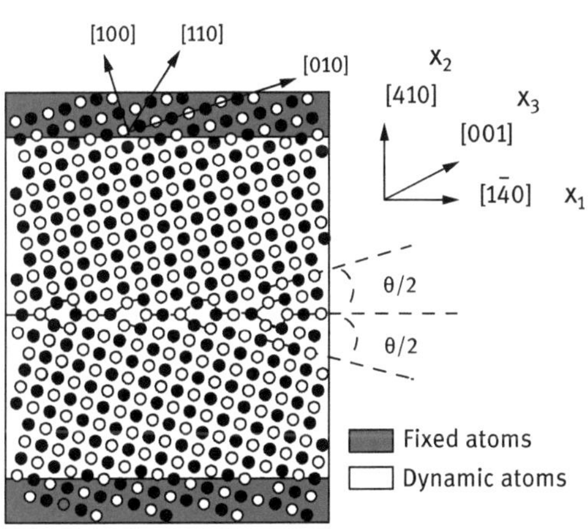

Fig. 4.2: Description of the simulation block for the $\Sigma 17(410)$ GB with a misorientation angle of $\theta = 28.07°$. Open and filled circles represent atoms in alternated planes along x_3-direction. Structural units forming the GB are outlined. Grey regions contain fixed atoms.

Tab. 4.1: Characteristics of [001] symmetrical tilt GBs in Ni. The ideal coupling factor predicted by geometric approaches and from MD simulations is unitless. The effective elastic shear modulus obtained by the transformation field analysis (TFA) and MD results is in GPa.

Boundary	θ [°]	Mode	β_{ideal}	β_{MD}	$\mu^{\mathrm{eff}}_{\mathrm{TFA}}$	$\mu^{\mathrm{eff}}_{\mathrm{MD}}$
Σ25(710)	16.26	⟨110⟩	0.286	0.305	117.51	119.40
Σ37(610)	18.92	⟨110⟩	0.333	0.372	115.03	122.02
Σ13(510)	22.62	⟨110⟩	0.400	0.423	111.15	113.30
Σ17(410)	28.07	⟨110⟩	0.500	0.507	104.71	107.76
Σ53(720)	31.89	⟨110⟩	−1.111	−1.160	99.83	95.85
Σ5(310)	36.87	⟨110⟩	−1.000	−1.004	91.42	93.27
Σ5(210)	53.13	⟨110⟩	−0.667	−0.673	72.60	66.45
Σ17(530)	61.93	⟨110⟩	−0.500	−0.545	63.43	57.88
Σ13(320)	67.38	⟨110⟩	−0.400	−0.430	58.81	53.10
Σ25(430)	73.74	⟨110⟩	−0.286	−0.259	54.59	47.12
Σ61(650)	79.61	⟨110⟩	−0.182	−0.186	51.88	46.68

volume and constant temperature) is used to perform the simulations. The time step is 0.2 fs. The shear rate applied in the top slab is given by $v_{\parallel} = \dot{E}L$, where $\dot{E}$ is the shear strain rate (10^8 s^{-1}) and L is the length of the box containing unconstrained atoms. The imposed grain translation velocity results about 1 m/s, well above the shear velocities under real conditions due to the time scale limitations of classical MD. The total stress tensor is averaged over all dynamic atoms using the standard virial expression and is constantly monitored during the simulation. GB position is tracked using common neighbor analysis [37].

4.3.2 Shear-coupled motion at low temperatures

Coupled motion is studied near 0 K: the upper slab is rigidly displaced with small increments, and after each one the system is fully relaxed while maintaining the constraint. Fig. 4.3 reports the shear stress and GB displacement variation as a function of the strain for Σ17(410). The GB does not move until the critical stress τ_c^0 is reached, being each peak of stress exactly correlated with an increment of the GB motion.

Every loading elastic step is associated with the same shear modulus (μ), obtained from the slope of the stress-strain curves. As discussed later, μ is a relevant input in KMC simulations, and it is noteworthy to check if the interatomic potential is reliable for the computation of the elastic constants of the bicrytal structures. As a consequence, this magnitude has been also calculated theoretically using the transformation field analysis (TFA), first proposed by Dvorak [38] and well described by Franciosi and Berbenni [39], see Tab. 4.1.

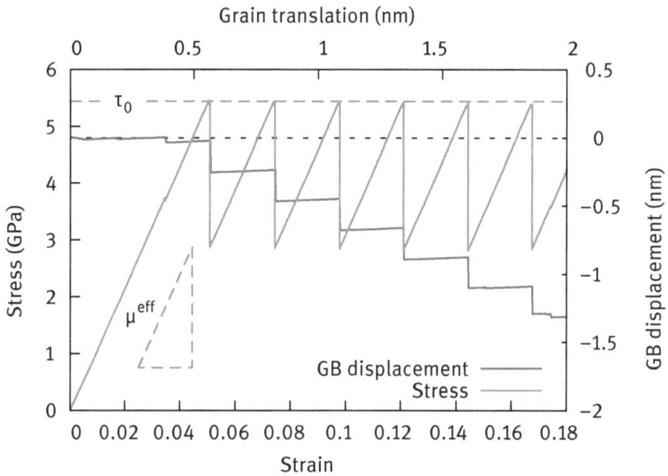

Fig. 4.3: Stress-Strain and GB displacement during coupled motion of the $\Sigma17(410)$ at 0 K with $v_{\parallel} = 1$ m/s. The perfect coupled motion is clearly observed in the absence of temperature. The critical stress and the effective shear modulus are identified, being 5.45 GPa and 107.76 GPa, respectively.

According to this approach (see Appendix 4.A), the effective shear modulus is determined to be:

$$\mu^{\text{eff}} = \frac{1}{2}\left(C_{66}^{\text{I}} - \frac{(C_{62}^{\text{I}})^2}{C_{22}^{\text{I}}} \right) + \frac{1}{2}\left(C_{66}^{\text{II}} - \frac{(C_{62}^{\text{II}})^2}{C_{22}^{\text{II}}} \right), \tag{4.5}$$

being C_{66}^{I}, C_{62}^{I} and C_{22}^{I}, and C_{66}^{II}, C_{62}^{II} and C_{22}^{II} the anisotropic elastic constants of constituent crystals I and II, respectively, obtained according to Eqs. (4.13) and (4.14).

Turning to the coupling modes, Fig. 4.4 displays the dependence of β with the misorientation angle. Values of β_{MD} are calculated as the ratio between S/H. At 0 K all the GBs move according to the $\langle 110 \rangle$ mode as expected, which could be explained by considering the discrete dislocation model of coupling at low temperatures, where the critical stress for the GB motion is related to the Peierls-Nabarro stress for the glide of the array of parallel edge dislocations [40]. Peierls-Nabarro stress for the two set of dislocations could be compared by checking the γ surfaces corresponding to each slip plane in the absence of temperature. A γ surface represents the extra crystal energy per unit area of a generalized stacking fault, obtained by rigidly displace the two halves of a crystal by a vector parallel to such relevant slip plane [40]. Studies on Cu bicrystals reported by Cahn et al. [17] determined that higher energy barriers lead to more compact dislocation core and larger Peierls-Nabarro stress. Energy for $\{110\}\langle 110 \rangle$ and $\{100\}\langle 110 \rangle$ slip systems have been calculated in Ni by rigidly displacing the two halves of a crystal on such slip planes, resulting in values of 1.15 and 2.12 J/m^2, respectively. These γ energies should be used as a prediction since complex atomic movements may not be included in the Peierls-Nabarro model, as is the case of

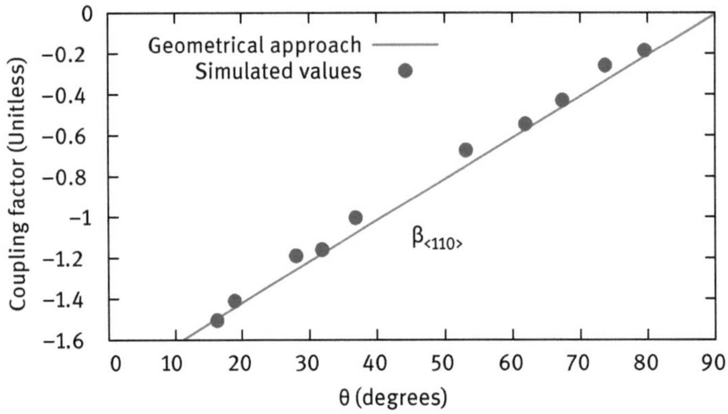

Fig. 4.4: Coupling factor obtained near 0 K by MD simulations is identified with red dots, and the branch of the ⟨110⟩ mode given by Eq. (4.4) is represented by a blue line.

the dissociation of the [100] dislocations at this temperature regime, involving screw and edge components.

4.3.3 Shear coupled motion at medium temperatures

At low and medium temperatures, the stop-and-go character is still the dominant response of the GB under simple shear. Fig. 4.5 illustrates the stress behaviour during coupling motion at 500 K for $\Sigma 17(410)$, showing the decreasing of the critical stress due to thermal fluctuations. Before doing the final movement, intermediate jumps are also observed, those are understood as a weak stability of the GB at this temperature [31]. The frequency of these reverse jumps increases with temperature, and GBs move in a more stochastic manner.

As temperature continues to increase, the shear stress does not display a clear stick-slip behaviour anymore, as illustrated in Fig. 4.6 for $\Sigma 17(410)$ at different temperatures. The stress is replaced by random noise with some average value, $\overline{\tau}_c$. This behaviour indicates a transition to a different response of the GB. It is interesting to examine the mechanisms existing at these temperatures. Despite describing such mechanisms is out of the scope of the present paper, modeling the coupling effect by the KMC method requires the knowledge of the temperature regimes at which the pure coupling is not observed anymore.

Fig. 4.7 illustrates the GB displacement as function of the strain for 600 and 800 K. Red line shows an overlap area in which both modes are observed. As shown in Fig. 4.8, after n-migration events according to the ⟨110⟩ mode, there is a switch to the ⟨110⟩ mode related to a change in the dislocation content, observed as an abrupt transition. This behaviour is only observed for $\theta < 30°$, within a temperature range between 600–900 K, specific to each GB. All the GBs identified as ⟨110⟩ in Tab. 4.1

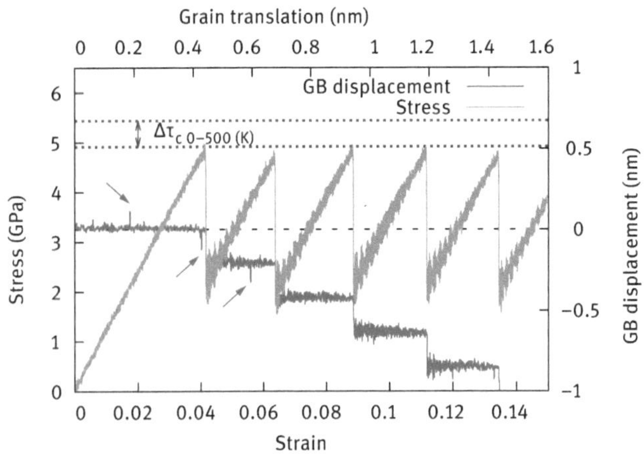

Fig. 4.5: Stress-strain and GB displacement during coupling motion of $\Sigma17(410)$ at 500 K with $v_{\|} = 1$ m/s. Dashed lines represent the decrease of the critical stress with temperature respect to τ_c^0. Arrows indicate reverse jumps of the GB.

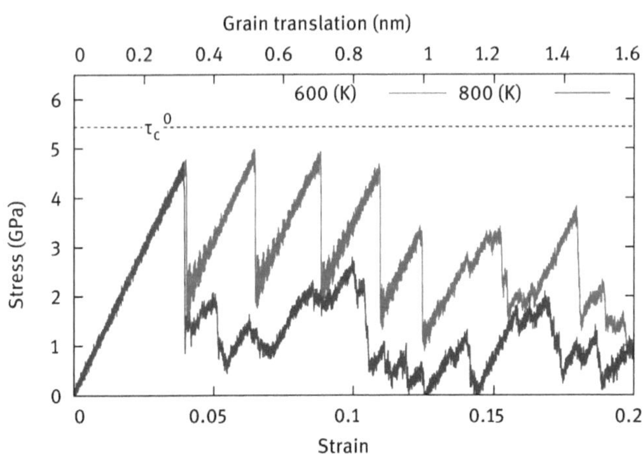

Fig. 4.6: Stress-strain curves during the coupled motion of $\Sigma17(410)$ at 600 K and 800 K, identified with red and blue lines respectively. The critical stress at 0 K is identified by a dashed line.

moves according to this mode once the activation stress is reached. Being known such GBs, as well as the temperature at which the second mode is activated, the Burgers vector to be introduced into KMC is defined as shown in the following sections.

Returning to Fig. 4.7, at 800 K GB coupled motion begins to be occasionally interrupted by sliding events, identified by the blue line. Grain translation is not accompanied by normal motion of the GB. Between sliding events, the GB continues to move in a coupled manner, with $\beta = 0.52$, in good agreement with predicted value shown in Fig. 4.8, $\beta = 0.50$ for $\Sigma17(410)$ $\theta = 28.07°$. In this case, the coupling factor is obtained

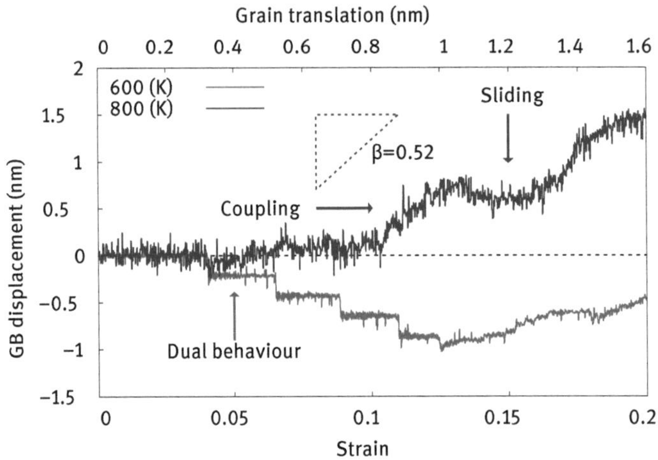

Fig. 4.7: GB displacement during the coupled motion of $\Sigma 17(410)$ at 600 K and 800 K, identified with red and blue lines respectively.

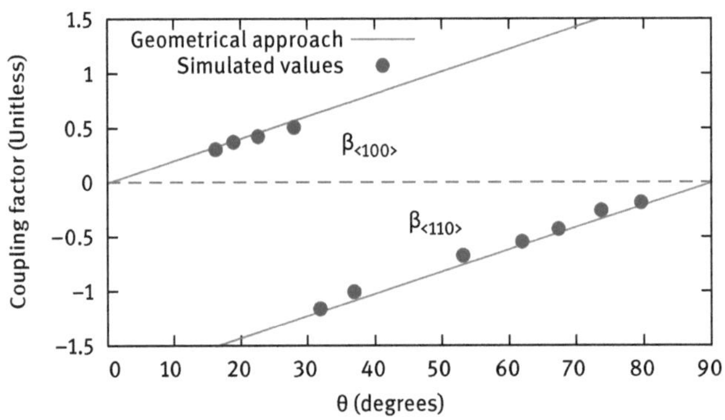

Fig. 4.8: Coupling factor obtained at 900 K as function of the tilt angle. Blue lines represent the two branches given by Eqs. (4.3) and (4.4).

from the slope of the correlation between grain translation and normal motion of the GB, since at this temperature, it is not possible to measure the ratio between S and H. The frequency of such sliding events increases with temperature, and the average normal velocity decreases. At a given temperature, approaching the melting point, $T_m = 1728$ K [41], GB shear-coupled motion is expected to be completely destroyed.

To understand the nature of the transition between these different mechanisms, $\Sigma 17(410)$ simulation results have been fit to Eq. (4.1), proposed by Ivanov and Mishin [28] to describe the temperature dependence of the peak stress within the stick-slip regime. Fig. 4.10 shows the linear dependence on $T^{2/3}$ of the stress averaged over the entire simulation time. The linearity indicates that the stick-slip behaviour is the dominant mechanism in this temperature regime. It should be noticed that Eq. (4.1) has been obtained assuming τ_c tends to τ_c^0. However, at temperatures 600 K and higher, $\tau \nrightarrow \tau_c^0$, and the average stress deviates from the linear fit. Such deviation indicates the transition to other mechanical response of the GB.

4.3.4 Nudged elastic band calculations

As described in Sec. 4.2, GB migration is a thermal activated process. GBs are described in terms of the energetic characteristics of the elementary steps that take place during the coupled motion. The nudged elastic band method (NEB) is used to compute the energy barrier to overcome for the GB motion between stable positions in the absence of temperature. The initial and final configurations, before and after the jump respectively, are known. The NEB uses N-replicas of the system obtained by linear interpolation between the initial and final state and tracks the atomic structure evolution through the sampling of a minimum energy path (MEP) [42]. The MEP describes the energy variation of the system along a collective reaction coordinate (RC) and provides the saddle point, which corresponds to the climbing replica with the highest energy.

In the present work, 32 replicas of the system are constructed. From the computed MEP for $\Sigma 17(410)$, the saddle point gives an energy barrier of 2.01 eV.

Assuming the GB shear-coupled motion is due to the collective glide of the array of dislocations, according to Hull and Bacon [40] the energy to be provided for the array motion has two contributions: the mechanical work done by the applied load and the thermal activation. As the latter increases, the mechanical contribution decreases since thermal fluctuations assist the array to overcome the barrier, and so the critical stress is reduced as explained in Sec. 4.2. The activation energy in the absence of temperature is linear in τ_c^0, and is given by:

$$\Delta E_0 = \tau_c^0 \boldsymbol{b}^2 l , \tag{4.6}$$

being τ_c^0 the required stress to move the array at zero temperature identified as the Peierls-Nabarro stress, $\boldsymbol{b}$ the Burgers vector, and l the length of the line. Eq. (4.6) estimates the $\Sigma 17(410)$ barrier as 2.20 eV, assuming τ_c^0 equal to 5.45 GPa (see Fig. 4.3). This value well matches the NEB results of 2.01 eV. It should be pointed out the relevance of the latter as input in the KMC modeling of this phenomena in next sections.

4.4 Kinetic Monte Carlo

While using MD, most of the time is spent in describing atomic vibrations, KMC characterizes the system by occasional transitions from state to state. KMC overcomes the time-scale problem by simulating the time evolution of a process occuring at a given known rate.

Coupled motion is a physical process in which the KMC method is applicable. GBs are not considered as a front, instead, the coupled motion is modeled in terms of the single constituent dislocations forming each boundary. By considering individual cores with an associate rate, KMC models the actual MEP: the motion begins with a step of a single unit or a group of them, initiating a disconnection, the first step spreads over the GB and the transition is completed [23, 43, 44].

4.4.1 Simulation methodology

In the present model, the KMC procedure is based on a stochastic algorithm that simulates the evolution of the system from state to state [45]. For a given configuration of the GB system, all possible processes are determined, e.g. the j-core jumps, and a list of rates, r_j, is built. The rates are calculated according to the transition state theory (TST) as:

$$r_j = r_0 \exp\left(-\frac{\Delta G}{k_B T}\right), \tag{4.7}$$

being r_0 the physical attempt frequency calibrated to be 10^{14} s^{-1}, ΔG the activation free energy, and having $k_B T$ the usual meaning. The total rate of events is calculated as:

$$R_i = \sum_{j=1}^{i} r_j , \tag{4.8}$$

for $i = 1, \ldots, N$, being N the total number of transitions. The actual escape takes place along one of these pathways. By generating a random number n within the range $[0, 1)$, the event to perform is selected, e.g. the j-core will jump. As a next step such event is performed, the disconnection is initiated, and the time is increased by:

$$\Delta t = \frac{1}{R_N} . \tag{4.9}$$

Once the affected rates are recalculated, the algorithm goes back to the computation of the cumulative function, Eq. (4.8).

As input of the KMC method, the transition rates are to be known in advance. Such rates are calculated based on the thermodynamics of dislocation motion reported by Hull and Bacon [40] and previously described in Sec. 4.3.4. The energy barrier that each unit sees when it moves is given by:

$$\Delta G = \Delta E_0 - \tau V , \tag{4.10}$$

being ΔE_0 equal to $\Delta E_{NEB} = 2.01$ eV, V the activation volume for the process calculated in the absence of temperature as $\Delta E_0/\tau$ and equal to 59.1 Å^3, and τV the mechanical contribution.

According to Tab. 4.2, ΔG in obtained knowing the stress at the initial step, t. As explained above, the event to perform is the discrete transition of a single core, that

Tab. 4.2: KMC algorithm to model the shear-coupled motion of GBs.

Calculate $\Delta G = \Delta E_0 - \tau V$
$\nu = \nu_0 \exp\left(-\frac{\Delta G}{k_B T}\right)$
$n \in [0, 1)$
Event j to perform: $R_{j-1} < rR_N < R_j$.
core position += b.
for (j < Ncores)
if (core position += b)
u −= Dissipated energy
$\tau = \sqrt{2\mu^{\mathrm{eff}} u}$
go to 1.

takes place in $\boldsymbol{b}$ units. When such transition occurs an amount of energy per unit volume is dissipated, and as a consequence there is a change in the total internal energy of the system which is related with the stress as:

$$u = \frac{1}{2}\frac{\tau^2}{\mu^{\mathrm{eff}}} \,. \tag{4.11}$$

By using the Eq. (4.11), the value of τ at $t + \Delta t$ is obtained and the algorithm goes back to the computation of ΔG.

4.4.2 Simulation results and discussion

The scaled-up parameters from MD to KMC are μ^{eff}, τ_c^0, $\boldsymbol{b}$, and ΔE_{NEB}, see Tab. 4.3. The dimensions and the number of cores of each GB are also introduced as an input. $\Sigma 17(410)$ is used as a model. The stress dependence on strain is computed at different temperatures and compared to MD results as displayed in Fig. 4.9.

Tab. 4.3: Input parameters incorporated to KMC obtained by MD simulations for $\Sigma 17(410)$.

$\mu^{\mathrm{eff}}_{\mathrm{MD}}$ (GPa)	τ_c^0 (GPa)	$b_{\langle 110\rangle}$ (Å)	$b_{\langle 100\rangle}$ (Å)	$\Delta E^0_{\mathrm{NEB}}$ (eV)
107.76	5.45	2.49	3.52	2.01

Results obtained via both techniques are in good agreement within a range of temperatures in which the stick-slip behaviour is the dominant response of the GB, seen in Fig. 4.9. As expected, the critical stress decreases with temperature. Thermal fluctuations reduce the mechanical contribution, allowing the GB to move before τ_c^0 is reached, identified with a dashed-blue line. For 800 K, the match between MD and KMC is not as clear as at lower temperatures. On the one hand, at such elevated tem-

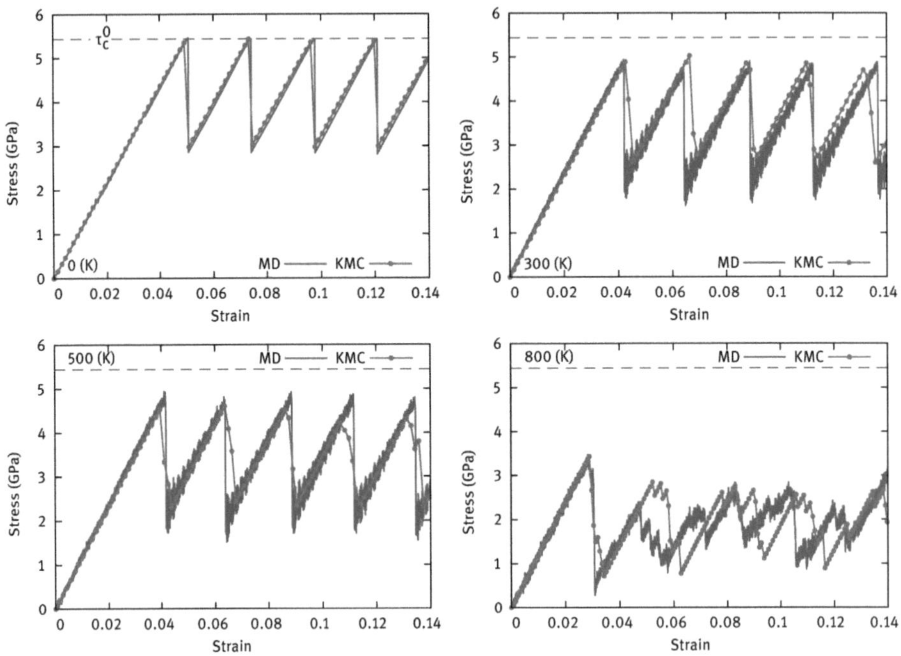

Fig. 4.9: MD and KMC modeling of GB shear-coupled motion at different temperatures for $\Sigma 17(410)$. Critical stress at zero temperature is identified with a dashed-blue line.

perature, there is random component in the process that should be taken into account, since there is a finite probability of spontaneous jumps to occur. On the other hand, from MD simulations, $\Sigma 17(410)$ coupled motion was observed to be occasionally interrupted by sliding events, and the noisy behaviour of the stress indicates the transition to other mode, see Fig. 4.6.

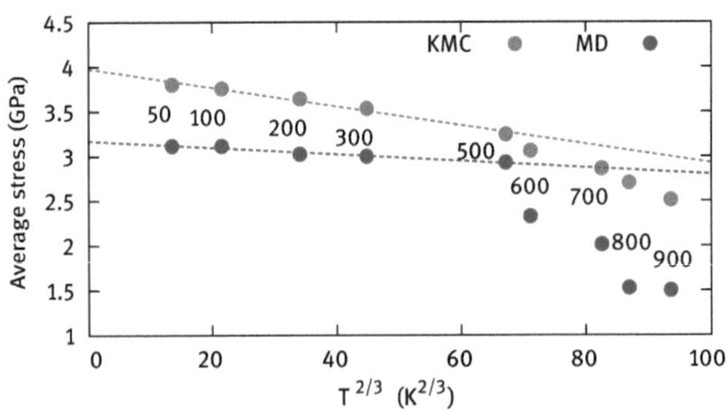

Fig. 4.10: Averaged stress as function of $T^{2/3}$ obtained by MD and KMC simulations. Blue and red lines represent the fit of KMC and MD results, respectively to Eq. 4.1 for temperatures below 500 K.

The presence of other mechanisms not modeled by KMC is thought to be responsible of such differences. Moreover, it should be taken into account that the $\Sigma 17(410)$ GB switches to the $\langle 110 \rangle$ mode at a temperature of 600 K, see Fig. 4.7. To model the GB motion in terms of the constituent dislocations, the change in the slip plane and the Burgers vector have to be included into the KMC model.

As a test of the agreement in the transition temperature in both models, the average stress obtained by KMC at different temperatures is fit to Eq. 4.1. Fig. 4.10 illustrates how the temperature at which the stick-slip behaviour breaks well matches the one obtained via MD. The deviation of the slope between both fittings is evident, however it should be noticed that here the average stress over the whole simulation time is considered, and according to Fig. 4.9 there is not a perfect match in the stress drop leading to different averaged stress. This mismatch in the stress drop is a limitation of the proposed KMC model, related with the fact that MD addresses the interaction between cores.

From Fig. 4.10, it could be confirmed that KMC restricts the modeling of GB migration to the case of stick-slip regime. This fact may reduce the accuracy of the description of the GB response in the whole range of temperatures. However the implementation of other mechanisms is an existing challenge for future work.

It should be mentioned at this point what is the relevance of KMC in terms of the computational time. MD simulations described in Sec. 4.3 take about 14 h, while modeling the same GB response using KMC takes only few seconds. The appeal of the KMC method is that it allows us to simulate larger sample sizes and longer time scales, allowing to overcome the limitations of regular MD, which implies that the imposed velocities are orders of magnitude higher than experiments. By using KMC, the shear velocity range has been greatly expanded, approaching to realistic deformation rates. The KMC model is applied to test the velocity dependence of the critical stress. At a fixed temperature of 400 K, the peak stress is averaged over the whole range of time and plotted versus the imposed velocities within a range from 60 µm/s to 2 m/s, as shown in Fig. 4.11. Simulated values are fit to Eq. 4.2, being η, τ_c^0 and E_0 adjustable parameters. The last two are compared to values obtained in previous sections in order to test the accuracy of the KMC method at modeling the coupling effect. The corresponding values from the fit in Fig. 4.11 are $\tau_c^0 = 5.63$ GPa and $E_0 = 2.01$ eV, both in good agreement with $\tau_c^0 = 5.45$ GPa obtained by MD simulations, and $E_0 = 2.01$ eV from NEB calculations, respectively.

After exploring the dynamics of coupled motion of GBs by means of MD and extending the model into KMC, a relation between the imposed velocity and the peak stress is proposed as:

$$v_{\parallel} = \gamma \boldsymbol{b} \exp\left(-\frac{\Delta E_0 - \tau V}{k_B T}\right), \tag{4.12}$$

being γ the attempt frequency related with vibration events, $\boldsymbol{b}$ the Burgers vector, ΔE_0 the energy barrier in the absence of temperature, V the activation volume for the process, and having $k_B T$ the usual meaning. This type of equations, introduced by Kocks

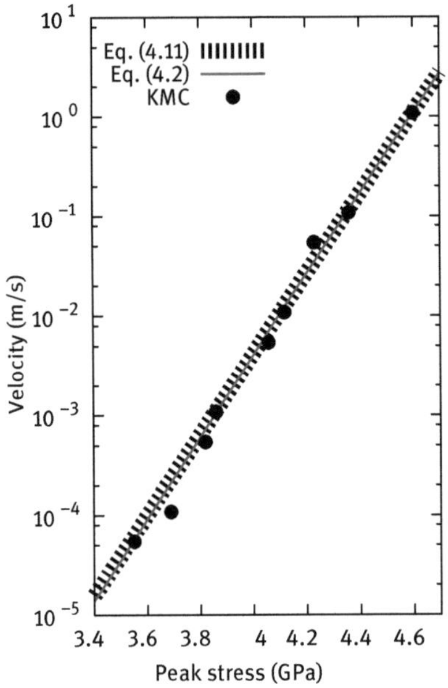

Fig. 4.11: $\Sigma17(410)$ velocity in logarithmic scale as function of the averaged peak stress at 400 K. Red line corresponds to the fit to Eq. (4.2) and black-dashed line to the fit to Eq. (4.12).

et al. [46] are commonly used in continuum crystal plasticity to link the macroscopic response to thermal activated processes occuring at the atomic scale. The KMC model accounts the thermal activated process assuming discrete dislocation cores. If the total number of cores tends to infinite, the response of the GB system fulfills Eq. (4.12).

To test this simple relation against atomistic simulations, points obtained at different imposed velocities at a temperature of 400 K are fit to Eq. (4.12), as shown in Fig. 4.11. The adjustable parameters obtained are $V = 52.8$ Å^3 and $\Delta E_0 = 1.99$ eV, both in good agreement with theoretical and NEB values of 59.1 Å^3 and 2.01 eV, respectively.

4.5 Concluding remarks

A bottom-up approach is presented to describe the shear coupled motion in Ni bicrystals. To that end two different atomistic simulation techniques are used: MD and KMC. Although MD is a useful tool in understanding the elementary atomic mechanism that govern the GB migration, simulating realistic sample sizes, boundary conditions and deformation rates are subjected to the limitations of the technique. Using a statistical approach involving KMC methods, the modeling is extended closer to reality by incorporating the output obtained via MD simulations.

At low temperatures, the GB migration exhibits the so-called stick-slip behaviour, characterized by the typical saw-tooth strain dependence of the stress. The stop-and-go motion of the eleven symmetrical tilt GBs has been described in terms of the ratio between the normal motion of the GB and the rigid translation of the adjacent grains.

The motion of the GBs has been also examined at high temperatures, to shed light into the possible changes of the dynamics of GB motion. Identifying the temperature at which the presence of competing mechanisms are observed, is the key to model the pure coupling response in KMC.

Exploring the dynamics of GB shear-coupled motion in terms of the energy leads to identify the mechanism as a thermal activated process [31]. The GB moves between equivalent stable positions separated by an energy barrier, which is calculated using the NEB method.

Once the GB structures are fully described focusing on both the energetic and structural characteristics, the output is extended into a KMC model. KMC describes the evolution of the system through occasional transitions from one state to another.

Furthermore, the connection between both techniques has allowed to extend the physical mechanisms of the GB shear-coupled motion into a statistical approach. It should be mentioned how the use of KMC involves an important saving of computational time. The simulation time is reduced from 14 h spent in MD simulations to 2 s in KMC. As a consequence, KMC allows to extend the deformation rates by more than 4 orders of magnitude, and imposing grain translation velocities during the shear closer to experimental velocities.

Acknowledgment: The authors acknowledge partial funding by Abengoa Research. I. M.-B. wants also to acknowledge partial funding from the “Subprograma Ramón y Cajal” fellowship RYC-2012-10639 by the Spanish Ministry of Economy and Competitiveness.

4.A Effective shear modulus for planar GBs: Application to [001] STGB contained in bicrystal structures

Considering the schematic structure of the bicrystal represented in Fig. 4.12, the effective elastic tensor is calculated according to [39]. In first place, the anisotropic elastic constants of each consituent crystal are obtained by the following transformation rules [47]:

$$C_{ijkl} = C_{12}\delta_{ij}\delta_{kl} + C_{44}(\delta_{ik}\delta_{jl} + \delta_{il}\delta_{jk}) + C_0 \sum_{s\neq 1}^{3} e_i^{(s)} e_j^{(s)} e_k^{(s)} e_l^{(s)}\,, \tag{4.13}$$

where $C_0 = C_{11} - C_{12} - 2C_{44}$, being C_0, C_{11}, and C_{14} the elastic constant for pure Ni [34]. φ is the rotation angle of [100] around the tilt axis, then $\varphi^{\mathrm{I}} = \theta/2$ and $\varphi^{\mathrm{II}} = -\theta/2$. The units vectors of the cubic lattice system are:

$$e_i^{(1)} = \begin{pmatrix} \cos\varphi \\ \sin\varphi \\ 0 \end{pmatrix}, \quad e_i^{(2)} = \begin{pmatrix} -\sin\varphi \\ \cos\varphi \\ 0 \end{pmatrix}, \quad e_i^{(3)} = \begin{pmatrix} 0 \\ 0 \\ 1 \end{pmatrix}. \tag{4.14}$$

Assuming the volume fraction is equal for both constituent crystals, $f^{\mathrm{I}} = f^{\mathrm{II}} = 0.5$, the transformation field analysis (TFA) is applied for this particular case. The convention proposed by Nye [48] is adopted,

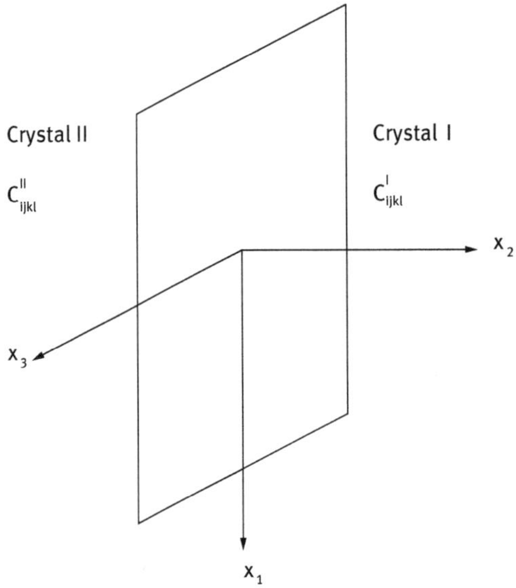

Fig. 4.12: Schematic structure of the bicrystal with the coordinate axes used to compute the elastic tensor.

where pairs of subscripts i, j and k, l are converted to single ones as: $11 \to 1$, $22 \to 2$, $33 \to 3$, 13 and $31 \to 5$, 12 and $21 \to 6$. Referred to the bicrystal structure in Fig. 4.12, the in-plane matrix components of the tensor are denoted $P = 1, 3, 5$ and the out-plane are $A = 2, 4, 6$. As a result, the effective elastic modulus C^{eff}_{ijkl}:

$$\begin{pmatrix} \Sigma_A \\ \Sigma_P \end{pmatrix} = \begin{pmatrix} C^{\text{eff}}_{AA} & C^{\text{eff}}_{AP} \\ C^{\text{eff}}_{PA} & C^{\text{eff}}_{PP} \end{pmatrix} \begin{pmatrix} E_A \\ E_P \end{pmatrix} \to C^{\text{eff}}_{ijkl} = \begin{pmatrix} C^{\text{eff}}_{AA} & C^{\text{eff}}_{AP} \\ C^{\text{eff}}_{PA} & C^{\text{eff}}_{PP} \end{pmatrix} . \tag{4.15}$$

By using the matrix forms of C^{eff}_{AA}, C^{eff}_{AP}, C^{eff}_{PA} and C^{eff}_{PP} in Ref. [39], the effective shear modulus is given by:

$$\mu^{\text{eff}} = \frac{1}{2}\left(C^{\text{I}}_{2121} - \frac{C^{\text{I},2}_{2122}}{C^{\text{I}}_{2222}} \right) + \frac{1}{2}\left(C^{\text{II}}_{2121} - \frac{C^{\text{II},2}_{2122}}{C^{\text{II}}_{2222}} \right) . \tag{4.16}$$

By using Nye convention, Eq. (4.16) is equal to Eq. (4.5) in Sec. 4.3.2.

References

[1] Hirth J. The influence of grain boundaries on mechanical properties. Metall Mater Trans B. 1972;3(12):3047.

[2] Wulfinghoff S, Bayerschen E, Böhlke T. A gradient plasticity grain boundary yield theory. Int J Plast. 2013;51:33.

[3] Farkas D. Atomistic theory and computer simulation of grain boundary structrure and diffusion. J Phys Conds Matter. 2000;12:497.

[4] Mishin Y, Asta M, Li J. Atomistic modeling of interfaces and their impact on microstructure and properties. Acta Mater. 2010;58:117–115.

[5] Bieler T, Eisenlohr P, Roters F, Kumar D, Mason D, Crimp M, et al. The role of heterogeneous deformation on damage nucleation at grain boundaries in single phase metals. Int J Plast. 2009;25:1665.

[6] Dahlberg C, Faleskog J, Niordson C, Legarth B. A deformation mechanisms map for polycrystals modeled using strain gradient plasticity and interfaces that slide separate. Int J Plast. 2013;43: 177.
[7] Du N, Qi Y, Krajewski P, Bower A. Aluminum grain boundary sliding enhanced by vacancy difussion. Acta Mater. 2010;58:4245–4252.
[8] Schäfer J, Albe K. Competing deformation mechanisms in nanocrystalline metals and alloys: Coupled motion versus grain boundary sliding. Acta Mater. 2012;60:6076.
[9] Warner D, Roters F, Molinari J. Atomistic based continuum investigations of plastic deformation in nanocrystalline copper. Int J Plast. 2006;22:754.
[10] Sansoz F, Molinari JF. Mechanical behaviour of Sigma tilt grain boundaries in nanoscale Cu and Al: A quasicontinuum study. Acta Mater. 2005;52:1931–1944.
[11] Tschopp M, McDowell D. Dislocation nucleation in $\Sigma 3$ assymmetric tilt grain boundaries. Int J Plast. 2008;24:191.
[12] Spearot D, Jacob K, McDowell D. Dislocation nucleation from bicrystal interfaces with dissociated structures. Int J Plast. 2007;23:143.
[13] Ohashi T, Barabash R, Pang J, Ice G, Barabash O. X-ray microdiffraction and strain gradient crystal plasticity studies on geometrically necessary dislocations near a ni bicrystal grain boundary. Int J Plast. 2009;25:920.
[14] Li C, Edwards E, Washburn J, Parker E. Stress-induced movements of crystal boundaries. Acta Metall. 1953;1:223.
[15] Read W, Shockley W. Dislocation models of crystal boundaries. Phys Rev B. 1950;78:275.
[16] Goryaka T, Modolov DA, Gottstein G. Stress-driven of symmetrical $\langle 110 \rangle$ tilt grain boundaries in al bicrystals. Acta Mater. 2009;57:5396–5405.
[17] Cahn JW, Mishin Y, Suzuki A. Coupling grain boundary motion to shear deformation. Acta Mater. 2006;54:4953–4975.
[18] Frolov T. Effect of interfacial structural phase transitions on the coupled motion of grain boundaries: A molecular dynamics study. Appl Phys Lett. 2014;104(211905).
[19] Huang B, Shang J, Liu Z, Chen Y. Atomic simulation of bcc niobium $\Sigma 5$ $\langle 110 \rangle$ $\{310\}$ grain boundary under shear deformation. Acta Mater. 2014;77:258.
[20] Cahn JW, Taylor J. A unified approach to motion of grain boundaries, relative tangential translation along grain boundaries, and grain rotation. Acta Mater. 2004;52(16):4887–4898.
[21] Homer E, Foiles S, Holm E, Olmsted D. Phenomenology of shear-coupled grain boundary motion in symmetric tilt and general grain boundaries. Acta Mater. 2013;61:1048.
[22] Tucker G, Zimmerman J, McDowell D. Shear deformation kinematics on bicrystalline grain boundaries in atomistic simulations. Modelling Simul Mater Sci Eng. 2010;18(0150002).
[23] Rajabzadeh A, Mompiou F, Lartigue-Korinek S, Combe N, Legros M, Modolov DA. The role of disconnections in deformation-coupled grain boundary migration. Acta Mater. 2014;77:223.
[24] Goryaka T, Modolov K, Modolov DA, Gottstein G. Concurrent grain boundary motion and grain boundary rotation under an applied stress. Acta Mater. 2011;59:5674.
[25] Berbenni S, Paliwal B, Cherkaoui M. Generalized Continua as Models for Materials, vol. 22. Berlin/Heidelberg: Springer; 2013.
[26] Berbenni S, Paliwal B, Cherkaoui M. A micromechanics-based model for the shear-coupled grain boundary migration in bicrystals. Int J Plast. 2013;44:68–94.
[27] Taupin V, Capolungo L, Fressengeas C. Disclination mediated plasticity in shear-coupled boundary migration. Int J Plast. 2014;53:179.
[28] Ivanov VA, Mishin Y. Dynamics of grain boundary motion coupled to shear deformation: An analytical model and its verification by molecular dynamics. Phys Rev B. 2008;78(064106).
[29] Sang Y, Dube M, Grant M. Thermal effects on atomic friction. Phys Rev Lett. 2001, 87.

[30] Garg A. Escape-field distribution for escape from a metastable potential well subjected to a steadily increasing bias field. Phys Rev B. 1995;51(15592).
[31] Mishin Y, Suzuki A, Uberuaga BP, Voter AF. Stick-slip behaviour of gbs studied by accelerated md. Phys Rev B. 2007;75(224101).
[32] Frank FC. The resultant content of dislocation in an arbitrary intercrystalline boundary. Symposium on the plastic deformation of crystalline solids. Pittsburgh: Carnegie Institute of Technology; 1953.
[33] Bilby BA. Types of dislocation sources. The Physical Society of London; 1955.
[34] Mishin Y, Farkas D, Mehl MJ, Papaconstantopoulos D. Interatomic potentials for monoatomic metals from experimental data and ab initio calculations. Phys Rev B. 1999;59(3393).
[35] Kronenberg M, Wilson F. Secondary recrystallization in copper. Trans Met Soc AIME. 1947;185: 501.
[36] Plimpton SJ. Fast parallel algorithms for short-range molecular dynamics. J Comp Phys. 1995; 117:1–19.
[37] Faken D, Jonsson H. Systematic analysis of local atomic structure combined with 3d computer graphics. Comput Mater Sci. 1994;2(279).
[38] Dvorak GJ. On uniform fields in heterogeneous media. In: Proc. R. Soc. Lond. A, vol. 431. pp. 89–110.
[39] Franciosi P, Berbenni S. Multi-laminate plastic strain organization for non-uniform tfa modeling of polycrystal regularized plastic flow. Int J Plast. 2008;24:1549–1580.
[40] Hull D, Bacon D. Introduction to dislocations. Pergamon Press; 1965.
[41] Turley J, Sines G. The anisotropy of young modulus, shear modulus and poisson ratio in cubic materials. J Phys D: Appl Phys. 1971, 4.
[42] Espinosa H, Bao G. Nano and cell mechanics: Fundamentals and frontiers. John Wiley and Sons; 2013.
[43] Rajabzadeh A, Legros M, Combe N, Mompiou F, Modolov DA. Evidence of gb dislocation step motion associated to shear coupled gb migration. Philos Mag. 2013;93(10–12):1299–1316.
[44] Rajabzadeh A, Mompiou F, Le M, Combe N. Elementary mechanisms of shear-coupled grain boundary migration. Phys Rev Lett. 2013;110(265507).
[45] Voter A. Introduction to kinetic monte carlo method. In: NATO Science Series, vol. 253.
[46] Kocks U, Argon A, Ashby F. Thermodynamics and kinetics of slip, vol. 19. Pergamon Press; 1975.
[47] Gemperlova J, Paidar V, Kroupa V. Compatibility stresses in deformed bicrystals. Czech J Phys B. 1989;39:427.
[48] Nye J. Physical properties of crystals. Oxford: Clarendon Press; 1957.

M. Labusch, M. Etier, D. Lupascu, J. Schröder, and M.-A. Keip

5 Product Properties of a two-phase magneto-electric composite

Synthesis and numerical modeling

Abstract: Magneto-electric (ME) materials are of high interest for a variety of advanced applications like in data storage and sensor technology. Due to the low ME coupling in natural materials, composite structures become relevant which generate the effective ME coupling as a strain-mediated product property. In this framework it seems to be possible to achieve effective ME coefficients that can be exploited technologically. The present contribution investigates the realization of particulate ME composites with a focus on their experimental and computational characterization. We will show that different states of pre-polarizations of the ferroelectric material have a decisive influence on the overall obtainable ME coefficient. Details on the synthesis of two-phase composite microstructures consisting of a barium titanate matrix and cobalt ferrite inclusions will be discussed. Subsequently we will employ computational homogenization in order to determine the effective properties of the experimental composite numerically. We investigate the influence of different states of pre-polarization on the resulting ME-coefficients. For the numerical incorporation of the pre-polarization we use a heuristic method.

Keywords: modulus and composites, homogenization, materials processing, product property, magneto-electric composite

5.1 Introduction

In the last decades there has been a steady increase in attention to smart functional materials that couple different physical quantities to one another like, for example, polarization and strain or magnetization and strain. By using such materials smart applications in sensor and actuator technology can be realized. An important coupling phenomenon that attracts particular attention recently is the *magneto-electric* (ME) coupling property. Materials exhibiting pronounced couplings between magnetic and electric quantities could be used in a number of applications as, for example, in electrical magnetic-field sensors or electric-write magnetic-read memories, see e.g. [1]. The latter could be constructed as a so-called Magneto-Electric Random Access Memory

---

Note: Contribution for the Special Issues on Modeling and Simulation of Advanced Manufacturing Processes – T. I. Zohdi, Handling Editor.

(MERAM), which would be realized by using materials that possess both *spontaneous magnetization and spontaneous polarization coupled to one another*, see, for example, [2]. A material that exhibits these two properties at the same time is termed *magneto-electric multiferroic*, where the term *multiferroic* defines a material that has at least two distinct ferroic states. Consequently, an ME multiferroic is ferroelectric and ferromagnetic (which is often accompanied by a third ferroic state given by ferroelasticity). ME multiferroics have been investigated intensively, see for example [3–8], [9, 10]. However, since the ME coupling in natural materials is found to be very weak, see [11] who determined an upper bound for the possible ME coupling coefficient of single-phase materials, the development of *ME composite materials* consisting of electro-active and magneto-active phases becomes relevant. ME composites produce the desired ME coefficients as a *strain-mediated product property*, see, for example, [12–18]. A product property of a composite is defined as a property that is not present in each of its constituents, but appears effectively through their interaction; a general treatment on possible product properties is given in [19]. The ME coupling in a composite is generated through electrically or magnetically induced strain, where one distinguishes between the direct and the converse ME effect. The direct effect characterizes magnetically induced polarization: an applied magnetic field yields a deformation of the magneto-active phase that is transferred to the electro-active phase. As a result, this delivers *strain-induced polarization* in the electric phase. On the other hand, the converse effect characterizes electrically activated magnetization: an applied electric field yields a deformation of the electro-active phase which is transferred to the magneto-active material. This deformation then results in *strain-induced magnetization*. Several experiments on composite ME multiferroics showed remarkable ME coefficients that are orders of magnitudes higher than those of single-phase materials, see e.g. [20].

From a theoretical viewpoint the macroscopic characterization of ME composites – and therewith the *determination of their effective behavior* – is of particular interest. As a consequence, the development of suitable homogenization methods becomes important. In recent years, numerous analytical methods for the determination of the effective response of ME composites have been developed, see, for example, [21–27]. However, analytical schemes are often limited to specific microscopic morphologies, so that computational schemes are becoming relevant. One such method based on a Finite-Element (FE) discretization has been developed in [28] in order to determine the effective properties of piezomagnetic/piezoelectric ME composites.

In the present contribution we employ an extension of the work [29] to the case of magneto-electro-mechanical coupling, see [30] for details. We will focus on the *computational characterization of realistic, experimentally measured ME microstructures* and investigate implications that originate from specific assumptions of microscopic properties. The composite under consideration is depicted on the left hand side of Fig. 5.1. Here we see the morphology of a 0-3 particulate composite, where the brighter regions show the electro-active barium titanate ($BaTiO_3$) and the darker regions show

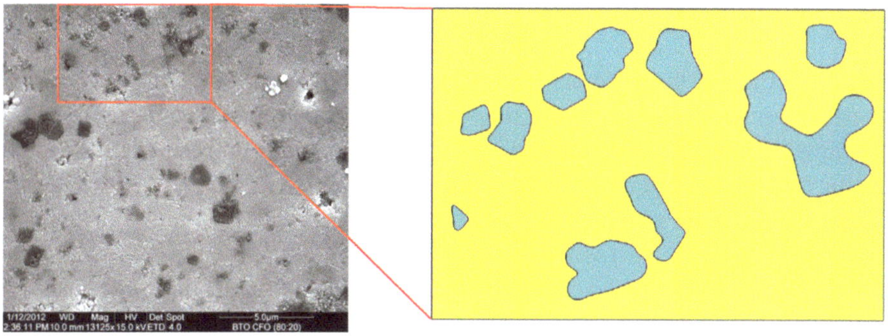

Fig. 5.1: Real composite microstructure with framed selected part.

the magneto-active cobalt ferrite ($CoFe_2O_4$). As implicitly defined above, we will denote the type of the composite by x-y with $\{x, y\} \in [1, 2, 3]$, where the first number indicates the spatial connectivity of the magneto-active phase and the second number indicates the connectivity of the electro-active phase.

In detail, the present paper reports on the investigation of experimental manufacturing and characterization techniques of ME composites as well as the associated implications for computational homogenization. Based on the homogenization procedure we compute the overall ME coefficient of two-phase composites consisting of piezoelectric and piezomagnetic phases. On the one hand, we apply the method to the simulation of idealized microstructures. Here, we compare our results to results taken from the literature [28] in order to validate the implemented model. On the other hand, we will apply the method to the computational characterization of a realistic microstructure, which we created using organosol crystallization. In this connection, we will discuss the influence of constitutive assumptions made in the numerical model and its implications on the predictable effective ME coefficient. The outline of the paper is as follows. In Sec. 5.2 the theoretical framework will be provided where the procedure for the determination of the effective properties will be discussed in detail. Sec. 5.3 gives a brief overview on the experimental manufacturing of the composite sample and describes the used measuring technique. Afterwards in Sec. 5.4, several numerical examples with idealized and realistic two-phase composite microstructures will be performed. The effective properties of the real microstructure are then compared to experimental measurements. Different considerations of the polarization directions on the microscale demonstrate their influence on the effective ME coupling coefficient. Sec. 5.5 closes the paper with a short conclusion.

5.2 Theoretical framework

In the following the magneto-electro-mechanically coupled boundary value problem is briefly described. We set up the fundamental continuum balance equations and introduce the basic kinematic, electric and magnetic quantities. Furthermore, we outline the constitutive equations for the piezoelectric and piezomagnetic phase on the microscale. The main goal of this work is to characterize the constitutive behavior of ME composites at the macroscopic scale in consideration of the composition of the microstructure. Due to the leading part of the morphology of the microstructure with respect to the overall (macroscopic) ME coupling coefficient we use a scale-transition between both scales, the macro- and the microscale. The scales are connected by a so-called localization and a homogenization step, where the latter one yields the effective properties of the composite, especially the overall ME-coefficient. In order to ease the readability of the following sections, we summarize the basic magneto-electro-mechanical quantities in Tab. 5.1.

Tab. 5.1: Magneto-electro-mechanical quantities and corresponding SI-Units.

Symbol	Continuum mechanical description	SI-Unit
$\boldsymbol{u}$	displacement vector	m
$\boldsymbol{\varepsilon}$	linear strain tensor	1
$\boldsymbol{\sigma}$	Cauchy stress tensor	N/m^2
$\boldsymbol{t}$	traction vector	N/m^2
$\boldsymbol{f}$	mechanical body forces	N/m^3
ϕ_e	electric potential	V
$\boldsymbol{E}$	electric field vector	V/m
$\boldsymbol{D}$	electric displacement vector	C/m^2
Q_e	electric surface flux density	C/m^2
ρ	density of free charge carriers	C/m^3
ϕ_m	magnetic potential	A
$\boldsymbol{H}$	magnetic field	A/m
$\boldsymbol{B}$	magnetic flux density	Vs/m^2
Q_m	magnetic surface flux density	Vs/m^2

5.2.1 Magneto-electro-mechanical boundary value problem

The considered micro-heterogeneous body $\mathcal{B} \subset \mathbb{R}^3$, characterized by a representative volume element $\mathcal{RVE} \subset \mathcal{B} \subset \mathbb{R}^3$ on the microscale, is parameterized with the coordinates $\boldsymbol{x}$. In order to capture the influence of the microstructure on the macroscopic behavior, we have to choose a suitable representative volume element which has to reflect the main mechanical, electrical, and magnetical characteristics of the microstructure. A periodic multiplication of the chosen $\mathcal{RVE}$ in all spatial directions is then imagined as an approximation of the real microstructure. The basic variables on the microscale, i.e., the linear strain tensor $\boldsymbol{\varepsilon}$, the electric field vector $\boldsymbol{E}$ and the magnetic field vector $\boldsymbol{H}$, are given by

$$\boldsymbol{\varepsilon} = \frac{1}{2}(\nabla \boldsymbol{u} + \nabla^T \boldsymbol{u}), \quad \boldsymbol{E} = -\nabla\phi_{\mathrm{e}}, \quad \text{and} \quad \boldsymbol{H} = -\nabla\phi_{\mathrm{m}}. \tag{5.1}$$

The underlying fundamental balance equations are given by the balance of linear momentum, Gauß's law of electrostatics and Gauß's law of magnetostatics

$$\nabla \cdot \boldsymbol{\sigma} + \boldsymbol{f} = \boldsymbol{0}, \quad \nabla \cdot \boldsymbol{D} = \rho, \quad \text{and} \quad \nabla \cdot \boldsymbol{B} = 0 \quad \text{in } \mathcal{RVE}. \tag{5.2}$$

The associated macroscopic quantities are (under some technical assumptions) defined in terms of suitable volume averages. For simplicity, we restrict ourselves to the following definitions

$$\overline{\boldsymbol{\xi}} = \frac{1}{\mathrm{vol}(\mathcal{RVE})} \int_{\mathcal{RVE}} \boldsymbol{\xi} \, \mathrm{d}v \quad \text{with } \{\boldsymbol{\xi} := \boldsymbol{\sigma}, \boldsymbol{\varepsilon}, \boldsymbol{D}, \boldsymbol{E}, \boldsymbol{B}, \boldsymbol{H}\}. \tag{5.3}$$

For a more general definition of the macroscopic quantities in terms of suitable surface integrals we refer to [31].

In order to derive energetically consistent boundary conditions in an algorithmically attractive representation we decompose the microscopic fields into the constant macroscopic part $\overline{\boldsymbol{\xi}}$ and a fluctuation part $\widetilde{\boldsymbol{\xi}}$, i.e.,

$$\boldsymbol{\xi} = \overline{\boldsymbol{\xi}} + \widetilde{\boldsymbol{\xi}} \quad \text{with } \{\boldsymbol{\xi} := \boldsymbol{\sigma}, \boldsymbol{\varepsilon}, \boldsymbol{D}, \boldsymbol{E}, \boldsymbol{B}, \boldsymbol{H}\}. \tag{5.4}$$

For the construction of energetically consistent boundary conditions we postulate a generalized macrohomogeneity condition, which states the equality of the macroscopic and averaged microscopic power [32]

$$\overline{\boldsymbol{\sigma}} : \dot{\overline{\boldsymbol{\varepsilon}}} - \overline{\boldsymbol{D}} \cdot \dot{\overline{\boldsymbol{E}}} - \overline{\boldsymbol{B}} \cdot \dot{\overline{\boldsymbol{H}}} = \frac{1}{\mathrm{vol}(\mathcal{RVE})} \int_{\mathcal{RVE}} (\boldsymbol{\sigma} : \dot{\boldsymbol{\varepsilon}} - \boldsymbol{D} \cdot \dot{\boldsymbol{E}} - \boldsymbol{B} \cdot \dot{\boldsymbol{H}}) \, \mathrm{d}v, \tag{5.5}$$

see also [29, 31]. The macrohomogeneity condition is fulfilled by applying Dirichlet-, Neumann-, or periodic boundary conditions, for details see [30]. Furthermore, also the microscopic displacement field, the electric potential, and the magnetic potential are additively decomposed into affine and fluctuating parts

$$\boldsymbol{u} = \overline{\boldsymbol{\varepsilon}} \cdot \boldsymbol{x} + \widetilde{\boldsymbol{u}}, \quad \phi_{\mathrm{e}} = -\overline{\boldsymbol{E}} \cdot \boldsymbol{x} + \widetilde{\phi}_{\mathrm{e}}, \quad \phi_{\mathrm{m}} = -\overline{\boldsymbol{H}} \cdot \boldsymbol{x} + \widetilde{\phi}_{\mathrm{m}}, \tag{5.6}$$

respectively. Suitable periodic boundary conditions are

$$\begin{aligned} \widetilde{\boldsymbol{u}}(\boldsymbol{x}^+) &= \widetilde{\boldsymbol{u}}(\boldsymbol{x}^-) \quad &\text{and} \quad \boldsymbol{t}(\boldsymbol{x}^+) &= -\boldsymbol{t}(\boldsymbol{x}^-), \\ \widetilde{\phi}_{\mathrm{e}}(\boldsymbol{x}^+) &= \widetilde{\phi}_{\mathrm{e}}(\boldsymbol{x}^-) \quad &\text{and} \quad Q_{\mathrm{e}}(\boldsymbol{x}^+) &= -Q_{\mathrm{e}}(\boldsymbol{x}^-), \\ \widetilde{\phi}_{\mathrm{m}}(\boldsymbol{x}^+) &= \widetilde{\phi}_{\mathrm{m}}(\boldsymbol{x}^-) \quad &\text{and} \quad Q_{\mathrm{m}}(\boldsymbol{x}^+) &= -Q_{\mathrm{m}}(\boldsymbol{x}^-), \end{aligned} \tag{5.7}$$

where $\boldsymbol{x}^+$ and $\boldsymbol{x}^-$ denote associated points on the boundary of a periodic unit cell. After the solution of the boundary value problem, we can compute the homogenized macroscopic response by averaging over the microscopic fields over the $\mathcal{RVE}$.

5.2.2 Constitutive framework on the microscale

We analyze the magneto-electric coupling behavior of a two-phase composite containing a piezoelectric and a piezomagnetic phase. In our oversimplified model we assume a transversely isotropic linear material law for both phases on the microscale and make use of the constitutive framework proposed in [33], which can be analogously extended to magneto-electro-mechanically coupled problems. In detail each phase is characterized by a preferred direction $\boldsymbol{a}$ in which the piezoelectric and piezomagnetic coupling is active, respectively. This preferred direction can be imagined as the direction of remanent polarization and magnetization, respectively. On the microscale the behavior of the two phases can be described by thermodynamical functions, which are defined for the piezoelectric phase as

$$\psi_{\mathrm{e}} = \frac{1}{2}\boldsymbol{\varepsilon} : \mathbb{C}_{\mathrm{e}} : \boldsymbol{\varepsilon} - \boldsymbol{E} \cdot \boldsymbol{e} : \boldsymbol{\varepsilon} - \frac{1}{2}\boldsymbol{E} \cdot \boldsymbol{\epsilon}_{\mathrm{e}} \cdot \boldsymbol{E} - \frac{1}{2}\boldsymbol{H} \cdot \boldsymbol{\mu}_{\mathrm{e}} \cdot \boldsymbol{H} \tag{5.8}$$

and for the piezomagnetic phase as

$$\psi_{\mathrm{m}} = \frac{1}{2}\boldsymbol{\varepsilon} : \mathbb{C}_{\mathrm{m}} : \boldsymbol{\varepsilon} - \boldsymbol{H} \cdot \boldsymbol{q} : \boldsymbol{\varepsilon} - \frac{1}{2}\boldsymbol{H} \cdot \boldsymbol{\mu}_{\mathrm{m}} \cdot \boldsymbol{H} - \frac{1}{2}\boldsymbol{E} \cdot \boldsymbol{\epsilon}_{\mathrm{m}} \cdot \boldsymbol{E}\,. \tag{5.9}$$

$\mathbb{C}$, $\boldsymbol{\epsilon}$, $\boldsymbol{\mu}$, $\boldsymbol{e}$, and $\boldsymbol{q}$ denote the tensors of elasticity, dielectric permittivity, magnetic permeability, piezoelectric coupling, and piezomagnetic coupling of the individual electroactive and magnetoactive phases $\{e, m\}$, respectively. The function of the piezoelectric phase ψ_{e} includes a mechanical part, a coupling term between electric fields and mechanical strains as well as a purely electrical and a purely magnetical part. Analogously, the piezomagnetic energy ψ_{m} includes besides the mechanical part a coupling term between the mechanical strains and the magnetic field as well as a purely magnetical and a purely electrical part. For both phases we write down the incremental constitutive equations

$$\underbrace{\begin{bmatrix} \Delta\boldsymbol{\sigma}_{\mathrm{e,m}} \\ -\Delta\boldsymbol{D}_{\mathrm{e,m}} \\ -\Delta\boldsymbol{B}_{\mathrm{e,m}} \end{bmatrix}}_{\mathfrak{S}} = \underbrace{\begin{bmatrix} \mathbb{C}_{\mathrm{e,m}} & -\boldsymbol{e}^T_{\mathrm{e,m}} & -\boldsymbol{q}^T_{\mathrm{e,m}} \\ -\boldsymbol{e}_{\mathrm{e,m}} & -\boldsymbol{\epsilon}_{\mathrm{e,m}} & \boldsymbol{0} \\ -\boldsymbol{q}_{\mathrm{e,m}} & \boldsymbol{0} & -\boldsymbol{\mu}_{\mathrm{e,m}} \end{bmatrix}}_{\mathfrak{C}} \underbrace{\begin{bmatrix} \Delta\boldsymbol{\varepsilon}_{\mathrm{e,m}} \\ \Delta\boldsymbol{E}_{\mathrm{e,m}} \\ \Delta\boldsymbol{H}_{\mathrm{e,m}} \end{bmatrix}}_{\mathfrak{D}}\,. \tag{5.10}$$

Here it has to be mentioned, that the magneto-electric (ME) coupling modulus for each of the two phases on the microscale, denoted by $\boldsymbol{\alpha}$, is equal to zero:

$$\boldsymbol{\alpha}_{\mathrm{e,m}} \equiv \boldsymbol{0}\,. \tag{5.11}$$

However, the overall ME-coefficients are generally nonzero:

$$\overline{\boldsymbol{\alpha}} \neq \boldsymbol{0}\,. \tag{5.12}$$

One of the main goals of this contribution is to determine the effective ME coupling coefficient computationally and to compare the results with experimental observations.

5.2.3 Constitutive framework of ME composites on the macroscale

In order to approximate the magneto-electric coupling behavior on the macroscale, we assume the existence of an overall (macroscopic) thermodynamical function

$$\overline{\psi} = \frac{1}{2}\overline{\boldsymbol{\varepsilon}} : \overline{\mathbb{C}} : \overline{\boldsymbol{\varepsilon}} - \overline{\boldsymbol{H}} \cdot \overline{\boldsymbol{q}} : \overline{\boldsymbol{\varepsilon}} - \overline{\boldsymbol{E}} \cdot \overline{\boldsymbol{e}} : \overline{\boldsymbol{\varepsilon}} - \frac{1}{2}\overline{\boldsymbol{E}} \cdot \overline{\boldsymbol{\epsilon}} \cdot \overline{\boldsymbol{E}} - \frac{1}{2}\overline{\boldsymbol{H}} \cdot \overline{\boldsymbol{\mu}} \cdot \overline{\boldsymbol{H}} + \overline{\psi}_{\mathrm{ME}} \, . \tag{5.13}$$

Obviously, the overall potential (5.13) reflects the characteristics of the functions (5.8) and (5.9) and has an *additional* essential contribution describing the ME-coupling. We define

$$\overline{\psi}_{\mathrm{ME}} := -\overline{\boldsymbol{H}} \cdot \overline{\boldsymbol{\alpha}} \cdot \overline{\boldsymbol{E}} \, , \tag{5.14}$$

so that the effective ME coefficient appears as

$$\overline{\boldsymbol{\alpha}} = \frac{\partial \overline{\boldsymbol{B}}}{\partial \overline{\boldsymbol{E}}} = \left[\frac{\partial \overline{\boldsymbol{D}}}{\partial \overline{\boldsymbol{H}}}\right]^T \quad \text{with } \overline{\boldsymbol{D}} = -\frac{\partial \overline{\psi}}{\partial \overline{\boldsymbol{E}}} \quad \text{and} \quad \overline{\boldsymbol{B}} = -\frac{\partial \overline{\psi}}{\partial \overline{\boldsymbol{H}}} \, . \tag{5.15}$$

Clearly, the ME coupling property of the composite arises as a product of the interaction between the individual phases. The interaction between the two phases is driven by the mechanical strains: on the one hand, an applied electric field induces a deformation of the electroactive material which is transferred to the magnetoactive material. The deformation of the magnetoactive material then induces a magnetic response. On the other hand, an applied magnetic field deforms the magnetic phase which then induces a strain-driven polarization in the electric phase. The remaining effective coefficients, the piezoelectric and piezomagnetic coupling moduli, are defined as

$$\overline{\boldsymbol{e}} = \frac{\partial \overline{\boldsymbol{D}}}{\partial \overline{\boldsymbol{\varepsilon}}} = -\left[\frac{\partial \overline{\boldsymbol{\sigma}}}{\partial \overline{\boldsymbol{E}}}\right]^T \quad \text{and} \quad \overline{\boldsymbol{q}} = \frac{\partial \overline{\boldsymbol{B}}}{\partial \overline{\boldsymbol{\varepsilon}}} = -\left[\frac{\partial \overline{\boldsymbol{\sigma}}}{\partial \overline{\boldsymbol{H}}}\right]^T \quad \text{with } \overline{\boldsymbol{\sigma}} = \frac{\partial \overline{\psi}}{\partial \overline{\boldsymbol{\varepsilon}}} \, , \tag{5.16}$$

with $\{\overline{e}, \overline{q}\}^T_{ijk} = \{\overline{e}, \overline{q}\}_{kij}$. The effective elastic modulus, dielectric permittivity and magnetic permeability are determined by the derivatives

$$\overline{\mathbb{C}} = \frac{\partial \overline{\boldsymbol{\sigma}}}{\partial \overline{\boldsymbol{\varepsilon}}} \, , \quad \overline{\boldsymbol{\epsilon}} = \frac{\partial \overline{\boldsymbol{D}}}{\partial \overline{\boldsymbol{E}}} \, , \quad \text{and} \quad \overline{\boldsymbol{\mu}} = \frac{\partial \overline{\boldsymbol{B}}}{\partial \overline{\boldsymbol{H}}} \, . \tag{5.17}$$

In order to achieve a compact notation we write the macroscopic incremental constitutive equations in standard matrix form

$$\underbrace{\begin{bmatrix} \Delta\overline{\boldsymbol{\sigma}} \\ -\Delta\overline{\boldsymbol{D}} \\ -\Delta\overline{\boldsymbol{B}} \end{bmatrix}}_{\overline{\mathfrak{S}}} = \underbrace{\begin{bmatrix} \overline{\mathbb{C}} & -\overline{\boldsymbol{e}}^T & -\overline{\boldsymbol{q}}^T \\ -\overline{\boldsymbol{e}} & -\overline{\boldsymbol{\epsilon}} & -\overline{\boldsymbol{\alpha}}^T \\ -\overline{\boldsymbol{q}} & -\overline{\boldsymbol{\alpha}} & -\overline{\boldsymbol{\mu}} \end{bmatrix}}_{\overline{\mathfrak{C}}} \underbrace{\begin{bmatrix} \Delta\overline{\boldsymbol{\varepsilon}} \\ \Delta\overline{\boldsymbol{E}} \\ \Delta\overline{\boldsymbol{H}} \end{bmatrix}}_{\overline{\mathfrak{D}}} \, . \tag{5.18}$$

In order to solve Eq. (5.2) with the finite-element method, we formulate the associated weak form of balance equations and approximate the displacements $\boldsymbol{u}$, the electric potential ϕ_e and the magnetic potential ϕ_m with suitable shape functions. We summarize the approximations of the basic field variables (5.1) on the microscale within a typical finite element as

$$\underline{\mathfrak{D}} = \underline{\mathbb{B}}\,\underline{\boldsymbol{d}}\,, \tag{5.19}$$

where the matrix $\underline{\mathbb{B}}$ contains the derivatives of the shape functions and the element vector $\underline{\boldsymbol{d}}$ contains the degrees of freedom, i.e. the displacements as well as the electric and magnetic potential. Following [30], we can compute the discrete values of the overall moduli $\overline{\underline{\mathfrak{C}}}$ on the fly during the iterative solution of the algebraic system of equations. The algorithmic expression for the overall moduli is

$$\overline{\underline{\mathfrak{C}}} = \langle\underline{\mathfrak{C}}\rangle - \frac{1}{\mathrm{vol}\,(\mathcal{RVE})}\,\underline{\mathfrak{L}}^T\underline{\mathfrak{K}}^{-1}\underline{\mathfrak{L}}\,, \tag{5.20}$$

with the volume average of the moduli over the $\mathcal{RVE}$

$$\langle\underline{\mathfrak{C}}\rangle = \frac{1}{\mathrm{vol}\,(\mathcal{RVE})}\int_{\mathcal{RVE}}\underline{\mathfrak{C}}\,\mathrm{d}v \tag{5.21}$$

and the FEM-matrices for the “generalized right-hand-sides” $\underline{\mathfrak{L}}$ and the global stiffness matrix for the unit-cell problem $\underline{\mathfrak{K}}$, i.e.,

$$\underline{\mathfrak{L}} = \int_{\mathcal{RVE}}\underline{\mathbb{B}}^T\underline{\mathfrak{C}}\,\mathrm{d}v \quad\text{and}\quad \underline{\mathfrak{K}} = \int_{\mathcal{RVE}}\underline{\mathbb{B}}^T\underline{\mathfrak{C}}\,\underline{\mathbb{B}}\,\mathrm{d}v\,. \tag{5.22}$$

5.3 Synthesis and manufacturing of ME composites

5.3.1 Synthesis schemes

The manufacturing approaches to magneto-electrics are highly depending on the nature of the resultant morphology [25]. Laminates (2-2 composites) can be easily manufactured by multiple methods depending on the dimensions of the intended composite. Thin films are grown by different methods well established in thin film technology like chemical vapor deposition (CVD), atomic layer deposition (ALD), sputtering, or laser ablation. Macroscopic laminates are generally glued together. Modeling of laminates is typically easy [21], because the boundary conditions are well set through the laminate structure for all fields, be they mechanical, electrical, or magnetic in nature. Laminates are presently the most effective structures for sensors if used on mechanically resonant very low stiffness beams [34]. 1D-composites like rod structures (1-3 structure) are harder to manufacture. First, rods have to be made and then appropriately aligned. Some direct growth techniques are available. Many subtractive techniques stem from the semiconductor industry. Lithography is the most widely used one. Typical feature sizes are in nanometer to micrometer range. Recently, additive manufacturing has become a broadly used tool in the larger micron to millimeter range. Here different pastes of materials are printed onto substrates layer by layer under computer control establishing an additive 3D structure, the simplest one being a rod structure. Subtractive techniques on the micron to millimeter range include dicing and etching. The latter may result in damage in sub-layers and structures. If the starting point for manufacturing is fibers, their alignment is a critical step in the process. One approach uses tape casting, where the flow dynamics of the highly viscous slurry under the doctor blade determines the orientation and degree of alignment of the fibers [35]. 0-3 structures require one component to be completely embedded in another. This is typically appropriate, if two different powders are compacted, one dominating in volume content over the other, or if e.g. powders are embedded into polymers. Many different approaches have been taken. For the making of magneto-electric composites based on mechanical coupling, two issues are crucial: (i) good mechanical adhesion between both phases and (ii) good electrical resistivity, because otherwise the material will be short circuited electrically.

5.3.2 Synthesis results for 0-3 composites

One of the most fundamental issues in providing a suitable magneto-electric 0-3 composite experimentally is its macroscopic resistivity. If this value drops below a certain threshold, electric poling of the sample is no longer possible. In thin films, the growth techniques typically assure a dense microstructure also in the electrical sense proving to exhibit no percolating conductive component [36]. In ceramic processing, this kind of "electrical" microstructure can typically not be provided by classical synthesis techniques like solid state reaction [37]. We recently developed the organosol route [38]. This method was adopted to coat the $CoFe_2O_4$ nanoparticles already in suspension. Subsequent densification during sintering then proved to be sufficiently gentle with this structure to generate a micron sized microstructure of non percolating $CoFe_2O_4$ particles in the system. In order to assure a robust system, we chose 20/80 for the present study containing a much higher content of the insulating ferroelectric than the partly conductive magnetostrictive component. This yielded a very suitable resistivity of the material of $\rho = 60\,\mathrm{G\Omega\,cm}$ which permitted poling of the sample. The relative difference in elastic modulus is small and thus equal volume contents of both constituents should provide best coupling coefficient.

Experimentally, it is clear that $BaTiO_3$ slightly changes its character in the composite. While in the single crystal and in mono-phase bulk ceramics, $BaTiO_3$ exhibits a clear first order phase transition. This typically entails a peak in dielectric constant and a step in its inverse [39]. While in our samples the Curie temperature is unchanged with respect to the pure phase ceramics, a broader distribution of the peak value is found (Fig. 5.2). Also the Curie–Weiss-Plot for $BaTiO_3$ should exhibit a fairly sharp minimum in $1/\epsilon$ which is broadened here. This is typical in nanograin materials [40], while micron size pure $BaTiO_3$ displays sharp maxima in ϵ even though there is a strong dependence of the explicit values of the dielectric constant at the phase transition points [41]. Thus, the grain size of our $BaTiO_3$-component in the magnetoelectric composite ceramic does exhibit the expected behavior for its grain size. No unexpected modification of the ferroelectric phase is observed. We also found no anomaly of the dielectric constant as reported in [42] which is likely an effect of high conductivity above the Curie point suggesting acceptor doping of $BaTiO_3$ in their samples [43]. Actually, our route obviously never showed any occurrence of such doping in the $BaTiO_3$-phase and the concurrent increase in conductivity of the ferroelectric part of the composite in our route. The somewhat increasing conductivity (apparent increase in $\overline{\epsilon}$) at low frequency and high temperature indicates a classical thermally activated transport process at medium distances, thus percolating conductive grains, but sufficiently small in cluster size not to short-circuit the entire sample.

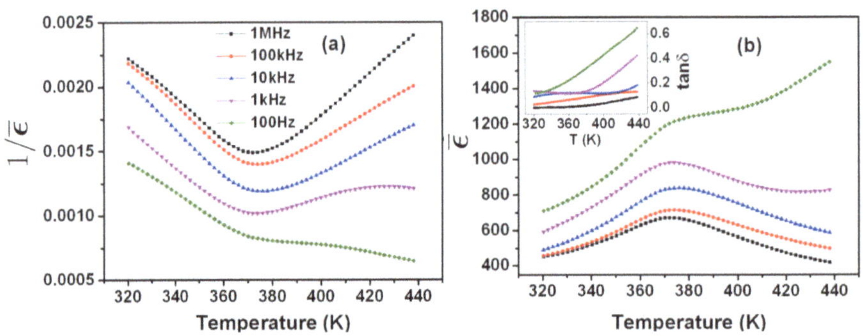

Fig. 5.2: Dielectric constant of 20/80 wt.% $CoFe_2O_4$-$BaTiO_3$ (0-3) ceramic: (a) Curie–Weiss-Plot and (b) experimental raw data and losses.

5.3.3 Experimental details

The $CoFe_2O_4/BaTiO_3$ nanoparticles with core/shell structure were synthesized via organosol crystallization, see [44, 45]. Initially, co-precipitation was used to synthesize $CoFe_2O_4$ particles with a mean size < 40 nm as described in detail in [45]. A stable ferrofluid of $CoFe_2O_4$ nanopowder was added to a tetramethyl-ammonium hydroxide (TMAH) solution containing the amorphous barium titanate precursor. The relative ratio of weights in the final products was adjusted by the amounts of the cobalt ferrite powder and barium titanate in the precursors yielding 20/80 by weight $CoFe_2O_4/BaTiO_3$. In order to achieve high electrical resistivity of the overall sample and optimal coupling in between the two phases, the cobalt iron oxide particles must be well distributed in the microstructure. This was assured by rotation milling for about 12 hours. The powder was then washed several times with ethanol, dried at room temperature under a fume hood, and ultimately calcinated at 750 °C for 15 minutes. Disks were pressed with a hydraulic press into disk-shaped pellets with diameter of 8 mm and thickness of 0.6 mm under 5 tons load. The discs were sintered into ceramics at 1200 °C for 2 hours. For comparison, pure cobalt ferrite and barium titanate ceramic samples were sintered alike. X-ray diffraction determined the phase content of the composites (Siemens D-5000 with Cu-Kα radiation at steps of $\delta(2\Theta) = 0.01°$ with a time constant of 1 s). Scanning and transmission electron microscopy (SEM and TEM) confirmed the morphology and structure of the specimens (SEM quanta 400 FEG and TEM TECNAI F20). Before SEM measurements, the ceramic samples were well polished and chemically etched. The particle/grain size distribution was analyzed by the analySIS software (Soft Imaging Systems). A self-built Sawyer-Tower circuit at a frequency of 250 Hz served to determine the electric field dependence of polarization. For electrical measurements silver electrodes were fired onto both sides of the samples at 500 °C. The magnetic properties as determined by SQUID magnetometry in the temperature interval from 5 to 300 K at magnetic fields up to 1 T can be found in [46]. Magneto-electric measurements were performed using a modified SQUID ac susceptometer in the temperature range 200–300 K at electric fields up to 1 kV/cm and a magnetic field of 0.15 T. The method is described in principles in [47]. In general, an ac electric field, $E = E_{ac} \cos(\omega t)$ generates an induced magnetic signal. The first harmonic of this magneto-electrically induced ac magnetic moment, $m' = m_{ME} \cos(\omega t)$, is detected using an internal lock-in amplifier. The converse magneto-electric coefficient, α_C, can be estimated from the ac electric field dependence of the induced magnetization, $\alpha_C = M_{ME}\mu_0/E_{ac}$, where $M_{ME} = m_{ME}/V_s$ is the magnetoelectrically induced magnetization, V_s is the sample volume, and μ_0 is the permeability of free space. The advantage of this method is its high sensitivity. The smallest measurable ME coefficient ranges around $0.01 \cdot 10^{-12}$ s/m.

5.4 Computational determination of magneto-electro-mechanical properties of ME composites

For the simulation of magneto-electric composites we consider different types of microstructures. First, in order to validate the implemented model, we consider a boundary value problem from the work [28]. After that, we compute the effective material behavior of an experimental ME microstructure consisting of piezoelectric $BaTiO_3$ matrix and piezomagnetic $CoFe_2O_4$ inclusions discussed in the previous paragraphs. The computationally determined ME coefficients are then compared to the experimentally measured data. In all calculations we consider periodic boundary conditions on the microscale.

Tab. 5.2: Material parameters.

Parameter	Unit	$BaTiO_3$	$CoFe_2O_4$
$\mathbb{C}_{11}$	N/mm^2	$16.6 \cdot 10^4$	$28.60 \cdot 10^4$
$\mathbb{C}_{12}$	N/mm^2	$7.7 \cdot 10^4$	$17.30 \cdot 10^4$
$\mathbb{C}_{13}$	N/mm^2	$7.8 \cdot 10^4$	$17.05 \cdot 10^4$
$\mathbb{C}_{33}$	N/mm^2	$16.2 \cdot 10^4$	$26.95 \cdot 10^4$
$\mathbb{C}_{44}$	N/mm^2	$4.3 \cdot 10^4$	$4.53 \cdot 10^4$
ϵ_{11}	mC/(kV m)	$112 \cdot 10^{-4}$	$0.80 \cdot 10^{-4}$
ϵ_{33}	mC/(kV m)	$126 \cdot 10^{-4}$	$0.93 \cdot 10^{-4}$
μ_{11}	N/kA^2	5.0	157.0
μ_{33}	N/kA^2	10.0	157.0
e_{31}	C/m^2	−4.4	0.0
e_{33}	C/m^2	18.6	0.0
e_{15}	C/m^2	11.6	0.0
q_{31}	N/(kA mm)	0.0	580.3
q_{33}	N/(kA mm)	0.0	-699.7
q_{15}	N/(kA mm)	0.0	550.0

The material parameters of the two phases are adopted from the work [28] and are listed in Tab. 5.2. They, however, differ for cobalt ferrite from the parameters used in the reference publication in such a way that we set $\mu_{11} := 157\,N/kA^2$, see also [48]. Second, the piezomagnetic coefficient q_{33} is set to $q_{33} := -699.7\,N/(kA\,mm)$, cf. [24][1].

5.4.1 Computational characterization of the magneto-electro-mechanical properties of an ideal microstructure

In order to validate the implemented material model we perform simulations of an ideal three-dimensional $\mathcal{RVE}$ taken from [28]. The $\mathcal{RVE}$ consists of a cubic piezomagnetic $CoFe_2O_4$ matrix and a cylindrical piezoelectric $BaTiO_3$ inclusion oriented in vertical direction. In analogy to the reference publication, the preferred directions of the two phases are assumed to point in positive vertical direction. For the description of both materials we use the set of material parameters given in [28]. This means that in this computation we set $\mu_{11} := -590\,N/kA^2$ and $q_{33} := +699.7\,N/(kA\,mm)$. We perform calculations with five different $\mathcal{RVE}$s characterized by volume fractions of 30 %, 40 %, 50 %, 60 %, and 70 % of the inclusion. For an illustration of the $\mathcal{RVE}$ with 40 % volume fraction of the inclusion see Fig. 5.3 (a).

1 In the work [28] these parameters are given by $\mu_{11} = -590\,N/kA^2$ and $q_{33} = +699.7\,N/(kA\,mm)$.

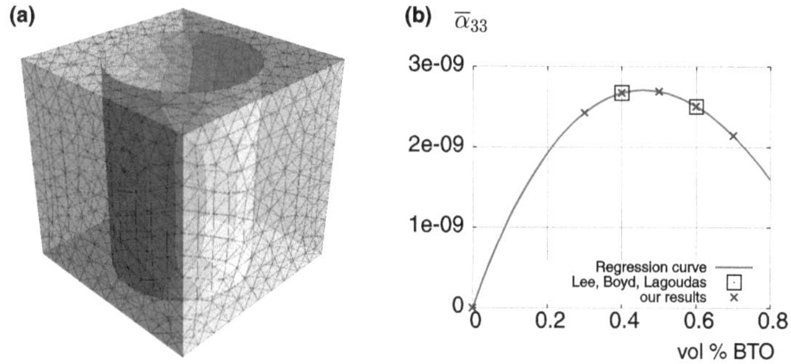

Fig. 5.3: (a) $\mathcal{RVE}$ consisting of piezomagnetic cube with piezoelectric cylindrical inclusion and (b) comparison of the determined ME-coefficient with results from [28]; please note, that the magnetic permeability is set to $\mu_{11} = -590\,\mathrm{N/kA^2}$ and the piezomagnetic coefficient q_{33} is set to $q_{33} = +699.7\,\mathrm{N/(kA\,mm)}$.

We apply a macroscopic electric field $\overline{\boldsymbol{E}} := [0, 0, \overline{E}_3]^T$. This field is transferred to the microscopic $\mathcal{RVE}$ using periodic boundary conditions. Due to the preferred orientation of the piezoelectric phase and the applied field, the inclusion reacts with an elongation in vertical direction. This deformation is transferred to the piezomagnetic phase, which leads to magnetization of the matrix. This strain-mediated coupling gives rise to a magneto-electric coupling coefficient that can be determined formally by

$$\overline{\boldsymbol{\alpha}} = \frac{\partial \overline{\boldsymbol{B}}}{\partial \overline{\boldsymbol{E}}} = \left[\frac{\partial \overline{\boldsymbol{D}}}{\partial \overline{\boldsymbol{H}}}\right]^T, \tag{5.23}$$

and which is evaluated using Eq. (5.20). The resulting ME coefficients are calculated for the different volume fractions of $BaTiO_3$ and are in perfect agreement with the results from [28], see Fig. 5.3 (b). For completeness the following graphs show the computed effective elastic, dielectric, magnetic, piezoelectric, piezomagnetic and ME coefficients of the composite.

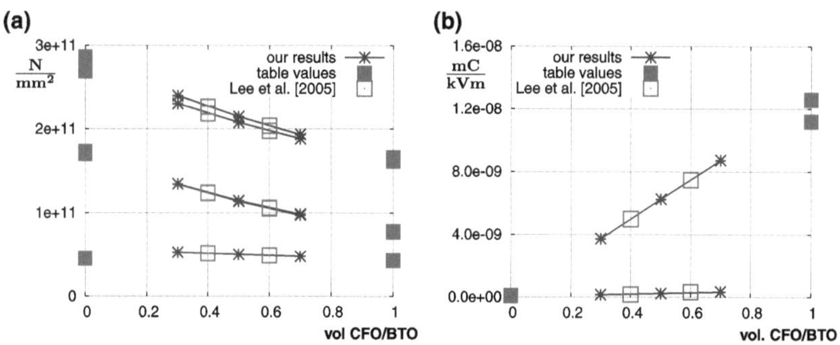

Fig. 5.4: (a) Elastic moduli, (b) dielectric permittivity depending on vol % BTO.

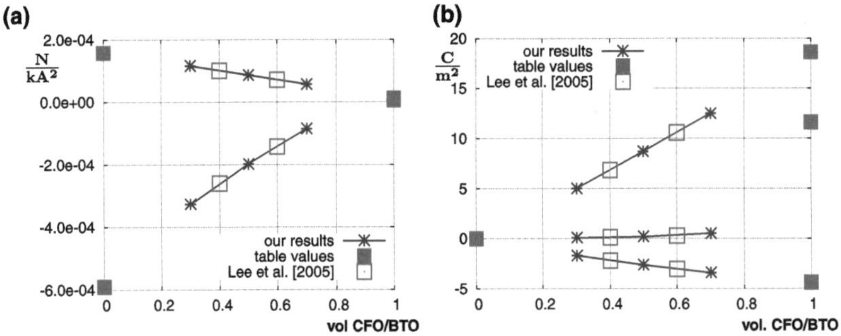

Fig. 5.5: (a) Magnetic permeability, (b) piezoelectric constants depending on vol % BTO.

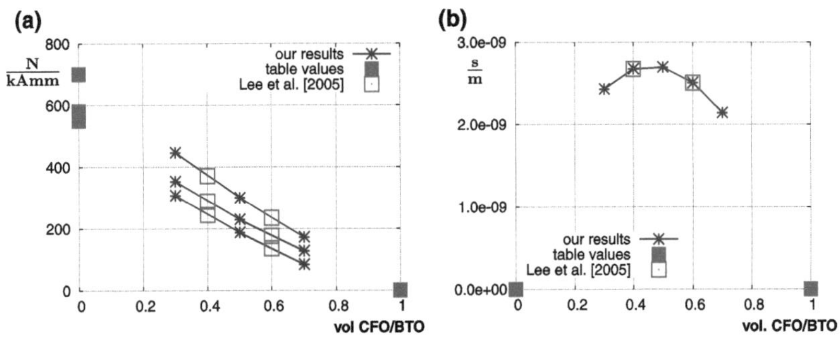

Fig. 5.6: (a) Piezomagnetic constants, (b) ME-coefficient depending on vol % BTO.

5.4.2 Computational characterization of the magneto-electro-mechanical properties of a real microstructure

Now we apply the computational method to the determination of the ME properties of the real composite microstructure discussed in Sec. 5.3. In Fig. 5.7 an electron microscopy image of the composite is shown, where the different phases are visible due to the slightly distinct local conductivity. The brighter areas reflect the piezoelectric barium titanate and the darker areas show the piezomagnetic cobalt ferrite.

The mass fractions of barium titanate and cobalt ferrite are 80 % and 20 %, respectively. Taking into account the densities of the two phases we use a sample with 16.5 % surface fraction of the inclusion. The two-dimensional microstructure is then discretized with 5324 quadratic triangular finite elements.

In a particulate ME composite the quality of magneto-electric coupling strongly depends on the pre-polarization and pre-magnetization of the individual phases. In order to analyze the effect of ferroelectric pre-polarization we will take into account different scenarios for the remanent pre-polarization. We incorporate the pre-polarization in the following way. As a first step, we apply a macroscopic electric

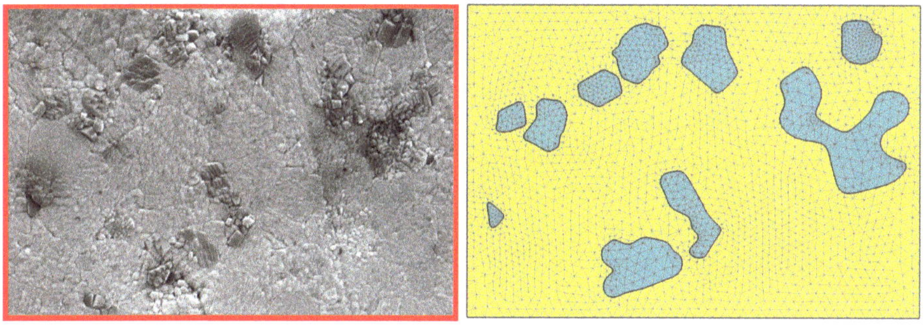

Fig. 5.7: Electron microscopy image of the composite microstructure (left) and finite element discretization (right), see also Fig. 5.1.

field in vertical direction on the initially unpolarized microstructure. That means we set $\overline{\boldsymbol{E}} = [0, \overline{E}_2, 0]^T$, $\overline{\boldsymbol{\varepsilon}} = \mathbf{0}$ and $\overline{\boldsymbol{H}} = \mathbf{0}$ in Eq. (5.6). In this first step the piezoelectric coupling parameters on the microstructure are set to zero. As a result of the macroscopic loading and the inhomogeneous distribution of the composite properties an inhomogeneous distribution of the electric field on the microscale is obtained. We use these electric field vectors to define the pre-polarization state of the piezoelectric matrix: first, the direction of the microscopic electric field is used to define the preferred direction $\boldsymbol{a}$ of the transverely isotropic model, i.e. we set $\boldsymbol{a} := \boldsymbol{E}/\|\boldsymbol{E}\|$ in each integration point on the microscale. Second, the norm of the electric field is used to define the amount of piezoelectric coupling at each particular integration point. The scaling is described by a factor p_s defined by a hyperbolic tangent function

$$p_s = \tanh\left(c \cdot \|\boldsymbol{E}\|\right). \tag{5.24}$$

where the factor c defines the slope of the hyperbolic tangent. In the next examples the value of c is set to 2 mm/kV in order to approximate the polarization behavior of barium titanate. By using this scaling factor, areas with high electric fields are characterized by higher piezoelectric coupling. For convenience, the algorithmic treatment is summarized in Tab. 5.3.

In the following simulations we will apply different macroscopic electric fields in the range between 0.01 and 3 kV/mm, which will lead to different pre-polarization states on the microstructure. We will then analyze the influence of the pre-polarization states on the overall ME coupling. The pre-magnetization directions of the piezomagnetic inclusions are assumed to point perfectly in vertical direction with fully activated coupling parameters.

Fig. 5.8 shows the pre-polarization directions as well as the distribution of the amount of pre-polarization for two different applied macroscopic electric fields of magnitude $\overline{E}_2 = 0.5$ kV/mm and $\overline{E}_2 = 1$ kV/mm.

Tab. 5.3: Determination of piezoelectric coupling properties on the microscale.

1. Set initial preferred direction $\boldsymbol{a}_0 = [0, 1, 0]$
2. Neglect constitutive couplings: set piezoelectric moduli equal zero $\boldsymbol{e} \equiv \boldsymbol{0}$
3. Apply macroscopic electric field $\overline{\boldsymbol{E}} = c_1^{+}\boldsymbol{a}$; compute local distribution of $\boldsymbol{E}$
4. Choose preferred directions $\boldsymbol{a}$ (for each Gauss point): $\boldsymbol{a} = \boldsymbol{E}/\|\boldsymbol{E}\|$
5. Estimate relative amplitude of remanent polarization $p_s = \tanh(c \cdot \|\boldsymbol{E}\|) \in [0, 1)$
6. Determine piezoelectric moduli

$$\boldsymbol{e} = p_s(-\beta_1 \boldsymbol{a} \otimes \boldsymbol{1} - \beta_2 \boldsymbol{a} \otimes \boldsymbol{a} \otimes \boldsymbol{a} - \beta_3 \hat{\boldsymbol{e}}) \quad \text{with } \{\hat{e}\}_{kij} := \frac{1}{2}[a_i \delta_{kj} + a_j \delta_{ki}]$$

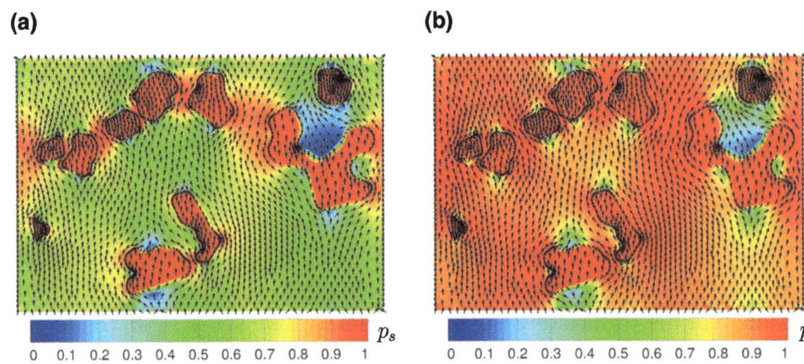

Fig. 5.8: Distribution of p_s (contour) with preferred directions $\boldsymbol{a}$ (vectors) for applied electric fields $\overline{E}_2 = 0.5$ kV/mm (a) and $\overline{E}_2 = 1$ kV/mm (b).

As can be seen the electric field in the matrix concentrates between the inclusions with maximum values in the areas with high surface fraction of the inclusion material. This is due to the lower dielectric permittivity of the inclusions. In these concentrated areas the electric fields are associated to a saturation of the hyperbolic function in Eq. (5.24). Thus, full pre-polarization of the matrix material is obtained.

Now having incorporated the pre-polarization of the ferroelecric matrix we can analyze the magneto-electro-mechanical behavior of the composite. In order to do so we load the pre-polarized specimen again with an electric field and study the resulting magneto-electric interactions on the microscale. In the following Fig. 5.9 (a) we see the distributions of microscopic electric potential and electric field that arise as a consequence of an applied macroscopic electric field of $\overline{E}_2 = 1$ kV/mm. Due to the piezoelectric coupling of the matrix this electric field yields a deformation that is mediated to the inclusions. Since we assumed piezomagnetic coupling in the inclusions we arrive at a magnetic reaction. In Fig. 5.9 (b) the strain-induced distribution of the magnetic potential as well as the vectors of magnetic flux density are shown.

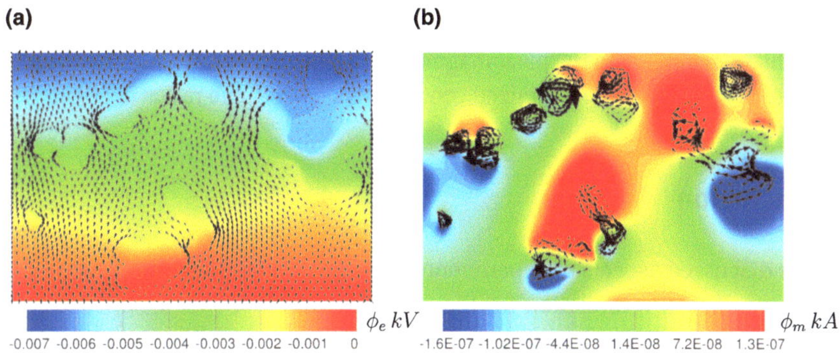

Fig. 5.9: (a) Electric potential ϕ_e with electric field vectors $\boldsymbol{E}$ in the piezoelectric matrix and (b) magnetic potential ϕ_m with magnetic flux density vectors $\boldsymbol{B}$ in the inclusions for $\overline{E}_2 = 1$ kV/mm.

As mentioned earlier, we applied different macroscopic electric pre-polarization fields in order to arrive at different states of pre-polarization on the microlevel. As can be seen in Fig. 5.10 the intensity of the pre-polarization field has a strong influence on the overall ME coupling.

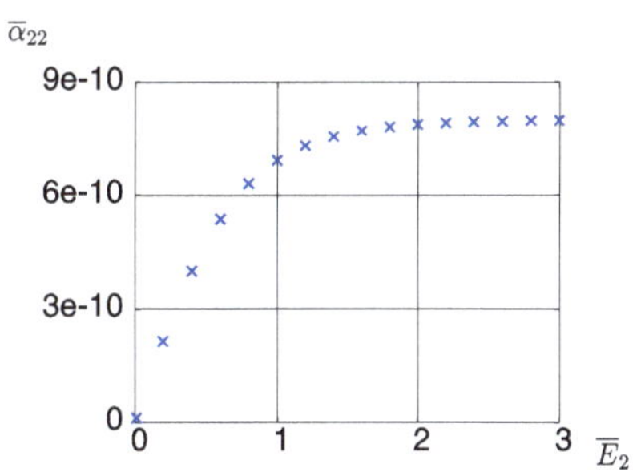

Fig. 5.10: Magneto-electric coupling coefficient $\overline{\alpha}_{22}$ in s/m for different electric fields $\overline{E}_2$ in kV/mm.

For high pre-polarization fields the ME-coefficient saturates to the value of $\overline{\alpha}_{22} \approx 8.04 \cdot 10^{-10}$ s/m. For comparison, [49] measured an effective ME-coefficient of $\overline{\alpha}_{22} \approx 4.4 \cdot 10^{-12}$ s/m of the sample. This rather large deviation allows for mainly two possible conclusions. On the one hand, on the modeling side, the method for the incorporation of ferroelectric pre-polarization was very simplified in nature and will be enhanced in future developments. Furthermore, the assumption of perfect pre-magnetization in the inclusions was very optimistic. In addition to that, the two-dimensional approximation of the particualte microstructure introduces another source for deviations. On the other hand, the simulations clearly indicate that there is still the need for improvement for the experimental preparation since an optimal pre-polarization and/or pre-magnetization state has not been obtained. In any case there will also be additional properties on the microscale that we have to address in future developments. One such property is the experimentally very relevant electric conductivity. Thus, the challenge for the experimental side now resides in assuring high resistivity in compositions of higher contents in the piezomagnetic phase which by nature of the coupling through strain should yield maximized coefficients near volume ratios of 50/50.

5.5 Conclusion

The paper discussed aspects of experimental manufacturing of multiferroic composites and a numerical formulation for the characterization of such composites. Due to the fact that the effective macroscopic magneto-electric coupling significantly depends on the composition and morphology of the microstructure, we used a computational homogenization procedure based on the FE^2 method which can be applied to arbitrary microstructural morphologies. This method was applied to two different kinds of microstructures. On the one hand, idealized microstructures taken from the literature were used to validate the implemented model. On the other hand, the method was applied to the characterization of a realistic microstructure, which was manufactured and measured in experiments. Due to the fact that the magneto-electric coupling in a real sample has to be activated through pre-polarization of the electroactive phase we considered different pre-polarization states of the piezoelectric matrix. The comparison between the experimental and computational results showed a rather large deviation. This could be explained by the partly oversimplified modeling assumptions and the experimentally not yet obtained optimal pre-polarization and -magnetization state.

Acknowledgment: We gratefully acknowledge the financial support by the "Deutsche Forschungsgemeinschaft" (DFG), research group "Ferroische Funktionsmaterialien – Mehrskalige Modellierung und experimentelle Charakterisierung", project 1 (SCHR 570/12-1) and project 2 (LU 729/12). Morad Etier acknowledges support by the DAAD-GRISEC program (Grant 50750877). Furthermore, we acknowledge the comprehensive discussions and suggestions by Dr. Vladimir V. Shvartsman.

References

[1] Eerenstein W, Wiora M, Prieto J, Scott JF, Mathur N. Giant sharp and persistent converse magnetoelectric effects in multiferroic epitaxial heterostructures. Nature Materials. 2007;6(5): 348–351.
[2] Bibes M, Barthélémy A. Multiferroics: Towards a magnetoelectric memory. Nature Materials. 2008;7(6):425–426.
[3] Schmid H. Multi-ferroic magnetoelectrics. Ferroelectrics. 1994;162:317–338.
[4] Hill N. Why are there so few magnetic ferroelectrics? Journal of Physical Chemistry B. 2000; 104:6694–6709.

[5] Cheong SW, Mostovoy M. Multiferroics: a magnetic twist for ferroelectricity. Nature Materials. 2007;6(1):13–20.
[6] Ramesh R, Spaldin N. Multiferroics: progress and prospects in thin films. Nature Materials. 2007;6(1):21–29.
[7] Khomskii D. Classifying multiferroics: Mechanisms and effects. Physics. 2009, 2.
[8] Martin L, Chu YH, Ramesh R. Advances in the growth and characterization of magnetic, ferroelectric, and multiferroic oxide thin films. Materials Science and Engineering: R: Reports. 2010; 68(4–6):89–133.
[9] Zohdi T. Simulation of coupled microscale multiphysical-fields in particulate-doped dielectrics with staggered adaptive FDTD. Computational Methods in Applied Mechanics and Engineering. 2010;199:3250–3269.
[10] Zohdi T. Electromagnetic Properties of Multiphase Dielectrics. Springer; 2012.
[11] Brown WF, Hornreich R, Shtrikman S. Upper bound on the magnetoelectric susceptibility. Physical Review. 1968;168(2):574–577.
[12] Nan CW. Magnetoelectric effect in composites of piezoelectric and piezomagnetic phases. Phys Rev B. 1994;50:6082–6088.
[13] Ryu J, Priya S, Uchino K, Kim HE. Magnetoelectric effect in composites of magnetostrictive and piezoelectric materials. Journal of Electroceramics. 2002;8(2):107–119.
[14] Eerenstein W, Mathur ND, Scott JF. Multiferroic and magnetoelectric materials. Nature. 2006; 442(7104):759–765.
[15] Priya S, Islam R, Dong S, Viehland D. Recent advancements in magnetoelectric particulate and laminate composites. Journal of Electroceramics. 2007;19(1):147–164.
[16] Nan CW, Bichurin MI, Dong S, Viehland D, Srinivasan G. Multiferroic magnetoelectric composites: Historical perspective, status, and future directions. Journal of Applied Physics. 2008; 103(3):031101 (pages 35).
[17] Srinivasan G. Magnetoelectric composites. Annual Review of Materials Research. 2010;40: 1–26. Available from: http://arjournals.annualreviews.org/doi/pdf/10.1146/annurev-matsci-070909-104459.
[18] Wang Y, Hu J, Lin Y, Nan CW. Multiferroic magnetoelectric composite nanostructures. Nature Asia-Pacific, Asia Materials. 2010;2:61–68.
[19] van Suchtelen J. Product properties: A new application of composite materials. Philips Research Reports. 1972;27:28–37.
[20] Fiebig M. Revival of the magnetoelectric effect. Journal of Physics D: Applied Physics. 2005; 38(8):R123–R152.
[21] Harshé G, Dougherty J, Newnham RE. Theoretical modeling of multilayer magnetoelectric composites. International Journal of Applied Electromagnetics in Materials. 1993;4:145–159.
[22] Harshé G, Dougherty J, Newnham RE. Theoretical modeling of 3-0/0-3 magnetoelectric composites. International Journal of Applied Electromagnetics in Materials. 1993;4:161–171.
[23] Huang JH, Kuo WS. The analysis of piezoelectric/piezomagnetic composite materials containing ellipsoidal inclusions. Journal of Applied Physics. 1997;81(3):1378–1386.
[24] Bichurin M, Petrov V, Srinivasan G. Theory of low-frequency magnetoelectric coupling in magnetostrictive-piezoelectric bilayers. Physical Reviews B. 2003;68(5):054402.
[25] Nan CW, Liu G, Lin Y, Chen H. Magnetic-field-induced electric polarization in multiferroic nanostructures. Physical Review Letters. 2005;94(19):197203–1–4.
[26] Srinivas S, Li JY, Zhou YC, Soh AK. The effective magnetoelectroelastic moduli of matrix-based multiferroic composites. Journal of Applied Physics. 2006;99(4):043905.
[27] Bichurin M, Petrov V, Averkin S, Liverts E. Present status of theoretical modeling the magnetoelectric effect in magnetostrictive-piezoelectric nanostructures. Part I: Low frequency and electromechanical resonance ranges. Journal of Applied Physics. 2010;107(5):053904.

[28] Lee JS, Boyd JG, Lagoudas DC. Effective properties of three-phase electro-magneto-elastic composites. International Journal Of Engineering Science. 2005;43(10):790–825.
[29] Schröder J, Keip MA. Two-scale homogenization of electromechanically coupled boundary value problems. Computational Mechanics. 2012;50:229–244.
[30] Schröder J, Keip MA, Labusch M. Algorithmic two-scale transition for magneto-electro-mechanically coupled problems – localization and homogenization; 2014. In preparation.
[31] Schröder J. Derivation of the localization and homogenization conditions for electro-mechanically coupled problems. Computational Materials Science. 2009;46(3):595–599.
[32] Hill R. Elastic properties of reinforced solids – some theoretical principles. Journal of the Mechanics and Physics of Solids. 1963;11:357–372.
[33] Schröder J, Gross D. Invariant formulation of the electromechanical enthalpy function of transversely isotropic piezoelectric materials. Archive of Applied Mechanics. 2004;73:533–552.
[34] Jahns R, Piorra A, Lage E, Kirchhof C, Meyners D, Gugat J, et al. Giant magnetoelectric effect in thin-film composites. Journal of the American Ceramic Society. 2013;96:1673–1681.
[35] Rauscher M, Roosen A. Effect of particle shape on anisotropic packing and shrinkage behavior of tape-cast glass-ceramic composites. International Journal of Applied Ceramic Technology. 2009;6:24–34.
[36] Bai F, Zhang H, Li J, Viehland D. Magnetic and magnetoelectric properties of as deposited and annealed batio3-cofe2o4 nanocomposite thin films. Journal of Physics D: Applied Physics. 2010, 43.
[37] Hanumaiah A, Bhimasankaram T, Suryanarayana S, Kumar G. Dielectric behavior and magnetoelectric effect in cobalt ferrite-barium titanate composites. Bulletin of Materials Science. 1994; 17:405–409.
[38] Gao Y, Shvartsman V, Elsukova A, Lupascu D. Low-temperature synthesis of crystalline $BaTiO_3$ nanoparticles by one-step "Organosol"-precipitation. Journal of Materials Chemistry. 2012;22: 17573–17583.
[39] Jaffe B, Jr WC, Jaffe H. Piezoelectric Ceramics. Academic Press; 1971.
[40] Wang T, Wang X, Yao G, Zhang Y, Liao X, Yang G, et al. Effects of processing on microstructure and dielectric properties of ultrafine-grained barium titanate based ceramics for BME-MLCC applications. Key Engineering Materials. 2008;368–372:43–46.
[41] Kinoshita K, Yamaji A. Grain-size effects on dielectric properties in barium titanate ceramics. Journal of Applied Physics. 1976;47:371–373.
[42] Hemeda O, Tawfik A, Amer M, Kamal B, Refaay DE. Structural, spectral and dielectric properties of piezoelectric-piezomagnetic composites. Journal of Magnetism and Magnetic Materials. 2012;324:3229–3237.
[43] Waser R, Baiatu T, Härdtl KH. dc electrical degradation of perovskite-type titanates. Journal of the American Ceramic Society. 1990;73:1645–1653.
[44] Etier M, Gao Y, Shvartsman V, Elsukova A, Landers J, Wende H, et al. Cobalt ferrite/barium titanate core/shell nanoparticles. Ferroelectrics. 2012;438:115–122.
[45] Etier M, Shvartsman V, Stromberg F, Landers J, Wende H, Lupascu D. Synthesis and magnetic properties of cobalt ferrite nanoparticles. Materials Research Society Symposium Proceedings. 2012, 1398.
[46] Etier M, Shvartsman V, Gao Y, Landers J, Wende H, Lupascu D. Magnetoelectric effect in (0-3) $CoFe_2O_4$-$BaTiO_3$ (20/80) composite ceramics prepared by the organosol route. Ferroelectrics. 2013;448:77–85.
[47] Borisov P, Hochstrat A, Shvartsman V, Kleemann W. Superconducting quantum interference device setup for magnetoelectric measurements. Review of Scientific Instruments. 2007;78: 106105–106108.

[48] Srinivas S, Li JY. The effective magnetoelectric coefficients of polycrystalline multiferroic composites. Acta Mechanica. 2005;53(15):4135–4142.
[49] Etier M, Gao Y, Shvartsman VV, Lupascu DC, Landers J, Wende H. Magnetoelectric properties of $0.2CoFe_2O_4$-$0.8BaTiO_3$ composite prepared by organic method. Proceedings of the European Conference on the Applications of Polar Dielectrics. 2012;:1–4.

K. Chockalingam, R. Janisch, and A. Hartmaier*

6 Coupled atomistic-continuum study of the effects of C atoms at α-Fe dislocation cores

Abstract: The influence of carbon at dislocation cores in α-Fe is studied to determine the Peierls stress, i.e the critical stress to move the dislocation at 0K. The effect of carbon on both edge and screw dislocations is investigated. A coupled molecular statics (MS) and extended finite element method (XFEM) is employed for this study, where the dislocation core is modeled atomistically. The results on pure Fe are found to be in good agreement with a fully atomistic study. The coupled approach captures the right core behavior and significantly reduces the size of the atomistic region, while describing the behavior of a single dislocation in an infinite anisotropic elastic medium. Furthermore, mechanical boundary conditions can be applied consistently. It was found that the influence of carbon on edge dislocations is much stronger than that on screw dislocations, and that carbon causes a directionally dependent Peierls stress in the case of a screw dislocation. Even though the increase of the Peierls stress is much more pronounced for edge dislocations, the total value does not reach the level of the Peierls stress for screw dislocations, either with or without carbon at the core. Hence, we conclude that the motion of screw dislocations remains the rate limiting factor for plastic deformation of α-Fe.

Keywords: dislocations, Peierls stress, molecular statics, XFEM, iron, carbon

6.1 Introduction

The alloying of crystalline metals with interstitial or substitutional elements leads to solid solution strengthening due to the interaction of dislocations with such point defects. Especially in the case of interstitials, their limited solubility and high diffusivity leads to segregation and accumulation of the atoms at defects. In the Fe-C system this leads to the formation of Cottrell clouds at dislocations, which drastically enhances solid solution strengthening and leads to a pronounced yield point phenomenon. Technologically, this effect is exploited in the so-called bake-hardening process to strengthen steel sheets after deep drawing.

While the solubility of C atoms in α-Fe with a perfect body-centered cubic (bcc) crystal structure is vanishingly small at room temperature, their solubility close to dislocations (and other crystal defects) is considerably higher. Thermodynamically this can be explained by a lowering of the chemical potential of C interstitials, either by hydrostatic tensile stresses close to edge dislocations or by shear stresses causing tetragonal distortions around screw dislocations. Thus carbon interstitials pin to the dislocations and thus lower the total energy of the system compared to a random carbon distribution. If a dislocation moves, it has to break out from this energetically favorable local structure. This gives rise to an energy barrier for dislocation motion and to an increase of the critical stress to move the dislocation; on a macroscopic level, this enhances the yield strength.

Models of solid solution hardening often consist of a purely elastic framework, describing the pinning effect via the interaction of the respective strain fields (see e.g. [1]). However, such models neglect the non-elastic behavior of the dislocation core, and the influence of interstitial or substitutional atoms on the core structure. Moreover, the presence of interstitial atoms changes the local electronic configuration in the crystal and, hence, its local elastic properties. It is not clear a priori how strong those effects are, and hence they should be quantified by a fully atomistic description. Molecular dynamics simulations of dislocations using embedded-atom method (EAM) type potentials are nowadays state of the art. For instance, to understand the role of carbon at dislocations and how it influences hardening, Tapasa et al. [2] made dislocations overcome a row of carbon atoms in α-Fe and determined the critical stress at different temperatures. Very recently, the influence of hydrogen on dislocations in α-Fe has been investigated by a fully atomistic model with semi-empirical potentials [3]. Clouet et al. [4] modeled the pinning of dislocations by carbon atomistically and compared the results to elasticity theory. There was good agreement in the binding energy between dislocations and carbon atoms in the atomistic simulation and the theoretical prediction when anisotropy was taken into account. However, such models cannot capture core effects. Diffusion of carbon in α-Fe and its interaction with dislocations was analyzed by Becquart et al. [5]. The evolution of the Fe lattice parameter with carbon concentration was verified with experimental data. Although these studies provided important insights, the transferrability of the EAM potentials is sometimes questionable [6] and a more rigorous model of the atomic interactions is desirable. Ab-initio density functional theory has been applied to study dislocation core structures [7] and Peierls barriers [8], but it is computationally too expensive to carry out simulations to determine the Peierls stress, which require a much larger supercell.

The objective of this paper is to implement a framework that allows characterization of carbon interstitial atoms at Fe dislocation cores and is not limited by the choice of method used to model interatomic interactions. To circumvent this limitation in modeling dislocations fully atomistically we make use of continuum elasticity theory. Continuum far field solutions [9] are available for edge and screw dislocation, but they break down at the dislocation core, producing infinitely high stresses. Hence, to capture the correct core behavior, the vicinity of the dislocation is modeled fully atomistically and coupled to a continuum region in the far field. There is an extensive amount of literature available to couple atomistic and continuum methods, and fourteen of the prominent methods are summarized by Miller and Tadmor [10] for different test cases. It is shown that coupled atomistics and discrete dislocations (CADD) as well as the finite element atomistics method (FEAt) perform reasonably well in terms of accuracy and efficiency. Such coupled schemes are ideally suited to the problem at hand. FEAt was developed by [11] and uses non-local elasticity, whereas CADD developed by Shilkrot et al. [12] uses linear elasticity, but allows for discrete dislocations in the continuum region. A coupled atomistic and continuum scheme with the atomistic domain modeled using density functional theory (DFT) can be found in [13]. A similar approach for an atomistic region containing a dislocation and described by DFT has been applied to study solid solution strengthening in aluminum alloys from first principles [14, 15]. However, in this work the mechanical boundary conditions were imposed by a lattice Green's function method proposed by [16, 17]. These coupled approaches significantly reduce the size of the computationally expensive atomistic domain to a small region around a dislocation core, yet describing the situation of a single defect embedded in an infinitely large elastic medium. We follow the FEAt scheme by Kohlhoff et al. [11], but add the extended finite element method description [18] to the originally proposed FEAt method, arriving at an "extended" FEAt, or XFEAt, method. With this, the correct slip caused by the dislocation is produced in continuum and atomistic regions at the same time. An EAM potential [19] is used here in the atomistic region for the purpose of demonstration. With this model we calculate the Peierls stress, i.e. the critical stress to move a dislocation at 0 K, with and without carbon interstitials being present at the dislocation core. We find that the results from our coupled atomistic-continuum system for pure α-iron are in good agreement with a fully atomistic system studied by Koester et al. [20]. The coupled atomistic-continuum XFEAt scheme is used to study the influence of carbon on both edge and screw dislocations and will be extended to more complex configurations in future work.

In Sec. 6.2, modeling aspects in atomistic and continuum domains are discussed, and the coupling scheme between both domains is outlined. In Sec. 6.3, the validation of the atomistic-continuum coupled system is demonstrated. Finally in Sec. 6.4, the influence of carbon on the critical stress for edge and screw dislocation is presented and discussed.

6.2 Coupling atomistic and continuum domains

For seamless coupling of atomistic and continuum domains two main factors are needed to be taken into consideration. First, the dislocation core should be correctly modeled in the atomistic domain and second, the continuum region surrounding the atomistic domain should be able to capture the correct far field dislocation behavior.

6.2.1 Atomistic domain

The core of the dislocation is modeled atomistically, so that accurate modeling of nonlinear effects on the core structure is ensured. Note that our scheme is flexible, and the description of the atomic interactions can be exchanged as desired. However, for testing purposes a Finnis Sinclair-type EAM potential [19] is used to characterize the Fe-Fe interactions. It captures lattice parameter, elastic constants, point-defect energies and other properties in good agreement with first-principle results. The Fe-C interactions were modeled using EAM Ruda potential [21] which is fitted to capture the dilute heat solution of carbon, the location of carbon interstitials and the migration energy of carbon in α-Fe. Both screw and edge dislocations are introduced into the atomistic domain by imposing the displacement field calculated by the Volterra construction on the pad atoms in the coupling region between atomistic and continuum regions. The role of pad atoms will be discussed in more detail in the subsequent section. The relaxation of the atomistic domain to its minimum potential energy configuration is carried out using the FIRE algorithm [22].

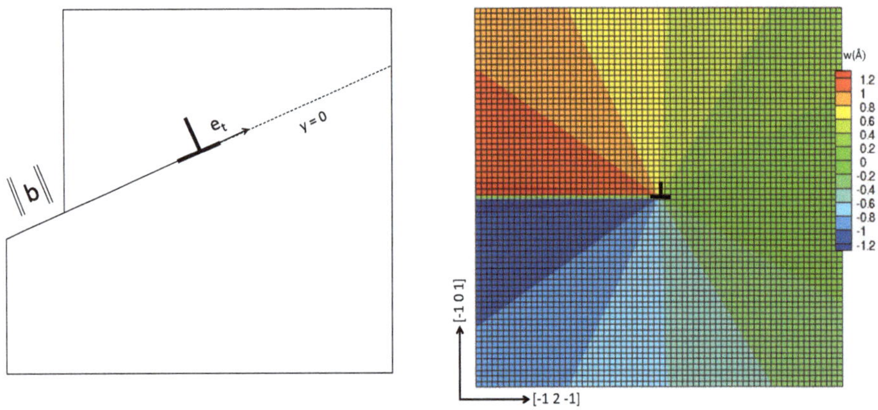

Fig. 6.1: Schematic of a glide plane is shown in (a), where the slip across the glide plane is one Burgers vector. XFEM modeling of a $\frac{1}{2}[111]$ screw dislocation in (b) is achieved using enriched elements.

6.2.2 Continuum domain

The continuum domain is modeled using the extended finite element method. The Volterra displacement field is also applied in the finite element model as boundary condition. It leads to a displacement jump of one Burgers vector perpendicular or parallel to the dislocation line depending on the type of dislocation. Imposing a displacement jump in the atomistic domain is straight forward given its discrete nature, but this not the case when using generic finite elements, which use continous shape functions. Instead displacement jumps can be handled using the extended finite element method (XFEM), since polynomial type shape functions are augmented with special enriched shape functions. The approach followed in this paper is based on the work by [23] to model dislocations using XFEM. To describe dislocations, the enriched part uses a Heaviside jump function that is commonly applied for strong discontinuities to capture the displacement jump. The first term on the right hand side of Eq. (6.1) is the standard finite element displacement approximation and the second term is the enriched part with the Heaviside step function.

$$\boldsymbol{r}(\boldsymbol{x}) = \sum_{i \in \zeta_I} N_i(\boldsymbol{x}) d_i + b\boldsymbol{e}_t \sum_{j \in \zeta_J} N_j(\boldsymbol{x}) H(f(\boldsymbol{x})) , \tag{6.1}$$

where $\boldsymbol{r}$ is the displacement vector field, $d_i(u, v, w)$ is the nodal displacement with corresponding three-dimensional directional components, ζ_I is the set of all nodes, ζ_J is the set of enriched nodes, N_i is the finite element shape functions, b is the Burgers vector, $\boldsymbol{e}_t$ is the tangential vector, $f(\boldsymbol{x})$ is a function that defines the glide plane and $H(\cdot)$ is the Heaviside step function, whose value is $-\frac{1}{2}$ above the glide plane and $\frac{1}{2}$ below the glide plane.

The second term is only active on the elements cut by the glide plane and ensures that the displacement jump across the enriched element has a magnitude of one Burgers vector as shown in Fig. 6.1 (a). Depending on the direction of the tangential vector $\boldsymbol{e}_t$ edge, screw, or mixed dislocations can be modeled. An in-house finite element program with hexahedron 3D elements is used to model the continuum domain. Fig. 6.1 (b) shows the w displacement plot along [111] with a $\frac{1}{2}[111]$ screw dislocation with boundary conditions corresponding to the Volterra construction. It can be observed that the displacement jump of one Burgers vector across the glide plane amounting to 2.417 Å is captured by using enriched elements. Note that the continuum elasticity solution is not valid at the dislocation core and will be replaced by the atomistic domain.

6.2.3 Coupling scheme

The approach used to couple the atomistic and the continuum domain is adopted from the work by [11]. The overall simulation domain is divided into four regions, namely zones I to IV, see Fig. 6.2. Zone I consists of a purely atomistic domain whereas zone IV represents the continuum domain surrounding the atomistic domain. Zone I and zone IV are bridged together by two transition regions, zone II and zone III. In the transition region both the atomistic and continuum domains overlap. In zone II, atomistic displacements are transferred from the atomistic domain to the continuum domain and in zone III, continuum displacements are transferred from the continuum domain to the atomistic domain. The atoms in zone I are labelled as core atoms. As the name implies, these atoms are used to model the dislocation core. The atoms in zone II and zone III are labelled as interfacial and pad atoms respectively. Both interfacial and pad atoms are coupled to the continuum finite element domain. The width of the pad atom region has to be at least that of the cut-off radius of the EAM potential used.

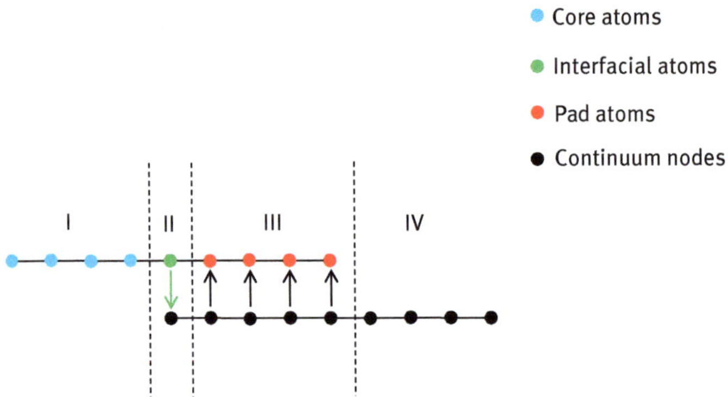

Fig. 6.2: Two-way transfer of displacements between the atomistic and continuum domain is illustrated. Atoms in zone I are marked as blue, atoms in zone II are marked as green, atoms in zone III are marked as red and the continuum nodes in zone IV are marked as black.

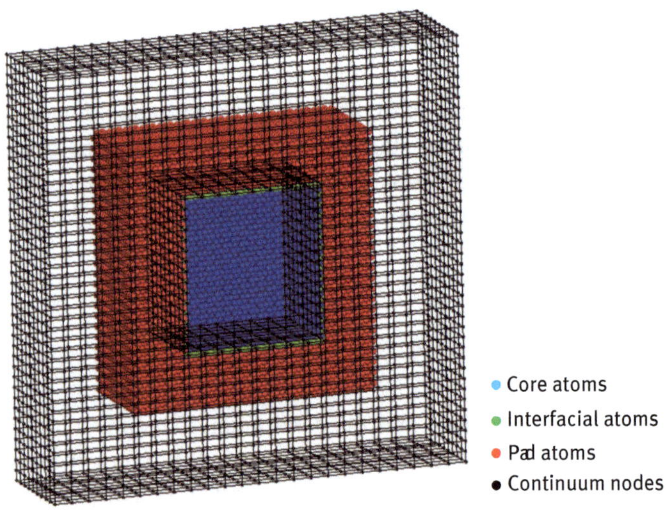

Fig. 6.3: A 3D perspective of the coupled atomistic and continuum domains with appropriately labeled lattice atoms and continuum nodes are presented.

The entire coupled 3D atomistic-continuum system with the short direction correspoding to the direction of the dislocation line is shown in Fig. 6.3 with appropriately labeled lattice atoms and continuum nodes. The steps involved in the coupling algorithm are as follows:

1. The Volterra construction is imposed on the atoms and as outer boundary conditions on the finite element domain.

2. Keeping the pad atoms fixed both the interfacial and core atoms are allowed to relax atomistically to the minimum energy configuration.
3. The displacements of the interfacial atoms are imposed on the finite element domain as inner boundary conditions.
4. The linear system of equilibrium equations are solved to obtain the finite element displacement field.
5. In turn the finite element displacement field is interpolated and imposed on the pad atoms.
6. The iteration is repeated from step 2 until solution global is obtained.

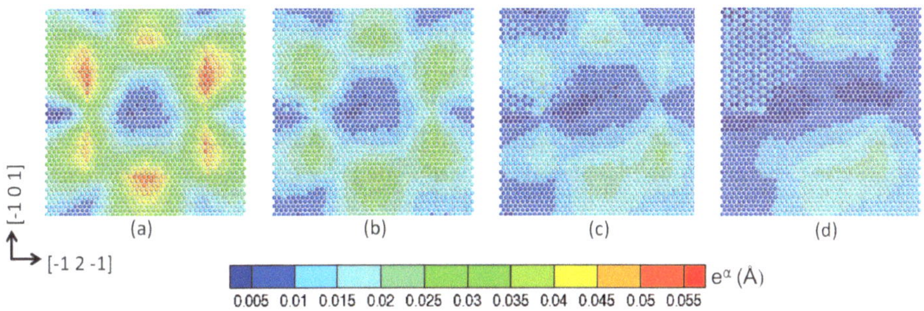

Fig. 6.4: L_2-norm for the $\frac{1}{2}[111]$ screw dislocation in the atomistic region of the coupled system, where a, b, c, and d correspond to the first, fifth, tenth and twentieth iteration respectively.

It is noted that since the atomistic and continuum domains are bridged in 3D, the structure is not restricted to modeling plane strain conditions. [11] verified force compatibility between the continuum and the atomistic region for a single crystal by applying normal and shear displancement boundary conditions in the continuum region and checking the transmission of the resulting stresses into the atomistic domain. In addition to this we show the continuity of the displacements for a screw dislocation in Fig. 6.6. More detailed analysis of this example is given in section 6.3. By using a large continuum region around the atomistic region, the core atoms get shielded from any boundary effects. A study of different domain sizes was performed to determine the optimum size of the pad and the core atoms. A core dimension of 44 Å × 42 Å and a pad of width 11 Å yielded the desired results. The number of atoms used in the coupled model is approximately a fraction of 1/50 of the fully atomistic model used by [20]. The finite element system of linear equations is solved using PARDISO [24, 25] direct solver. For larger system sizes, iterative solvers would be a favorable option compared to direct solvers to obtain maximum numerical efficiency. In the continuum finite element region a uniform hexahedral mesh size of 7 Å × 4 Å × 5 Å for the screw dislocation and 5 Å × 4 Å × 7 Å for the edge dislocation is used. The simulation can be further optimized by coarsening the mesh away from the dislocation core.

6.3 Verification by dislocation analysis

A $\frac{1}{2}[111]$ screw dislocation is introduced in the $(\bar{1}01)$ plane of the coupled system of dimension 500 Å × 500 Å with a thickness of 15 Å along the $[111]$ line direction. The atomistic domain is restricted to 98 Å × 101 Å with a core size of 56 Å × 60 Å. Anisotropic elastic constants corresponding to the Mendelev potential [19] with $C_{11} = 243.4$ GPa, $C_{12} = 145.0$ GPa and $C_{44} = 116.0$ GPa are used in the finite element domain. Periodic boundary conditions are applied both in the atomistic and finite element domain in the direction of the dislocation line. It is noted that the linear strain assumption and an anisotropic material model are used in the finite element domain.

In order to validate the coupled scheme the results are compared to a fully atomistic model with a dimension of 500 Å × 500 Å × 15 Å. To quantify the displacement error incurred in the coupled system compared to the fully atomistic model we follow the work of [10]. The displacement error per atom α of the coupled system is computed using the L_2-norm of the difference between the fully atomistic displacement vector $\tilde{\boldsymbol{u}}^{\alpha}_{\text{exact}}$ and the coupled displacement vector $\tilde{\boldsymbol{u}}^{\alpha}$. The error e^{α} is defined as follows,

$$e^{\alpha} = \|\tilde{\boldsymbol{u}}^{\alpha} - \tilde{\boldsymbol{u}}^{\alpha}_{\text{exact}}\| . \tag{6.2}$$

The global displacement error e normalized by the total number of atoms N_{A} is given by

$$e = \sqrt{\frac{\sum_{\alpha=1}^{N_{\text{A}}} (e^{\alpha})^2}{N_{\text{A}}}} \tag{6.3}$$

and the global percent error $e_{\%}$ is obtained by dividing e by the average of the atomic displacement $\tilde{u}_{\text{avg}}$:

$$e_{\%} = \frac{e}{\tilde{u}_{\text{avg}}} \cdot 100 , \tag{6.4}$$

where

$$\tilde{u}_{\text{avg}} = \frac{1}{N_{\text{A}}} \sum_{\alpha=1}^{N_{\text{A}}} \|\tilde{u} - \tilde{u}_{\text{exact}}\| . \tag{6.5}$$

As described in the previous section, the relaxation procedure in the coupled scheme is an iterative process, and Fig. 6.4 shows the error plots of the atomistic domain. It can be observed that initially much of the error is accumulated at the transition region and the error decreases in the subsequent iterations as the system converges. The $e_{\%}$ for the first, fifth, tenth and twentieth iteration are 4.65 %, 2.98 %, 2.07 %, and 1.78 % respectively.

Once convergence is obtained it can be seen from Fig. 6.5 that there is a good agreement between the coupled model shown on the left and the pure atomistic model shown on the right. From the line plot shown in Fig. 6.6 for the section marked by the black line shown in Fig. 6.5, it can be observed that there is continuity of displacement from the atomistic to the continuum region.

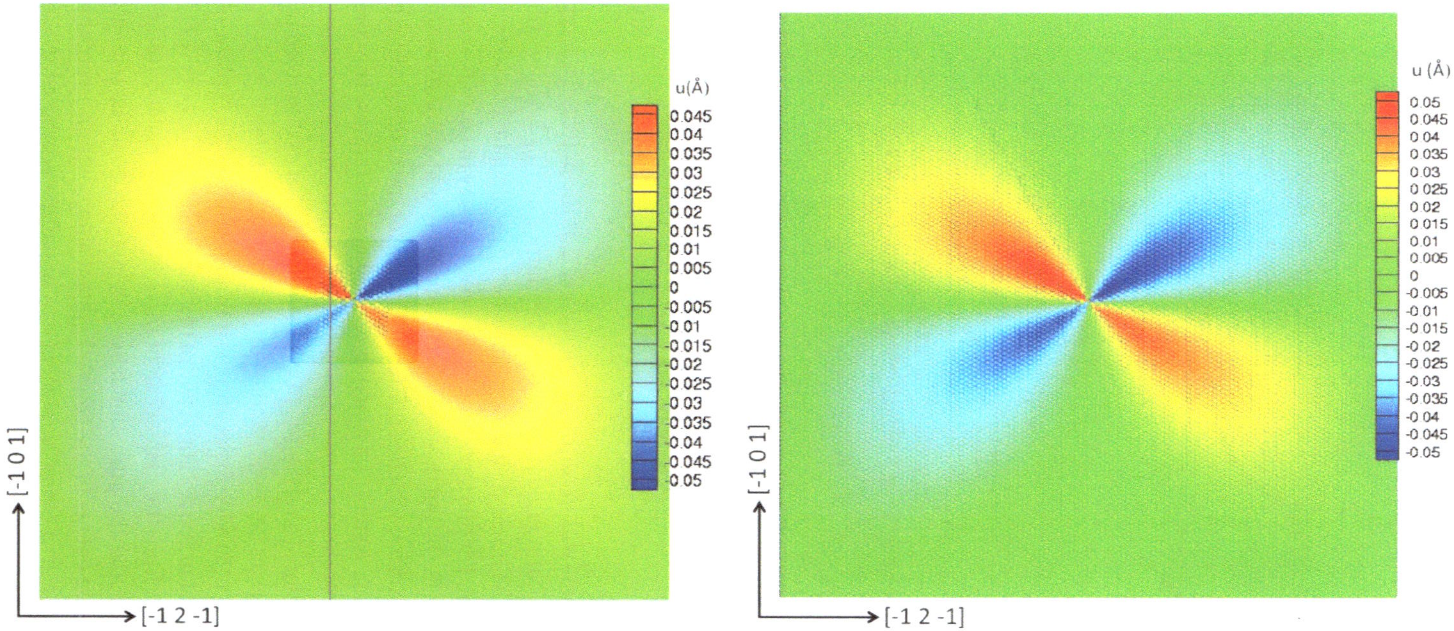

Fig. 6.5: Displacement plots of a $\frac{1}{2}[111]$ screw dislocation, where the coupled atomistic-continuum system is shown on the left and the fully atomistic model of the same size is shown on the right. The displacement along $[\bar{1}2\bar{1}]$ is labeled as u. The black line marks the intersection for which the continuity of displacements in shown in Fig. 6.6.

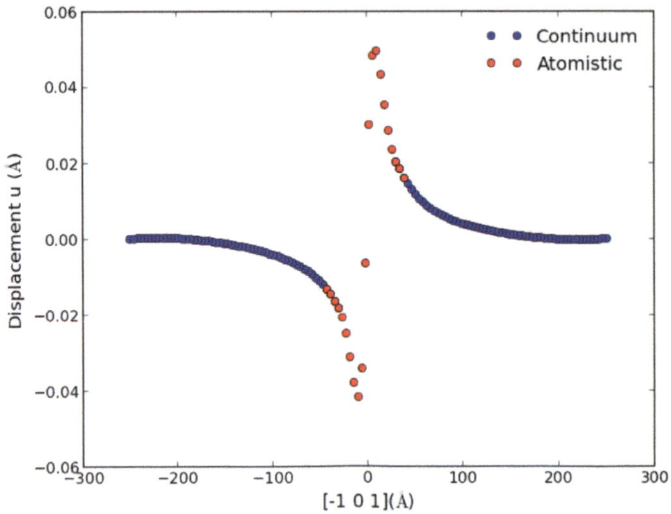

Fig. 6.6: Atomistic-continuum displacement line plot for the section marked by the black line shown in Fig. 6.5.

To further quantify the convergence behavior of the coupled system, we consider the evolution of the dislocation core energy during the iteration procedure. The dislocation core energy E_c per unit length is calculated as,

$$E_c = \frac{\sum_{i=1}^{n} E_i - nE_b}{L_z} \tag{6.6}$$

where n is the number of atoms within the dislocation core, E_i is the potential energy of atom i, E_b is the potential of a bulk atom in a perfect lattice and L_z is the width along the dislocation line. The screw dislocation core energy is calculated for atoms located within a radius of 15 Å. For this large radius, the excess energy in Eq. (6.6) most likely contains linear elastic effects, but these contributions to the total dislocation energy cancel out in the relative error, which is plotted in Fig. 6.8 for the coupled system compared to the pure atomistic system as a function of number of iterations. The core energy of the coupled system is shown to converge to that of the fully atomistic system. A differential displacement plot of a relaxed $\frac{1}{2}[111]$ screw dislocation as described by [26] is shown in Fig. 6.9 (a). It can be seen that the core is non-degenerate.

In a similar manner, a $\frac{1}{2}[111]$ edge dislocation is introduced in the $(\bar{1}2\bar{1})$ plane of the coupled system of dimension 500 Å × 500 Å with a thickness of 21 Å along the $[\bar{1}2\bar{1}]$ line direction. It can be seen from the displacement plots in Fig. 6.7 that the results between the coupled system and the pure atomistic system are in good agreement. In Fig. 6.8 the dislocation core energy of the coupled system agrees within 1 % of the pure atomistic system. The lack of better agreement can in part be attributed to the mismatch between the isotropic boundary conditions imposed and the anisotropic

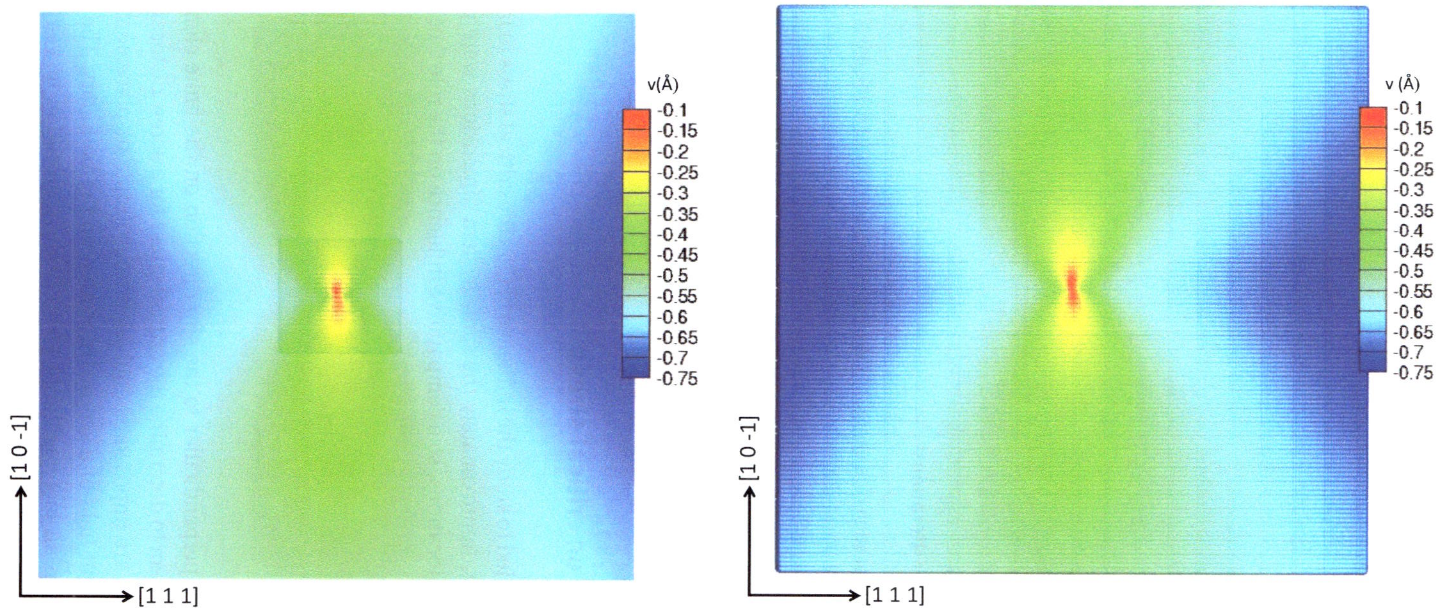

Fig. 6.7: Displacement plots of a $\frac{1}{2}[111]$ edge dislocation, where the coupled atomistic-continuum system is shown on the left and the fully atomistic model of the same size is shown on the right.

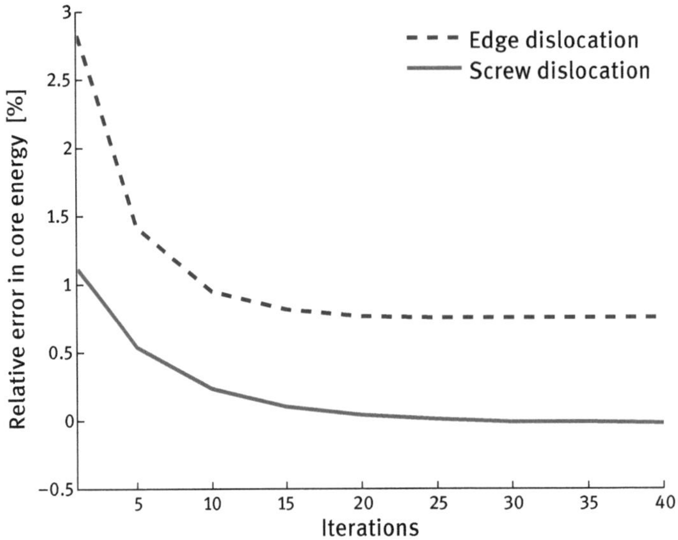

Fig. 6.8: The screw and edge dislocation core energy of the coupled system relative to the ones of a fully atomistic system.

behavior of the crystal. It is more pronounced for the edge dislocation because of its full strain tensor with non-zero hydrostatic and normal components, while the strain tensor of the screw dislocation has only shear components.

6.4 Carbon influence on critical stress

The coupled scheme is applied to study the influence of C on the critical stress to move edge and screw dislocation. Approximately the same system dimensions were used for both edge and screw dislocations, so a comparative influence on the role C can be obtained.

6.4.1 Screw dislocation

6.4.1.1 In pure Fe

Shear stresses are applied perpendicular to the slip direction on the finite element domain of the coupled system to determine the critical stress for a screw dislocation. The applied shear stresses are imposed as equivalent anisotropic displacements in addition to the displacement field of a screw dislocation on the boundary. The displacements were applied in an incremental of 1 Å until the dislocation moved and further increments were bisected to the desired accuracy. The corresponding stress

τ, is obtained using Hooke's law. The error bar corresponds to the least incremental used to determine dislocation movement. Having the x-axis along the direction of slip, y-axis normal to the slip plane and z-axis along the direction of the dislocation line the relevant components of the stress tensor σ, is $\tau = \sigma_{23} = \sigma_{32}$.

The critical stress of the screw dislocation in a coupled system of dimension 500 Å × 500 Å × 15 Å is found to be 1205 ± 25 MPa. The determined critical stress is compared to a pure atomistic study [20] of the same dimensions using EAM Mendelev potential is found to be within the error bar. The result also compares well to the study by [27] on glide of screw dislocations in Fe.

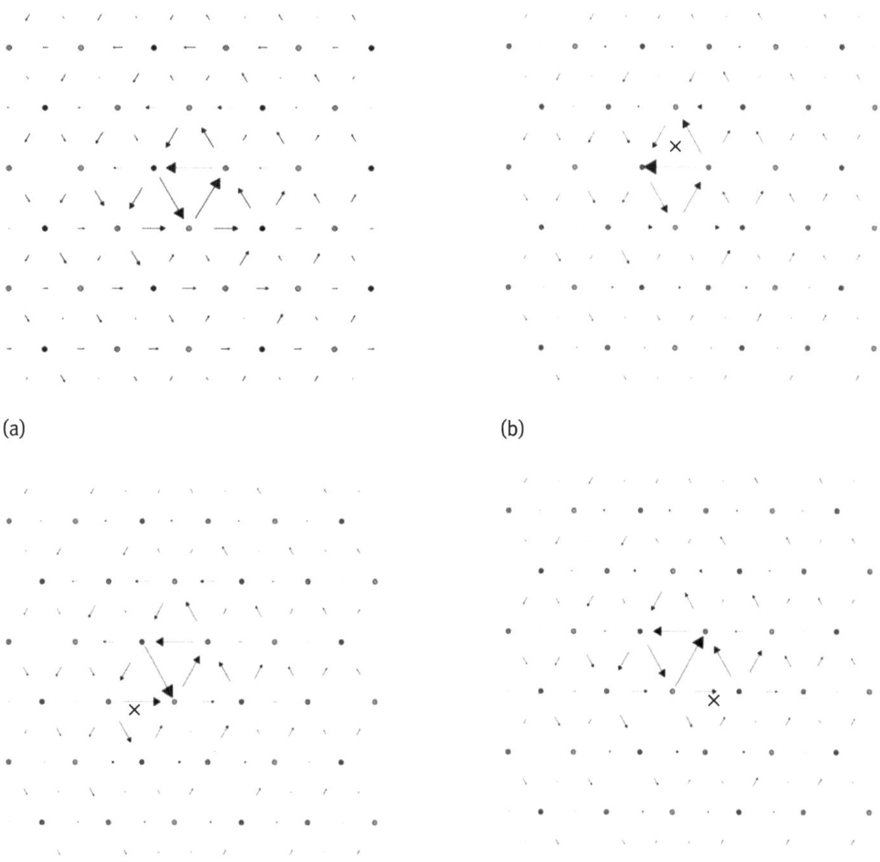

(a) (b) (c) (d)

Fig. 6.9: A differential displacement plot of a relaxed $\frac{1}{2}[111]$ screw dislocation in Fe. The × denotes the location of carbon in the screw dislocation.

6.4.1.2 Interstitial C, case (i)

A coupled system dimension of 500 Å × 500 Å × 45 Å is used for this study. The first position of the carbon atom is shown in Fig. 6.9 (b) and the change in the dislocation structure due to the presence of carbon can be observed. The segregation energy of carbon at the screw dislocation core is determined to be −0.59 eV per C atom. Three scenarios are studied, the first with three carbon atoms at a equal spacing of 15 Å, the second with two carbon atoms at a equal spacing of 22.5 Å and the third with one carbon atom along the dislocation line i.e. at a spacing of 45 Å due to periodic boundary conditions. The intention was to obtain the Peierls stress as a function of C concentration. However, in all three scenarios no bowing out of the dislocation line around the C atoms was observed, which means we did not reach a dilute limit. Moreover, the critical stress is found to be 2129 ± 50 MPa in all scenarios, which means the system is saturated with C already at a spacing of 45 Å. To observe any bowing out effect due to the presence of carbon, a larger system dimension along the dislocation line than the one used is required.

6.4.1.3 Interstitial C, case (ii)

In order to investigate any anisotropic behavior in the influence of carbon on the critical stress, two more additional carbon sites, namely position two and three as shown in Fig. 6.9 (c) and (d) respectively are considered. One carbon site is obtained from another by a rotation of 120° around the dislocation line as can be seen by the differential plot in Fig. 6.9 and correspond to symmetric position within the non-degenerate screw dislocation core. All positions are energetically equivalent. The dimensions, loading direction and other parameters used are same as in case (i). The critical stress for positions two and three was found to be 1580 ± 50 MPa. Both positions have the same critical stress because C in position three reshuffles to position two under the application of load. The critical stress is lower compared to the one for C in position one, and notably for these positions the dislocation starts to move on an inclined slip plane rather than on the horizontal plane. The anisotropy in the influence of carbon on screw dislocations, certainly deserves more attention in the future.

6.4.2 Edge dislocation

The critical stress to move a pure edge dislocation is calculated by applying shear displacements parallel to the slip direction. In this case the equivalent stress components are $\tau = \sigma_{23} = \sigma_{32}$.

The critical stress of an edge dislocation in pure Fe in a coupled system of dimension 500 Å × 500 Å × 42 Å is found to be 85 ± 12 MPa. The computed critical stress compares well to a pure atomistic study of the same dimensions. To determine the influence of carbon on the critical stress, a carbon atom is placed below the dislocation line at the energetically most favorable position. The segregation energy of carbon at the edge dislocation core is determined to be –0.75 eV. The critical stress to move the dislocation in the presence of a C atom is found to be 486 ± 30 MPa.

6.4.3 Discussion

For both edge and screw dislocations, carbon increases the critical stress giving evidence to the Cottrell effect. The increase in critical stress due to the presence of carbon for the screw dislocation is 77 %, and for the edge dislocation it is 470 %. As reported in the work by [4], since the maximum binding energy for C in edge dislocation cores is higher compared to screw dislocation cores, carbon pins edge dislocations more effectively. However, note that even in the presence of carbon the critical stress to move an edge dislocation is a factor of 3.5 smaller than to move a screw dislocation with or without carbon. Hence, the motion of screw dislocation remains the limiting factor in the deformations of ferritic steels. This is also supported by the lower segregation energy which we obtain for the edge dislocation. The critical stress values for screw and edge dislocation with and without carbon are summarized in Tab. 6.1.

Tab. 6.1: Critical stress of screw and edge dislocation with and without carbon.

	Critical stress (MPa)		
Dislocation	**Without carbon**	**With carbon**	**Relative increase**
Screw	1205 ± 25		
Position 1		2129 ± 50	77 %
Position 2		1580 ± 50	32 %
Position 3		1580 ± 50	32 %
Edge	85 ± 12	486 ± 30	470 %

6.5 Conclusion

XFEAt, a coupled extended finite element (XFE) and atomistic (At) framework, was formulated and implemented to study the effect of carbon on the motion of screw and edge dislocations in the Fe-C system. An embedded atom method (EAM) potential [19] is used to model the dislocation core in the inner atomistic domain and the extended finite element method is used in the outer continuum domain to model the slip along the glide plane and the correct mechanical boundary conditions. With this model it is possible to study the behavior of dislocations in an effectively infinite anisotropic elastic medium. This coupled XFEAt scheme is applied to study the critical stress to move edge and screw dislocations in α-Fe with and without interstitial C atoms being present at the dislocation core. We obtain good agreement with the results of a fully atomistic study of pure Fe [20]. The presence of carbon increases the critical stress of a screw dislocation by 77 % and that of an edge dislocation by 470 %, as carbon pins more effectively to edge dislocations compared to screw dislocations. However, the critical stress to move a screw dislocation is still a factor 3.5 higher, even in the presence of carbon. Thus, the motion of screw dislocations remains the rate limiting factor even in the Fe-C system. Furthermore, it is found that carbon placed at energetically equivalent sites around the screw dislocation leads to different behaviors. This phenomenon shows once more the importance of deriving critical stresses and flow rules used in continuum models from atomistic simulations. Our coupled system can be used to study such an influence of interstitials on dislocation cores and can be easily extended to other types of atomic interaction models.

Acknowledgment: The Authors acknowledge financial support through Thyssen-Krupp AG, Bayer MaterialScience AG, Salzgitter Mannesmann Forschung GmbH, Robert Bosch GmbH, Benteler Stahl/Rohr GmbH, Bayer Technology Services GmbH, and the state of North-Rhine Westphalia as well as the European Commission in the framework of the European Regional Development fund (ERDF).

References

[1] Hull D, Bacon D. Introduction to Dislocations. 4th ed. Oxford, UK: Butterworth-Heinemann; 2001.

[2] Tapasa K, Osetsky Y, Bacon D. Computer simulation of interaction of an edge dislocation with a carbon interstitial in ff-iron and effects on glide. Acta Mater. 2007;55:93–104.

[3] Song J, Curtin W. Mechanisms of hydrogen-enhanced localized plasticity: An atomistic study using α-Fe as a model system. Acta Mater. 2014;68:61–69.

[4] Clouet E, Garruchet S, Nguyen H, Perez M, Becquart C. Dislocation interaction with C in α-Fe: A comparison between atomic simulations and elasticity theory. Acta Mater. 2008;56:3450–3460.

[5] Becquart C, Raulot J, Bencteux G, Domain C, Perez M, Garruchet S, et al. Atomistic modeling of an Fe system with a small concentration of C. Computational Materials Science. 2007;40: 119–129.

[6] Hristova E, Janisch R, Drautz R, Hartmaier A. Solubility of carbon in alpha-iron under volumetric strain and close to the Σ5(310)[001] grain boundary: Comparison of DFT and empirical potential methods. Comp Mat Sci. 2011;50:1088–1096.

[7] Li H, Wurster S, Motz C, Romaner L, Ambrosch-Draxl C, Pippan R. Dislocation-core symmetry and slip planes in tungsten alloys: Ab initio calculations and microcantilever bending experiments. Acta Mater. 2012;60:748–758.

[8] Dezerald L, Ventelon L, Clouet E, Denoual C, Rodney D, Willaime F. *Ab initio* modeling of the two-dimensional energy landscape of screw dislocations in bcc transition metals. Phys Rev B. 2014;89(024104).

[9] Hirth J, Lothe J. Theory of Dislocations. 2nd ed. New York, NY: John Wiley & Sons; 1982.

[10] Miller RE, Tadmor EB. A unified framework and performance benchmark of fourteen multiscale atomistic/continuum coupling methods. Modelling Simul Mater Sci Eng. 2009;17(053001).

[11] Kohlhoff S, Gumbsch P, Fischmeister HF. Crack propagation in b.c.c. crystals studied with a combined finite-element and atomistic model. Philos Mag A. 1991;64:851–878.

[12] Shilkrot L, Miller RE, Curtin WA. Multiscale plasticity modeling: Coupled atomistics and discrete dislocation mechanics. Journal of the Mechanics and Physics of Solids. 2004;52:755–787.

[13] Nair A, Warner D, Hennig R. Coupled quantumcontinuum analysis of crack tip processes in aluminum. Journal of the Mechanics and Physics of Solids. 2011;59:2476–2487.

[14] Leyson GPM, Hector Jr LG, Curtin WA. Solute strengthening from first principles and application to aluminum alloys. Nat Mater. 2010;9:750–755.

[15] Leyson G, Hector Jr LG, Curtin W. Solute strengthening from first principles and application to aluminum alloys. Acta Mater. 2012;60:3873–3884.

[16] Woodward C, Rao SI. Flexible ab initio boundary conditions: Simulating isolated dislocations in bcc Mo and Ta. Phys Rev Lett. 2002;88(216402).

[17] Woodward C, Trinkle DR, Hector LG, Olmsted DL. Prediction of dislocation cores in aluminum from density functional theory. Phys Rev Lett. 2008;100(045507).

[18] Moës N, Dolbow J, Belytschko T. A finite element method for crack growth without remeshing. International Journal for Numerical Methods in Engineering. 1999;46:131–150.

[19] Mendelev MI, Han S, Srolovitz DJ, Ackland GJ, Sun D, Asta M. Development of new interatomic potentials appropriate for crystalline and liquid iron. Philos Mag. 2003;83:3977–3994.

[20] Koester A, Ma A, Hartmaier A. Atomistically informed crystal plasticity model for body-centered cubic iron. Acta Mater. 2012;60:3894–3901.

[21] Ruda M, Farkas D, Garcia G. Atomistic simulations in the FeC system. Computational Materials Science. 2009;45:550–560.

[22] Bitzek E, Koskinen P, Gähler F, Moseler M, Gumbsch P. Structural relaxation made simple. Phys Rev Lett. 2006;97:170–201.

[23] Gracie R, Ventura G, Belytschko T. A new fast finite element method for dislocations based on interior discontinuities. International Journal for Numerical Methods in Engineering. 2007;69: 423–441.

[24] Schenk O, Wächter A, Hagemann M. Matching-based preprocessing algorithms to the solution of saddle-point problems in large-scale nonconvex interior-point optimization. Computational Optimization and Applications. 2007;36:321–341.

[25] Schenk O, Bollhöfer M, Römer RA. On large-scale diagonalization techniques for the Anderson model of localization. SIAM Rev. 2008;50:91–112.

[26] Vitek V, Perrin RC, Bowen DK. Core structure of 1/2(111) screw dislocations in bcc crystals. Philos Mag. 1970;21(173).

[27] Chaussidon J, Fivel M, Rodney D. The glide of screw dislocations in bcc Fe: Atomistic static and dynamic simulations. Acta Mater. 2006;54:3407–3416.

Part II: **Multiscale simulations of plastic deformation and fracture**

A. Siddiq*, S. Schmauder, and M. Rühle

7 Niobium/alumina bicrystal interface fracture

A theoretical interlink between local adhesion capacity and macroscopic fracture energies

Abstract: In the presented work, an effort has been put to clear up the theoretical interlink between local adhesion capacity and macroscopic fracture energies by bridging different length scales, such as nano-, meso-, and macro-scale. Crystal plasticity theory along with a cohesive modelling approach has been used during this work. The influence of different cohesive law parameters (cohesive strength, work of adhesion) on the macroscopic fracture energies for three different orientations of niobium/alumina bicrystal specimens has been presented. It is found that cohesive strength has a stronger effect on macroscopic fracture energies as compared to work of adhesion. In the last part a generalized correlation among macroscopic fracture energy, cohesive strength, work of adhesion and yield stress is derived. The presented results can provide a great help to experimentalists in order to design better metal/ceramic interfaces.

Keywords: metal/ceramic interface, crystal plasticity, cohesive model, fracture mechanics, work of adhesion

7.1 Introduction

Interfaces between metal and ceramic play an important role in many applications, because they combine the properties of metals like ductility, high electrical and thermal conductivity and the properties of ceramics like high hardness, corrosion resistance and capacity of resistance to wear. Some of the applications of metal/ceramic joints include automotive industry, turbine blades, high pressure sodium lamps, squid magnetometers and dental implants. The fracture at or near such interfaces often limits the reliability of these joints. In the past, niobium/alumina interfaces have been studied experimentally [1–5] and numerically [6–8]. Single crystalline alumina/niobium/alumina joints were investigated in [1, 2] for different orientations of single crystalline niobium. The results showed that the orientations of single crystalline materials have significant effects on the energies of the niobium/alumina interface fracture. It has been observed experimentally [3, 4] that the amount of plasticity initiated in the metal during fracture of such interfaces is strongly influenced by both, the interface chemistry and the orientations of slip systems relative to the crack surface due to which the fracture energy of niobium/alumina bicrystal interfaces changes dramatically with relative crystallographic orientation of the two constituents, i.e. metal and ceramic. Isotropic polycrystalline niobium/alumina interfaces were studied experimentally as well as numerically in [5, 6]. It was shown that a low yield stress results in high plastic deformation in the metal part and consequently a higher plastic energy is dissipated before the critical stress value for fracture is reached. In order to study crystal orientation effects on crack initiation energies of bicrystalline niobium/alumina interfaces, crystal plasticity theory [7, 8] was used in [9, 10]. The results showed that by changing the orientation of the single crystalline material, one can change the amount of plasticity induced due to the activation of different slip systems, also, the level of stresses induced at and near the crack tip were found to be different. The reason of the variation in fracture energies for various orientations were explained on the basis of induced stresses and plasticity due to the activation of different slip systems around the crack tip. Interface fracture of a metal/ceramic bicrystalline interface (niobium(110)[001]/alumina($11\bar{2}0$)[0001]) using a cohesive modelling approach and conventional crystal plasticity theory was studied in [11]. The effect of different cohesive law parameters, such as, cohesive strength and work of adhesion on the macroscopic fracture energy was studied. It was found that the effect of cohesive strength on fracture energies is more profound as compared to the work of adhesion. A comparison between experimental and simulation results was presented and a theoretical interlink among cohesive strength, work of adhesion, yield stress and macroscopic fracture energy was presented.

The aim of the present work is to study the interface fracture behaviour of two other orientations (niobium(100)[001]/alumina($11\bar{2}0$)[0001] and niobium(111) [$\bar{1}\bar{1}2$]/alumina($11\bar{2}0$)[0001]) of niobium/alumina bicrystal specimens using crystal plasticity theory [7, 8] and a cohesive modelling approach [12]. The effect of different cohesive law parameters, such as cohesive strength and work of adhesion, on the fracture energies of the bicrystal niobium/alumina interfaces is studied for the different orientations. Cohesive model parameters are identified for different combinations of cohesive strength and work of adhesion by applying a scale bridging procedure as shown in Fig. 7.1. For each value of work of adhesion [13–15], the cohesive strength is varied and transferred to the bicrystal niobium/alumina finite element model as input parameters for the cohesive law to simulate the interface fracture. As discussed above, niobium single crystals are modelled using crystal plasticity theory in bicrystal niobium/alumina interfaces. The result of the simulation, i.e. the computed fracture energy of the system is compared with the experimental fracture energy value. When convergence is reached then the final set of local interface fracture parameters (cohesive strength, work of adhesion) and global fracture energy of the system is obtained, otherwise the simulation is rerun with the new guess of cohesive strength. It should be noted that, in general, the cohesive parameters can be directly computed from atomistic (ab initio) calculations. Although for the present niobium/alumina system only work of adhesion values were available, that is why different cohesive strength values have been used to validate the experimental and simulated results. A generalized correlation among cohesive strength, work of adhesion, yield stress (the stress level at which the single crystalline material starts to yield, i.e. primary slip system is activated in the single crystalline material; it should be noted that yielding in single crystalline material depends upon the orientation of the single crystalline material with respect to loading axis [10, 16, 17]) and macroscopic fracture energy will be deduced for all three orientations of niobium/alumina bicrystal specimens.

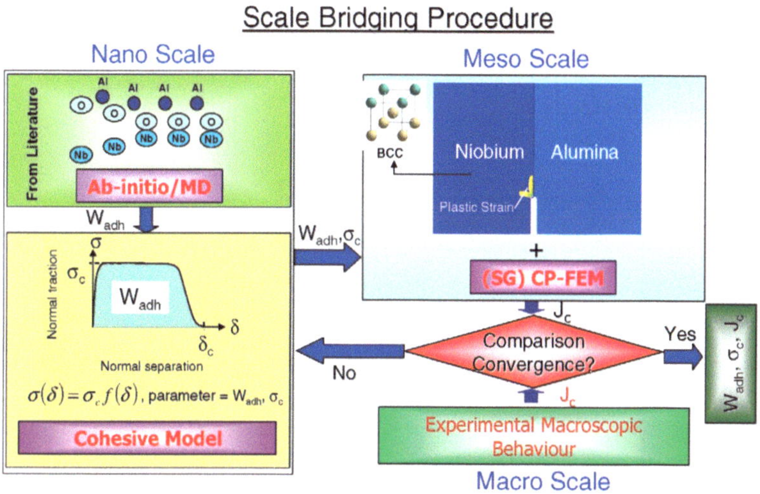

Fig. 7.1: Scale bridging procedure for metal/ceramic interface fracture.

7.2 Concept of modelling

The finite element model is based on the experiments performed in [4], in which the infuence of orientation and impurities on the fracture behaviour of niobium/alumina interfaces was studied using notched bending tests. The finite element model is similar to the one discussed in [10, 11]. The dimensions of the finite element model were $2 \times 4 \times 36\,\text{mm}^3$ with a notch length of 0.4 mm. A schematic of the specimen is shown in Fig. 7.2. The details of the specimen preparation are discussed in detail in [4]. Finite element models of alumina shanks and a polycrystalline niobium sheet were always joined with the corresponding niobium single crystal finite element model and alumina finite element model using tie constraints available in ABAQUS. For crack propagation, cohesive elements of zero thickness were used along the interface of the niobium/alumina bicrystal. The cohesive law used for this study is proposed in [12]. The finite element mesh consisted of 39 508 plane strain four-noded quadrilateral elements, as shown in Fig. 7.3. The specimens are loaded with a loading rate of 96.8 μm/min.

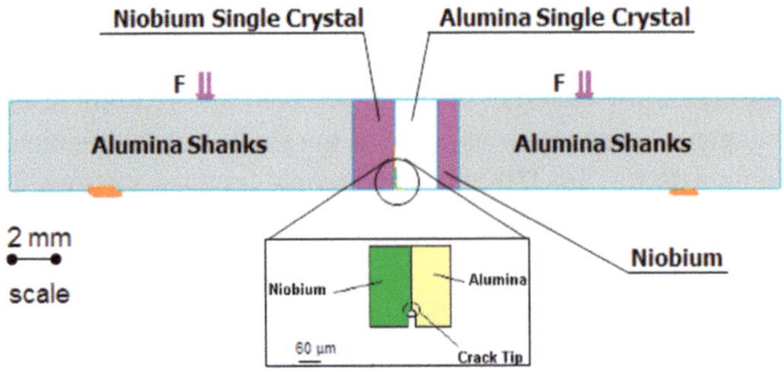

Fig. 7.2: Four point bending test specimen.

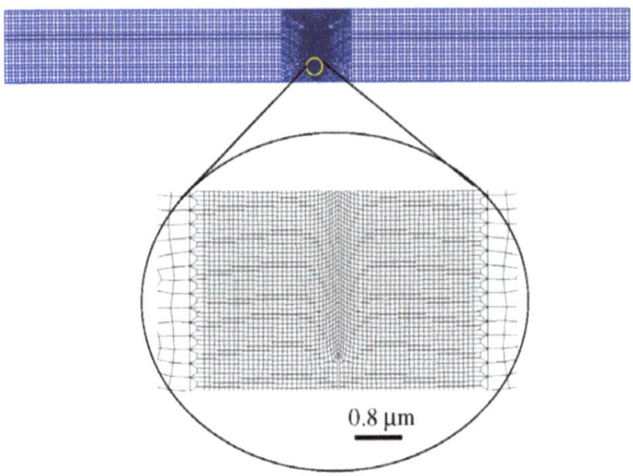

Fig. 7.3: Finite element mesh of four point bending test bicrystal specimens.

Throughout this work, both outer alumina shanks (ceramic) and the alumina single crystalline material, at the center of the specimen, were treated as purely elastic with a Young's modulus of 390 GPa and a Poisson's ratio of 0.27. The polycrystalline niobium sheet (which was used in experiments to join alumina shank with alumina single crystals) is always modelled with an elastic–plastic constitutive law. Young's modulus and Poisson's ratio for the polycrystalline niobium sheet were the same for all simulations ($E = 104.9$ GPa, $m = 0.397$). These elastic and plastic data are adjusted to alumina and niobium, respectively, given in [5, 6]. The plastic behaviour of the stress–strain curve of the polycrystalline niobium sheet is approximated by a Ramberg–Osgood relation [5], which is described in a one dimensional case by the following equation:

$$E_c \varepsilon = \sigma + \alpha \left(\frac{\sigma}{\sigma_0} \right)^{n-1} \sigma \tag{7.1}$$

This material model for deformation plasticity is also valid for multi-axial stress state. Here, n denotes the hardening exponent, α the yield offset and σ_0 the yield stress. The parameters of the above equation are adjusted to the niobium stress–strain curves in [5, 6]. The parameters used are $n = 6$, $\sigma_0 = 180$ MPa and $\alpha = 0.3$.

Single crystalline niobium is always modelled using the hardening law given in [18]. The Bassani and Wu model [18] is based upon the analytical characterization of the hardening moduli at any of the three stages during deformation. Their expression for self and latent hardening depends on the shear strain γ^α of all slip systems

$$\begin{aligned} h_{\alpha\alpha} &= \left\{\left(h_0 - h_s \operatorname{sech}^2\left[\frac{(h_0 - h_s)\,\gamma_\alpha}{\tau_s - \tau_0}\right] + h_s\right)\right\} G\left(\gamma^\beta; \beta \neq \alpha\right) \\ h_{\alpha\beta} &= q h_{\alpha\alpha}(\alpha \neq \beta) \\ &G(\gamma^\beta; \beta \neq \alpha) = 1 + \sum_{\beta\neq\alpha} f_{\alpha\beta} \tanh\left(\frac{\gamma^\beta}{\gamma_0}\right) \end{aligned} \tag{7.2}$$

where, h_0 is the initial hardening modulus, τ_0 the initial yield stress, τ_s the saturation stress, γ^α the total shear strain in slip system α, h_s the hardening modulus during stage I deformation, $f_{\alpha\beta}$ the interaction strength between slip systems α and β, and γ^β the total shear strain in slip system β while sech is the secant hyperbolic function.

The hardening moduli are described with an initial hardening (h_0) which saturates after reaching the resolved shear stress (τ_s). After the diminishing of the hyperbolic secant term, the saturation hardening term (h_s) specifies that each slip system has a finite hardening rate. The function G deals implicitly with cross-hardening that occurs between slip systems during stage II hardening.

Tab. 7.1: Hardening parameters for the Bassani and Wu hardening law.

	τ_0 (MPa)	τ_s (MPa)	h_0 (MPa)	h_s (MPa)	γ_0	γ_{0l}	$f_{\alpha\beta}$	$f_{\alpha\beta l}$	q	q_l
(110)[111]	13.5	16.4	292.262	1.4	0.25	0.25	10.0	9.9	0.2891	0.2315
(112)[111]	13.07	16.344	49.03325	39.2266	0.1	0.1	0.34	0.3	0.01	0.011

The hardening parameters for each slip system are derived in [19]. The hardening parameters used for each family of slip systems, i.e., (110)[111] and (112)[111], are given below in Tab. 7.1. Slip systems (123)[111] were not considered during the simulation, based on the experimental findings of Bowen [19]. His experimental results showed that (123)[111] slip systems are only activated when substitutional impurities are present in the single crystalline niobium.

For crack growth analyses, a cohesive law proposed by Scheider and Brocks [12] has been applied. In this cohesive law the initial stiffness of the cohesive element can be varied by changing δ_1, as shown in Fig. 7.4. If δ_1 is increased, the slope of the curve, before σ_c is reached, decreases and similarly the slope of the curve increases if δ_1 is

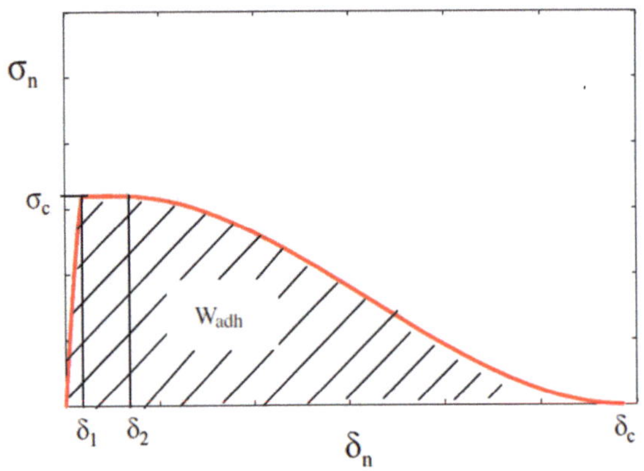

Fig. 7.4: Cohesive law proposed by Scheider and Brocks [12].

decreased. Also a region can be defined, where the traction in the cohesive element is kept constant. This has been achieved by using two additional parameters δ_1 and δ_2 (Fig. 7.4), leading to the following formulation for the function $\sigma(\delta)$:

$$\sigma(\delta) = \sigma_c \begin{cases} 2\left(\frac{\delta}{\delta_c}\right) - \left(\frac{\delta}{\delta_1}\right)^3 & \delta \le \delta_1 \\ 1 & \delta_1 \le \delta \le \delta_2 \\ 2\left(\frac{\delta-\delta_2}{\delta_c-\delta_2}\right)^3 - 3\left(\frac{\delta-\delta_c}{\delta_c-\delta_2}\right)^2 & \delta_2 \le \delta \le \delta_c \end{cases} \tag{7.3}$$

The cohesive models can be described by two independent parameters. These parameters may be two of the three parameters, namely the cohesive energy also known as work of adhesion (W_{adh}), and either of the cohesive strength (σ_c) or the separation length (δ_c). The cohesive strength (σ_c) is the maximum stress (traction) value at which damage initiates while the separation length (δ_c) is the amount of separation at which the interface completely fails.

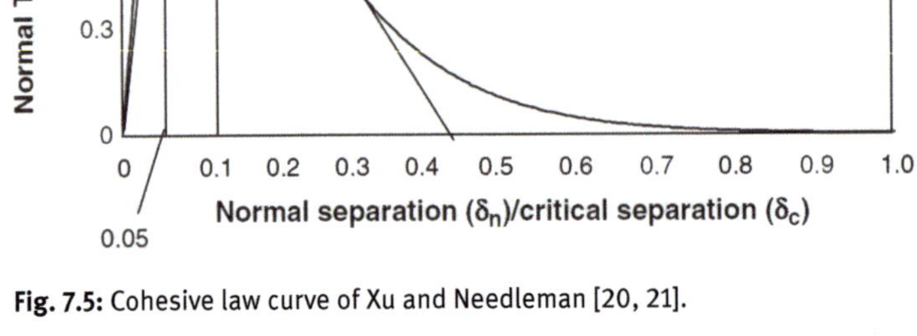

Fig. 7.5: Cohesive law curve of Xu and Needleman [20, 21].

Throughout this work δ_1 and δ_2 are chosen by comparing the slopes of the above discussed cohesive model with the cohesive law curves of Xu and Needleman [20, 21]. This is done in order to find the values of δ_1 and δ_2 so that the region of peak cohesive stress is similar to that used by Xu and Needleman [20, 21] in their cohesive law derived from atomistic calculations. δ_1 is obtained by comparing the initial slope of the cohesive law before damage is initiated (i.e. before σ_c is achieved) as shown in Fig. 7.5, while δ_2 is obtained by comparing the slope of the curve after the damage has been initiated (as shown in Fig. 7.5). The straight line, after the damage has been initiated, does not show the complete separation (δ_c), but it is just the tangent to the curve of Xu and Needleman cohesive model in order to reach to the cohesive strength value, which is then projected on the horizontal axis to obtain δ_2. This cohesive law has been used by Xu and Needleman [21] and Kysar [17, 22] for copper/alumina bicrystal interfaces and the values δ_1 and δ_2 are found to be $\delta_1 = 0.05\delta_c$ and $\delta_2 = 0.1\delta_c$.

All the simulations have been performed for three different orientations of niobium/alumina bicrystal interfaces, i.e. niobium(110)[001]/alumina(1$\bar{1}$20)[0001], niobium(100)[001]/alumina(1$\bar{1}$20)[0001] and niobium(111)[$\bar{1}\bar{1}$2]/alumina(1$\bar{1}$20) [0001].

7.3 Results and discussion

The effect of cohesive law parameters, such as cohesive strength and work of adhesion, on the fracture energies of niobium(110)[001]/alumina(1$\bar{1}$20)[0001] interfaces were studied in [10]. A correlation among macroscopic fracture energy, cohesive strength, work of adhesion and yield stress was also deduced for this orientation. The present work is an extension of the studies performed in [11]. In the following, the effect of cohesive strength and work of adhesion on macroscopic fracture energies are presented for three different orientations of niobium/alumina interfaces, i.e. niobium (110)[001]/alumina(1$\bar{1}$20)[0001], niobium(100)[001]/alumina(1$\bar{1}$20)[0001] and niobium(111)[$\bar{1}\bar{1}$2]/alumina(1$\bar{1}$20)[0001]. Also, a generalized correlation among macroscopic fracture energy, cohesive strength, work of adhesion and yield stress is deduced for these three orientations.

Three different values of work of adhesion (1, 4, 9.8 J/m^2) are used for the finite element simulations based on the reported values in [13–15, 22–25]. In the following, crack propagation analyses have been performed using three different sets of values of cohesive law parameters, such as cohesive strength and work of adhesion. Similar to [10], the effect of cohesive strength for a constant work of adhesion on the steady state fracture energy is studied and later the effect of work of adhesion on the steady state fracture energy is studied for constant cohesive strength values.

Fracture energies as a function of cohesive strength to yield stress ratio (σ_c/σ_y) have been plotted in Fig. 7.6 for three different work of adhesion values at steady state crack growth for niobium(110)[001]/alumina(1$\bar{1}$20)[0001] interfaces [11]. These

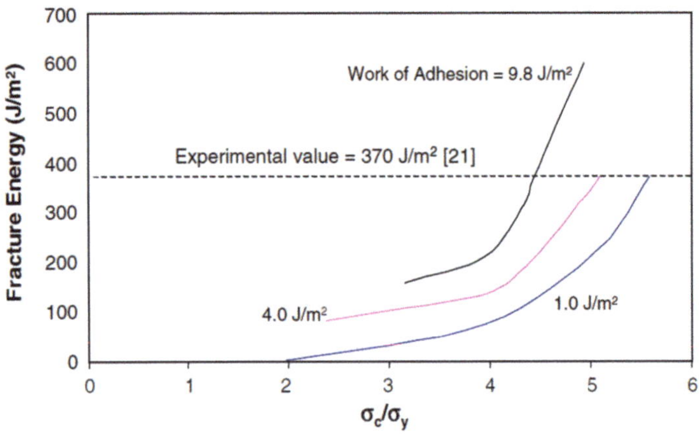

Fig. 7.6: Fracture energy vs cohesive strength (σ_c) to yield stress (σ_y) ratio at steady state crack growth for different work of adhesion, γ_0 (Nb(110)[001]||Al_2O_3(1120)[0001]) [11].

curves show that as the cohesive strength to yield stress ration (σ_c/σ_y) increases the fracture energy at the steady state crack growth also increases. This increase is gradual until the cohesive strength to yield stress ratio (σ_c/σ_y) is approx. 4.0. As soon as this value is reached the slope of the fracture energy curve increases which is due to the high plastic energy dissipation according to the higher number of activated slip systems. The reason for the higher number of activated slip systems is that, as the cohesive strength is increased the resolved shear stress of the slip systems with lower Schmidt factor also increases until it reaches the yield stress of such slip systems. As soon as the resolved shear stress becomes equal to the yield stress, the slip systems with lower Schmidt factor are activated. This is also shown in Fig. 7.7 [10], where slip systems $(1\bar{1}2)[\bar{1}11]$, $(\bar{2}11)[111]$, and $(211)[\bar{1}11]$ possess the highest Schmidt factors, namely the Schmidt factors −0.4714, −0.2357, and −0.408, respectively. Due to the highest Schmidt factor −0.4714 appearing for the $(1\bar{1}2)[\bar{1}11]$ system, it is the first slip system which is activated. As the deformation process still continues, stresses around the crack tip also increase, causing other slip systems to activate, e.g. the $(\bar{2}11)[111]$ and $(211)[\bar{1}11]$ slip systems. This activation of slip systems continues as the stresses around the crack tip increase.

In order to reach up to the experimental results, different work of adhesion values yield different cohesive strength to yield stress ratios (σ_c/σ_y). The values of cohesive strength to yield stress ratio (σ_c/σ_y) were found for three different work of adhesion values, 1, 4, and 9.8 J/m^2 to be 5.59, 5.095, and 4.443, respectively. In a similar fashion the fracture energy is plotted as a function of the work of adhesion for different values of cohesive strength at steady state crack growth, as shown in Fig. 7.8. It shows that a linear relation exists between fracture energy and work of adhesion. These results (in Figs. 7.6 and 7.8) show that the fracture energy depends more strongly on the cohesive strength than on the work of adhesion.

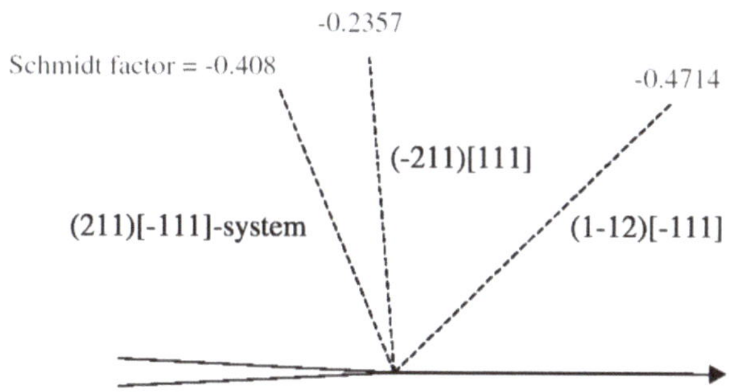

Fig. 7.7: Slip systems with highest Schmidt factors for niobium(110)[001]||alumina(11$\bar{2}$0)[0001] interfaces.

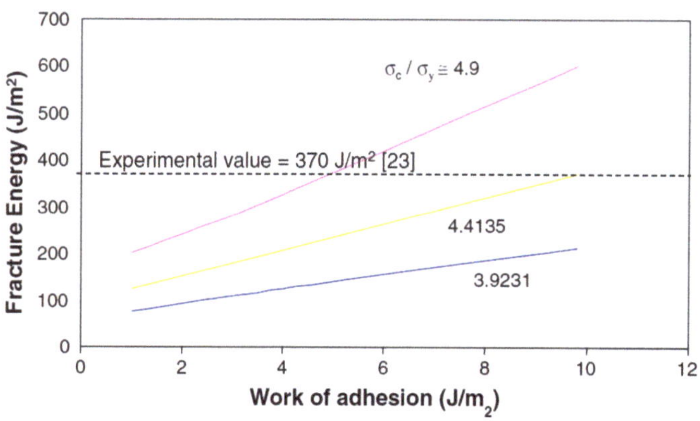

Fig. 7.8: Fracture energy vs. work of adhesion for the Nb(110)[001]||Al_2O_3(11$\bar{2}$0)[0001] interface crack [11].

Similar studies have been performed for the other two orientations, i.e., when niobium(100)[001] is bonded with alumina(11$\bar{2}$0)[0001] and when niobium(111)[$\bar{1}\bar{1}$2] is bonded with alumina(1120)[0001]. Figs. 7.9 and 7.10 show the plots of the fracture energies as a function of cohesive strength for three different values of work of adhesion at steady state crack growth. Fig. 7.9 shows the fracture energy results for a niobium (100)[001]|alumina(11$\bar{2}$0)[0001] interface and Fig. 7.10 for the niobium(111)[$\bar{1}\bar{1}$2]|alumina(11$\bar{2}$0)[0001] interface. The plot in Fig. 7.9 shows that as the cohesive strength increases the fracture energies are almost linearly dependent on σ_c/σ_y and when the value of the cohesive strength reaches the range of 200–210 MPa (i.e., $\sigma_c/\sigma_y \cong 5$) the slopes of the fracture energy curves increase. As discussed before, the reason is higher plastic energy dissipation due to the higher number of slip systems being activated. A similar (as seen in Fig. 7.6) trend is depicted from Fig. 7.10 for niobium(111)[$\bar{1}\bar{1}$2]|alumina(11$\bar{2}$0)[0001] interfaces and the change in the slope of the fracture energy curve is again due to the higher number of slip systems being

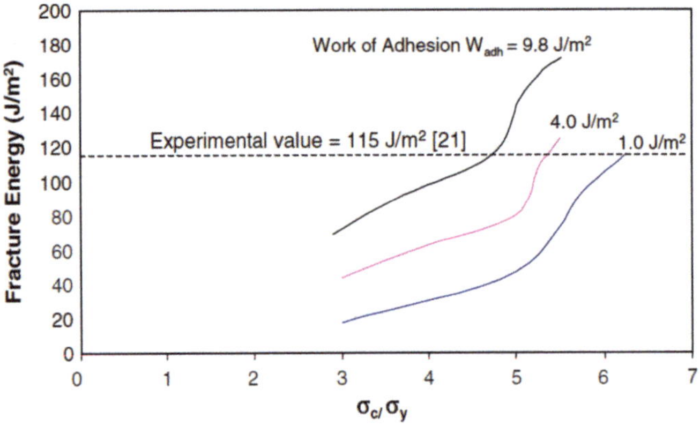

Fig. 7.9: Fracture energy vs. cohesive strength (σ_c) to yield stress (σ_y) ratio at steady state crack growth for different values of work of adhesion, W_{adh} for the Nb(100)[001]||$Al_2O_3(11\bar{2}0)[0001]$) interface crack.

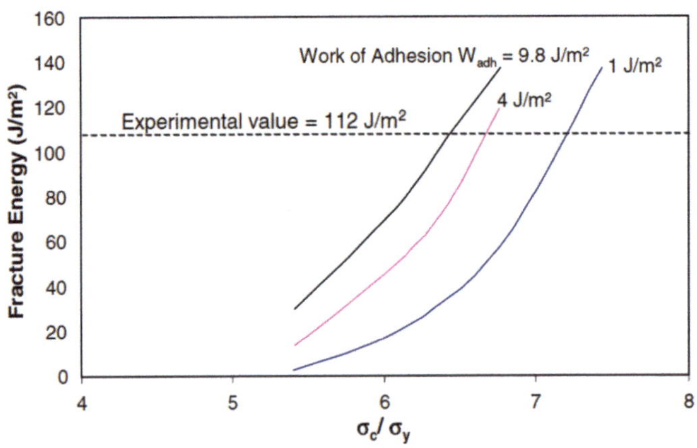

Fig. 7.10: Fracture energy vs. cohesive strength (σ_c) to yield stress (σ_y) ratio at steady state crack growth for different values of work of adhesion, W_{adh} for the Nb(111)[$\bar{1}\bar{1}2$]||$Al_2O_3(11\bar{2}0)[0001]$) interface crack.

activated at larger σ_c/σ_y-ratios. As discussed before, the reason for the higher number of activated slip systems is that, as the cohesive strength is increased the resolved shear stress of slip systems with lower Schmidt factor also increases until it reaches the yield stress of such slip systems.

The fracture energy is replotted as a function of the work of adhesion for different values of cohesive strength at steady state crack growth in Fig. 7.11 for the niobium (100)[001]|alumina$(11\bar{2}0)[0001]$ interface and in Fig. 7.12 for the niobium$(111)[\bar{1}\bar{1}2]$ |alumina$(11\bar{2}0)[0001]$ interface. As the work of adhesion increases, the fracture en-

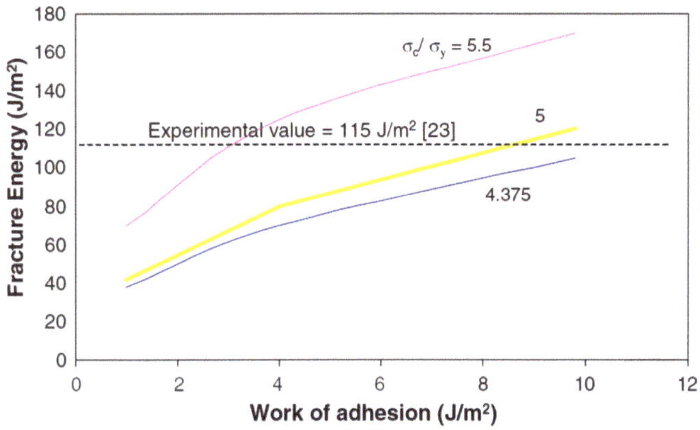

Fig. 7.11: Fracture energy vs. work of adhesion at steady state crack growth along the Nb(100)[001]||Al_2O_3(11$\bar{2}$0)[0001] interface.

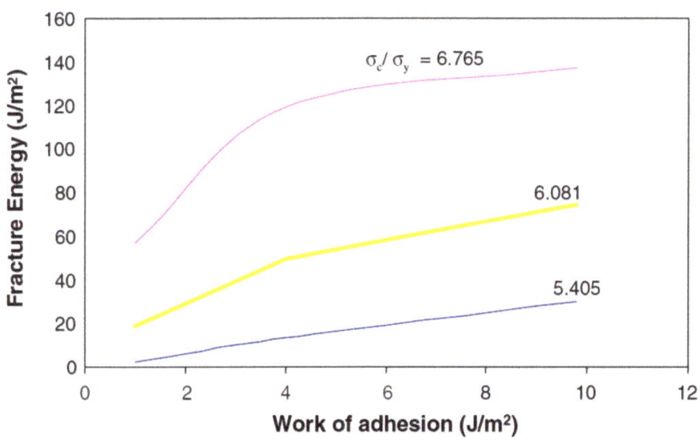

Fig. 7.12: Fracture energy vs. work of adhesion at steady state crack growth along the Nb(111)[$\overline{11}$2]||Al_2O_3(11$\bar{2}$0)[0001] interface.

ergy also increases but the change is gradual when compared to the increase in the fracture energy due to a change in cohesive strength.

These results show that the fracture energy is more sensitive to cohesive strength than to the work of adhesion. It is also found that as the cohesive strength increases, the fracture energy increases as well. It is also seen during the simulations that the slope of the fracture energy curve increases when the number of activated slip systems increase.

The cohesive law parameters identified for the three different values of work of adhesion for three different orientations of niobium/alumina interfaces are given in Tab. 7.2. It must be noted that the macroscopic yield stress (σ_y), is obtained during each simulation and is the stress value at which first slip system is activated. It is related to

Tab. 7.2: Cohesive model parameters identified for three different orientations of niobium/alumina interfaces.

Work of adhesion (J/m^2)	Cohesive Strength (MPa)	Steady state fracture energy J_c (J/m^2)
Nb(100)[001]‖Al_2O_3(11$\bar{2}$0)[0001] interfaces (σ_y = 40 MPa)		
1	250	116
4	215	115.8
9.8	190	114.8
Nb(110)[001]‖Al_2O_3(11$\bar{2}$0)[0001] interfaces (σ_y = 50.98 MPa)		
1	283	370.6
4	258	371
9.8	225	372
Nb(111)[$\bar{1}\bar{1}$2]‖Al_2O_3(11$\bar{2}$0)[0001] interfaces (σ_y = 29.6 MPa)		
1	214	112
4	198	111.8
9.8	193	113

the initial yield stress of each slip system (Tab. 7.1) by $\tau_0 = \mu\sigma_y$ [8], Schmidt factors (μ) for different slip systems for the three orientations in Tab. 7.2 were found to be in the range of 0.2357–0.4714 (see Fig. 7.7, for Nb(110)[001]‖Al_2O_3(11$\bar{2}$0)[0001] interfaces), which gives the macroscopic yield stress (σ_y) value to be in the range of 28–51 MPa.

The results in Tab. 7.2 show that as the work of adhesion increases the cohesive strength required to reach the experimental fracture energy value decreases. In the following, a correlation among cohesive strength (σ_c), work of adhesion (W_{adh}) and yield stress (σ_y) has been obtained by using least square surface fitting technique (available in MATLAB [26]) to estimate the cohesive strength (σ_c) for three differently oriented niobium/alumina interfaces. A surface has been plotted in Fig. 7.13 using cohesive strength (σ_c) and work of adhesion (W_{adh}) given in Tab. 7.2, and yield stress (σ_y) for each orientation, i.e. 40 MPa for Nb(100)[001]‖Al_2O_3(11$\bar{2}$0)[0001]), 50.98 for Nb(110)[001]‖Al_2O_3(11$\bar{2}$0)[0001], and 29.6 for Nb(111)[$\bar{1}\bar{1}$2]‖Al_2O_3(11$\bar{2}$0)[0001] interfaces.

A correlation among cohesive strength (σ_c), work of adhesion (W_{adh}) and yield stress (σ_y) has been derived for three different orientations of niobium/alumina interfaces, i.e. Nb(110)[001]‖Al_2O_3(11$\bar{2}$0)[0001], Nb(100)[001]‖Al_2O_3(11$\bar{2}$0)[0001]), and Nb(111)[$\bar{1}\bar{1}$2]‖Al_2O_3(11$\bar{2}$0)[0001], using least square surface fitting technique in MATLAB [26] and the correlation is given as:

$$\sigma_c = 6.112\left(\alpha_i \cdot \frac{W_{adh}}{W_0}\right)^{-0.0919} \cdot \sigma_y\,, \tag{7.4}$$

where W_0 is the reference work of adhesion taken to be 1 J/m^2 for this study. The effect of orientation of the bicrystal specimen is taken into account by introducing

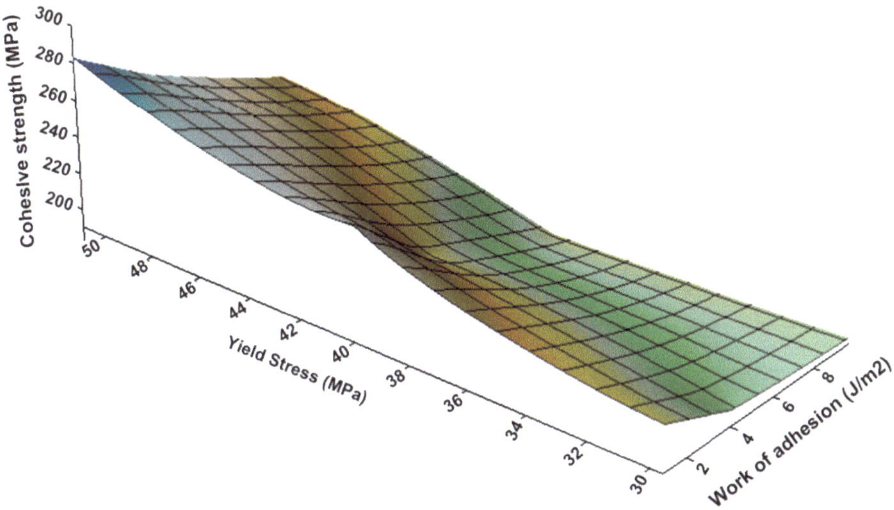

Fig. 7.13: Surface plot of fracture energy vs. work of adhesion and cohesive strength for three different orientations of niobium/alumina interfaces (see Tab. 7.2).

an additional parameter α_i which depends on the orientation of the bicrystal specimens. The value of α_i is found to be 1.0 for Nb(100)[001]||Al_2O_3(11$\bar{2}$0)[0001]), 1.35 for Nb(110)[001]||Al_2O_3(11$\bar{2}$0)[0001], and 0.853 for Nb(111)[$\overline{1}\overline{1}2$]||$Al_2O_3$(11$\bar{2}$0)[0001] interfaces.

This correlation can be used to estimate the cohesive strength (σ_c) for a given work of adhesion (W_{adh}) and yield stress (σ_y) for any of the three orientations of niobium/alumina interfaces. The reason is, this relation is derived from the identified set of cohesive law parameters (cohesive strength and work of adhesion) and of the know yield stress values for three different orientations of niobium/alumina bicrystal specimen (see Tab. 7.2 and Fig. 7.13).

In the following a generalized correlation among local adhesion capacity, cohesive strength, yield stress and macroscopic fracture energy for the analysed three different orientations of niobium single crystals in niobium/alumina bicrystal specimens is deduced. Such a correlation has already been deduced for Nb(110)[001]||Al_2O_3 (11$\bar{2}$0)[0001] interfaces in [11]. This correlation was based on the conclusion that a cubic relation exists between the cohesive strength (σ_c) to yield stress (σ_y) ratio and the fracture energy while results in [11] showed that a linear relation exists between work of adhesion (W_{adh}) and fracture energy (J_c). The correlation is given by

$$J_c = \left(a\left(\frac{\sigma_c}{\sigma_y}\right)^3 + b\left(\frac{\sigma_c}{\sigma_y}\right)^2 + c\left(\frac{\sigma_c}{\sigma_y}\right) + d \right) W_{adh} \,. \tag{7.5}$$

The coeffcients a, b, c, and d in Eq. (7.5) were identified by least square surface fitting technique with MATLAB [26]. In order to generalize the results presented above for three different orientations of niobium/alumina bicrystal specimens, a generalized relation has been deduced which can be used for any of the above three orientations of niobium/alumina bicrystal specimens. This has been done by introducing the parameter α_i already used in Eq. (7.4). The generalized relation is given by

$$J_c = \left(2.98\left(\alpha_i \cdot \frac{\sigma_c}{\sigma_y}\right)^3 + 34.8\left(\alpha_i \cdot \frac{\sigma_c}{\sigma_y}\right)^2 + 135.24\left(\alpha_i \cdot \frac{\sigma_c}{\sigma_y}\right) - 164.89\right) W_{adh} . \quad (7.6)$$

Tab. 7.3: Orientation parameter (α_i) for three orientation of bicrystal niobium/alumina interfaces.

Orientation	Orientation parameter α_1
$Nb(100)[001]\|\|Al_2O_3(11\bar{2}0)[0001]$	1.0
$Nb(110)[001]\|\|Al_2O_3(11\bar{2}0)[0001]$	1.35
$Nb(111)[\overline{11}2]\|\|Al_2O_3(11\bar{2}0)[0001]$	0.853

Using the least square surface fitting technique available in MATLAB [26], the orientation coefficients α_i have been identified for the three different orientations of niobium/alumina bicrystal interfaces which are given in Tab. 7.3 for three different orientations of niobium/alumina bicrystal interfaces.

A generalized correlation among cohesive strength, work of adhesion, yield stress and fracture energy for the three different orientations examined is thus deduced in the present work with orientation parameters α_1 0.8–1.35 for the orientations studied. In the above correlation (Eq. (7.6)) the cohesive strength can be estimated using the correlation given in Eq. (7.4).

The relation found in Eq. (7.6) can be used to predict the fracture energy of single crystalline niobium/alumina bimaterial systems, if the rest of the material parameters are available. The relation not only takes into account the dependence of the fracture energy on cohesive model parameters, such as cohesive strength and work of adhesion, but also on material properties, such as the yield stress which is strongly dependent on the crystal orientation [9, 10]. It should be noted that the three orientations of niobium/alumina bicrystal specimen showed similar hardening behaviour during deformation [9, 10] the only difference was the yield stress. Therefore, for other orientations of the same niobium/alumina bicrystal specimen with different hardening during deformation will have significant effect on the fracture energies.

7.4 Conclusions

Interface fracture analyses of bicrystal niobium/alumina specimens were presented in this work using a cohesive modelling approach. In the first part, the influence of cohesive law parameters, such as cohesive strength and work of adhesion were studied for different orientations of niobium single crystalline materials in niobium/alumina bicrystal specimens. It was found that cohesive strength has a strong influence on the fracture energy of the bicrystal niobium/alumina interface for the different niobium orientations studied. The relation between the cohesive strength and the fracture energy is found to be non-linear cubic. It is also found that as the work of adhesion increases the fracture energy also increases and both (work of adhesion and fracture energy) have almost a linear relation.

It is also found that set of cohesive law parameters found for niobium/alumina bicrystal specimens for three different orientation is not unique and several combinations of cohesive law parameters can validate experimental results. Therefore, ab initio calculations are inevitable to compute the work of adhesion of a realistic niobium/alumina interface, i.e. creating an interface which is not ideally clean. After which one can fix one parameter, i.e. work of adhesion, of the cohesive law and can find the other (cohesive strength).

In the last part a generalized correlation was derived among the fracture energy, cohesive strength, work of adhesion and yield stress. Orientation parameters for this correlation were identified for three different orientations of niobium single crystalline materials in the bicrystal niobium/alumina specimens. This correlation interlinks the local adhesion capacity, cohesive strength, yield stress and macroscopic fracture energy for niobium/alumina bicrystal specimen. As an outlook, the correlation which interlinks the local adhesion capacity, cohesive strength, yield stress and macroscopic fracture energy is phenomenological, however, derivation of such relation from basic mechanics is of utter importance and will be a topic of future research.

Acknowledgment: The presented work is funded by the Deutsche Forschungsgemeinschaft within the Graduiertenkolleg "Internal Interfaces in Crystalline Materials", which is gratefully acknowledged.

References

[1] Soyez G. Plastische Verformung und Rißbildung in druckbeanspruchten Niob/a-Al_2O_3-Verbunden. PhD thesis, Stuttgart; 1996.

[2] Soyez G, Elssner G, Rühle M, Raj R. Constrained yielding in niobium single crystals bonded to sapphire. Acta Mater. 1998;46:3571.

[3] Fischmeister HF, Mader W, Gibbesch B, Elssner G. Joining of ceramic, glass and metal. Mater Res Soc Symp Proc. 1988;122:529.

[4] Korn D, Elssner G, Cannon RM, Rühle M. Fracture properties of interfacially doped nb–al_2o_3 bicrystals: I. Fracture characteristics. Acta Mater. 2002;50:3881.

[5] Kohnle C, Mintchev O, Brunner D, Schmauder S. Fracture of metal/ceramic interfaces. Comp Mater Sci. 2000;19:261.

[6] Kohnle C, Mintchev O, Schmauder S. Elastic and plastic fracture energies of metal/ceramic joints. Comp Mater Sci. 2002;25:272.

[7] Hill R, Rice JR. Constitutive analysis of elastic–plastic crystals at arbitrary strain. j mech phys solid. 1972;20:719.

[8] Asaro RJ. Crystal plasticity. J Appl Mech. 1983;50:921.
[9] Siddiq A, Schmauder S. Crystal orientation effects on the fracture energies of a bimaterial interface, abstract no. 274. In: Proceedings of the McMAT 2005 mechanics and materials conference (CD-ROM).
[10] Siddiq A, Schmauder S. Modelling of crystal plasticity effects on the crack initiation energies of a bicrystal interface. Comp Assist Mech Engng Sci. 2007;14:67–78.
[11] Siddiq A, Schmauder S. Interface fracture analyses of a bicrystal niobium/alumina specimen using cohesive modelling approach. Modell Simul Mater Sci Engng. 2006;14:1015–30.
[12] Scheider I, Brocks W. Simulation of cup-cone fracture using the cohesive model. Engng Fract Mech. 2003;70:1943.
[13] Finnis MW, Kruse C, Schönberger U. Ab initio calculations of metal/ceramic interfaces: what have we learned, what can we learn? Nanostruct Mater. 1995;6:145.
[14] Baitrev IG, Alavi A, Finnis MW. First-principle calculations of the ideal cleavage energy of bulk niobium(111)/alpha-alumina(0001) interfaces. Phys Rev Lett. 1999;82:1510.
[15] Zhang W, Smith JR. Stoichiometry and adhesion of Nb/Al_2O_3. Phys Rev B. 2000;61:16883.
[16] Siddiq A, Schmauder S. Simulation of hardening in high purity niobium single crystals during deformation, steel grips. J Steel Relat Mater. 2005;3:281.
[17] Kysar JW. Continuum simulations of directional dependence of crack growth along a copper/sapphire bicrystal interface. Part I: Experiments and crystal plasticity background. J Mech Phys Solid. 2001;49:1099.
[18] Wu TY, Bassani JL. Latent hardening in single crystals. I: Theory and experiments. Math Phys Sci. 1993;435:21.
[19] Bowen DK, Christian JW, Taylor G. Deformation properties of niobium single crystals. Can J Phys. 1967;45:903.
[20] Xu XP, Needleman A. Numerical simulation of fast crack growth in brittle solids. J Mech Phys Solid. 1994;42:1397.
[21] Xu XP, Needleman A. Void nucleation by inclusion debonding in a crystal matrix. Modell Simul Mater Sci Engng. 1993;1:111.
[22] Kysar JW. Continuum simulations of directional dependence of crack growth along a copper/sapphire bicrystal interface. Part II: Crack tip stress and deformation analysis. J Mech Phys Solid. 2001;49:1129.
[23] Cannon RM, Korn D, Elssner G, Rühle M. Fracture properties of interfacially doped $Nb–Al_2O_3$ bicrystals. II: Relation of interfacial bonding, chemistry and local plasticity. Acta Mater. 2002;50:3903.
[24] PhD thesis; .
[25] Duffy DF, Hardning JH, Stoneham AM. A calculation of the structure and energy of the niobium/alumina interface. Acta Mater. 1996;44:3293.
[26] MATLAB Version 6.0. The Mathworks Inc., Massachusetts; 2004.

A. Koester, A. Ma, and A. Hartmaier*

8 Atomistically informed crystal plasticity model for body-centred cubic iron

Abstract: The glide of screw dislocations with non-planar dislocation cores dominates the plastic deformation behavior in body-centred cubic iron. This yields a strong strain rate and temperature dependence of the flow stress, the breakdown of Schmid's law and a dependence of dislocation mobility on stress components that do not contribute to the mechanical driving force of dislocation glide. We developed a constitutive plasticity model that takes all these effects into account. The model is based on the crystal plasticity approach and parameterized by performing molecular statics calculations using a semi-empirical potential. The atomistic studies yield quantitative relations between local stress tensor components and the mobility of dislocations. Together with experimental stress-strain curves obtained for two different orientations of iron single crystals taken from the literature, the constitutive law is completely parameterized. The model is validated by comparing numerical single crystal tension tests for a third orientation to the equivalent experimental data from the literature. We also provide results for the temperature and strain rate dependence of the new atomistically informed constitutive model.

Keywords: bcc iron, molecular statics simulations, crystal plasticity, non-Schmid, constitutive modeling

8.1 Introduction

The characteristic deformation behavior of body-centred cubic (bcc) metals can be ascribed to their screw dislocation core configuration. The glide of screw dislocations with a non-planar core dominates the plastic deformation within bcc metals. This core configuration leads to a high Peierls barrier and a strong dependence of the flow stress on temperature and strain rate [1, 2]. Experimental and theoretical studies revealed two pivotal phenomena, which characterize the behavior of bcc metals: The breakdown of Schmid's law and the dependence of the yield stress on the so-called non-glide stress components of the stress tensor. The non-glide components are the stresses, which do not move the dislocation, but change the dislocation core configuration and thus assist or retard the glide of the dislocation [3, 4].

Another characteristic feature of the bcc crystal structure is the absence of truly close packed planes such as the {111} planes in face centered cubic (fcc) crystals. This hinders the determination of active glide systems. On the one hand, based on some experimental studies [5] the choice of glide planes has been found to be temperature dependent. At low temperatures dislocation glide on the {110} planes is observed and at high temperatures on {112} planes. On the other hand, atomistic studies identify the {110} planes as the glide planes [4, 6, 7]. To clarify the question on what kind of planes slip takes place and whether the choice of glide planes is temperature dependent, recent experimental results have been published by Caillard [8, 9]. In these studies the glide in bcc iron at low and room temperature has been investigated. The results show that slip of screw dislocations takes place on {110} planes and this observation is independent of temperature within the investigated range.

Based on the experimental and theoretical observations with respect to the mechanical behavior of bcc iron, an atomistically informed constitutive model in the framework of the crystal plasticity approach is developed in this paper. In this model the yield law is based on the transition state theory and atomistic simulation results, which are sensitive to strain rate and temperature. Compared to the non-Schmid law reported in the literature [4], the yield law proposed in this paper additionally takes into account the influence of the normal stress components on screw dislocation mobility. With the help of this new constitutive model the mechanical behavior of bcc iron single crystals has been investigated carefully and the results are compared with experimental data from the literature.

8.2 Crystal plasticity approach

The constitutive theory of crystal plasticity follows the work of Asaro [10]. In this theory, the deformation gradient $\boldsymbol{F}$ can be multiplicatively decomposed as

$$\boldsymbol{F} = \boldsymbol{F}^{\mathrm{e}}\boldsymbol{F}^{\mathrm{p}} , \tag{8.1}$$

where $\boldsymbol{F}^{\mathrm{p}}$ is the plastic deformation gradient due to slip and $\boldsymbol{F}^{\mathrm{e}}$ the elastic deformation gradient including crystal lattice stretching and rotation. The elastic law is described with the help of the second Piola-Kirchhoff stress tensor $\boldsymbol{S}$ in the intermediate configuration and the right Cauchy-Green tensor $\boldsymbol{C}_e = \boldsymbol{F}^{\mathrm{eT}}\boldsymbol{F}^{\mathrm{e}}$. The Cauchy stress can be calculated as $\boldsymbol{\sigma} = (\det \boldsymbol{F}^{\mathrm{e}})^{-1}\boldsymbol{F}^{\mathrm{e}}\boldsymbol{S}\boldsymbol{F}^{\mathrm{eT}}$. Assuming the elastic deformation is small, $\|\boldsymbol{C}_e - \boldsymbol{I}\| \ll 1$, the resolved shear stress reads

$$\tau^\alpha = \boldsymbol{\sigma} : \boldsymbol{m}^\alpha \otimes \boldsymbol{n}^\alpha = \boldsymbol{C}_e\boldsymbol{S} : \boldsymbol{m}_0^\alpha \otimes \boldsymbol{n}_0^\alpha \approx \boldsymbol{S} : \boldsymbol{m}_0^\alpha \otimes \boldsymbol{n}_0^\alpha , \tag{8.2}$$

where $\boldsymbol{n}^\alpha$ and $\boldsymbol{m}^\alpha$ are unit vectors of slip plane normal and the slip direction in the current state and $\boldsymbol{n}_0^\alpha$ and $\boldsymbol{m}_0^\alpha$ the slip plane normal vector and slip direction vector in the reference configuration. The plastic deformation is described taking into account the sum of shear of all slip systems. Through determining the shear rate $\dot{\gamma}_\alpha$ with respect to the slip system $\boldsymbol{m}_0^\alpha \otimes \boldsymbol{n}_0^\alpha$ the plastic deformation velocity gradient amounts to

$$\boldsymbol{L}_p = \dot{\boldsymbol{F}}^{\mathrm{p}}\boldsymbol{F}^{\mathrm{p}-1} = \sum_\alpha \dot{\gamma}^\alpha \boldsymbol{m}_0^\alpha \otimes \boldsymbol{n}_0^\alpha \tag{8.3}$$

and the evolution of the plastic deformation can be formulated with the help of Eq. (8.3). Plasticity in metallic materials is the consequence of dislocation motion through the crystal. This process requires that a finite driving force, i.e. the resolved shear stress (Eq. (8.2)), is larger than the critical resolved shear stress (CRSS) of the crystal. There are two kinds of barriers against dislocation motion: Short-range barriers, which can be overcome by thermal activation, such as the Peierls resistance or local forest dislocations, which have to be cut; and long-range barriers, which are also called athermal resistance, such as grain boundaries, large incoherent precipitates and dislocation microstructure. The sum of the thermal resistance s_{t}^{α} and athermal resistance s_{a}^{α} to dislocation motion results in the total resistance to slip as the following:

$$s^{\alpha} = s_{\mathrm{t}}^{\alpha} + s_{\mathrm{a}}^{\alpha}. \tag{8.4}$$

Based on the transition state theory the average dislocation velocity $v_{\mathrm{avg}}^{\alpha}$ is given by

$$v_{\mathrm{avg}}^{\alpha} = v_{0}^{\alpha} \exp\left\{\frac{-\Delta G}{k_{\mathrm{B}} T}\right\}, \tag{8.5}$$

with ΔG being the activation enthalpy, v_0^{α} the reference velocity, k_{B} the Boltzmann constant and T the temperature. Following the Orowan relation, which connects the dislocation velocity with the slip rate, the slip rate can be written as

$$\dot{\gamma}^{\alpha} = \begin{cases} 0 & \text{if } \tau_{\mathrm{eff}}^{\alpha} \leq 0 \\ \dot{\gamma}_{0}^{\alpha} \exp\left\{\frac{-\Delta G}{k_{\mathrm{B}} T}\right\} & \text{if } \tau_{\mathrm{eff}}^{\alpha} = 0 \end{cases} \tag{8.6}$$

with $\dot{\gamma}_0^{\alpha}$ being the reference strain rate and $\tau_{\mathrm{eff}}^{\alpha} = |\tau^{\alpha}| - s_{\mathrm{a}}^{\alpha}$ the effective resolved shear stress acting on the slip system α. The activation enthalpy is calculated using the approach of Kocks [11]

$$\Delta G = G_0 \left[1 - \left(\frac{\tau_{\mathrm{eff}}^{\alpha}}{s_{\mathrm{t}}^{\alpha}}\right)^{p}\right]^{q}, \tag{8.7}$$

where G_0 is the activation free energy for dislocation motion. The parameters p and q are fitting parameters, which are responsible for the description of the activation profile. The thermal resistance s_{t}^{α} is calculated for a given strain rate $\dot{\gamma}$ and temperature with the help of Eq. (8.8) based on the results of [5], so that the material behavior becomes temperature and strain rate dependent.

$$\tau_t = \tau_{t0} \left(1 - \left(\frac{k_{\mathrm{B}} T}{G_0} \ln\left(\frac{\dot{\gamma}_0}{\dot{\gamma}}\right)\right)^{\frac{1}{p}}\right)^{\frac{1}{q}}, \tag{8.8}$$

where τ_{t0} is the stress at which the temperature tends towards 0 K. The approach introduced so far is a standard approach, which does not take into account the breakdown of Schmid's law or the influence of the non-glide stress components, which was observed in experimental and atomistic studies [3, 4]. The development of a flow rule based on the work of Groeger et al. [4, 6, 7] taking into consideration these effects shall be discussed in more detail in the next section. To complete the theory of the crystal plasticity, the hardening law used in this work is described in more detail.

$$\dot{s}_{\mathrm{a}}^{\alpha} = \sum_{\beta} h^{\alpha\beta} |\dot{\gamma}^{\beta}| = \sum_{\beta} q^{\alpha\beta} h^{\beta} |\dot{\gamma}^{\beta}| \,, \tag{8.9}$$

where $q^{\alpha\beta}$ is the hardening matrix, which takes into account the interaction of different active slip systems [12]. The diagonal values ($\alpha = \beta$) of the matrix $q^{\alpha\beta}$ describe self hardening and the values on the non-diagonal positions ($\alpha \neq \beta$) describe latent hardening. h^{β} denotes the self hardening rate of the slip system β and is defined as

$$h^{\beta} = h_0^{\beta} \left| 1 - \frac{s_{\mathrm{a}}^{\beta}}{s_{\mathrm{a,s}}^{\beta}} \right|^{r} \operatorname{sign}\left(1 - \frac{s_{\mathrm{a}}^{\beta}}{s_{\mathrm{a,s}}^{\beta}} \right), \tag{8.10}$$

with h_0^{β} being the initial hardening rate, $s_{\mathrm{a,s}}^{\beta}$ the saturation value of the athermal slip resistance and the exponent r being a constant material parameter [13].

8.3 Atomistic studies

The breakdown of Schmid's law and the influence of non-glide components of the stress tensor are embedded in the crystal plasticity approach via the flow rule. The formulation of the flow rule is based on the atomistic studies of molybdenum and tungsten performed by Groeger [4, 6, 7]. Within these atomistic studies the dependence of the critical stress to move a dislocation through the crystal lattice on the orientation of the maximum resolved shear stress plane (MRSSP) under pure shear parallel to the slip direction has been investigated. Furthermore the influence of shear stresses perpendicular to the slip direction has been studied. From the atomistic studies slip has only been detected within the $\{110\}\langle 111\rangle$ glide systems. Hence, the following flow rule for bcc materials has been introduced

$$\begin{aligned} \tau^{\alpha} = {} & \boldsymbol{\sigma} : \boldsymbol{m}^{\alpha} \otimes \boldsymbol{n}^{\alpha} \\ & + a_1 \boldsymbol{\sigma} : \boldsymbol{m}^{\alpha} \otimes \boldsymbol{n}_1^{\alpha} \\ & + a_2 \boldsymbol{\sigma} : (\boldsymbol{n}^{\alpha} \times \boldsymbol{m}^{\alpha}) \otimes \boldsymbol{n}^{\alpha} \\ & + a_3 \boldsymbol{\sigma} : (\boldsymbol{n}_1^{\alpha} \times \boldsymbol{m}^{\alpha}) \otimes \boldsymbol{n}_1^{\alpha} \le \tau_{\mathrm{cr}} \,. \end{aligned} \tag{8.11}$$

According to Eq. (8.11) the resolved shear stress τ^{α} on slip system α is a function of the local stress tensor $\boldsymbol{\sigma}$ and three material constants a_1, a_2, and a_3. This shear stress is the effective driving force on screw dislocations. Hence, when it reaches or exceeds the critical value τ_{cr} plastic deformation sets in. The first term in Eq. (8.11) is the well known Schmid's law. The other three terms, which contain the constants a_1, a_2, and a_3 take into account the non-Schmid effects. The second term integrates the effect known as twinning-antitwinning assymetry. The third and fourth term include the effect of shear stresses perpendicular to the slip direction. $\boldsymbol{n}_1^{\alpha}$ is the specific $\{110\}$ plane normal vector, which includes an angle of −60° with the reference slip plane defined by $\boldsymbol{n}^\alpha$. Since $\boldsymbol{n}_1^\alpha$ is different for opposed glide directions, it is necessary to calculate the non-Schmid terms of the flow rule for 24 glide systems. The parameters a_1, a_2, and a_3 are specific material constants and need to be fitted for the particular material from atomistic studies.

To determine the flow rule for bcc iron, we perform molecular statics calculations, using a Finnis Sinclair type EAM Potential [14]. For the simulations a box of atoms is set up, in which the atoms in the outer frame are at fixed positions, while the inner atoms are mobile. The setup of the simulation box is illustrated in Fig. 8.1.

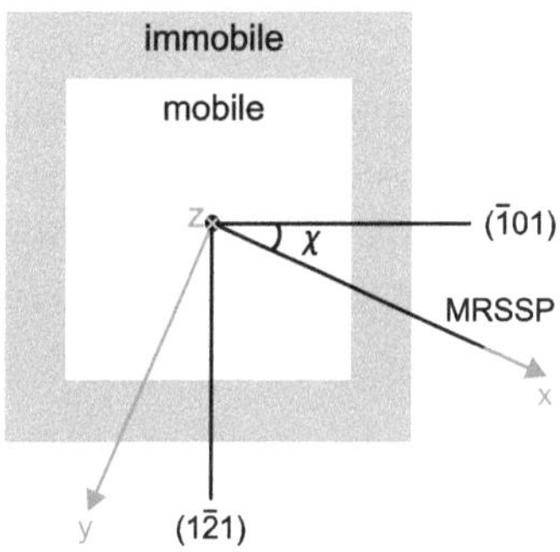

Fig. 8.1: Schematic drawing of the MD simulation setup. An immobile frame of atoms surrounds the mobile ones. The orientation of the maximum resolved shear stress plane (MRSSP) is defined by the angle χ.

To avoid boundary effects the dimensions of the box in the (111) plane are 500×500 Å. The height of the box in $[111]$ direction amounts to six atom layers and periodic boundary conditions are applied in this direction. The width of the immobile frame has to be at least two times the cut-off radius of the inter atomic potential. A $\frac{1}{2}[111]$ screw dislocation is inserted in the following way: The displacement field of the screw dislocation is calculated based on the Volterra construction and added to the positions of atoms initially sitting on the sites of a perfect bcc lattice. The atoms in the mobile region are subsequently relaxed using the EAM potential, while the outer frame of atoms is kept fixed. During the relaxation that is carried out with the FIRE algorithm described in [15], the atoms reach their equilibrium positions characterised by a minimum of their potential energies. The relaxed dislocation core shows a non-degenerated core structure. Fig. 8.2 shows the differential displacement map of the screw dislocation core as it is described in [16].

To investigate the influence of stresses parallel and perpendicular to the glide direction on the critical stress, external stresses are applied to the boundaries of the simulation box. For the application of external stresses, the corresponding elastic displacement field following the approch of anisotropic elasticity is added to the displacement field of the screw dislocation.

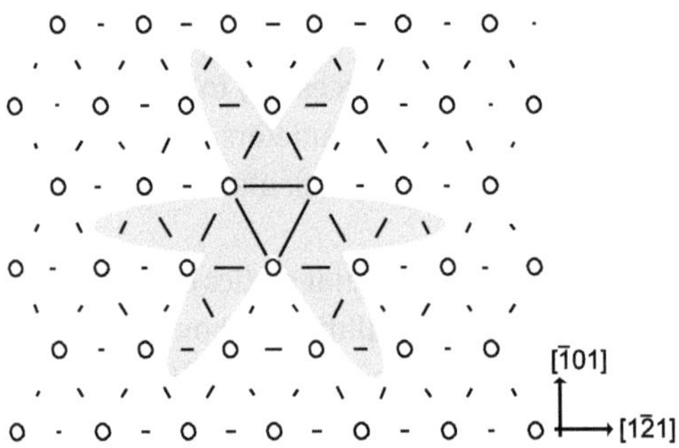

Fig. 8.2: Differential displacement map of a non-degenerated dislocation core of bcc-iron that has been generated using the simulation setup shown in Fig. 8.1. The grey shaded area highlights the dislocation core.

8.3.1 Orientation dependence of the critical stress

Shear stresses τ parallel to the slip direction and acting on planes inclined by an angle χ to the slip plane are applied to the simulation box to investigate the dependence of the critical stress on the MRSSP orientation. Under the assumption of the x-axis lying in the MRSSP, the y-axis being normal to the MRSSP, and the z-axis parallel to the dislocation line direction, the stress tensor takes the form

$$\boldsymbol{\sigma} = \begin{bmatrix} 0 & 0 & 0 \\ 0 & 0 & \tau \\ 0 & \tau & 0 \end{bmatrix} . \tag{8.12}$$

Due to the symmetry of the bcc lattice it is sufficient to investigate the orientation dependence in the range of $\pm 30°$ to the reference $(\bar{1}01)$ slip plane. The results of the lattice statics calculations are shown in Fig. 8.3 and compared to the results of Chaussidon [17], who also investigated the orientation dependence of the Peierls stress in bcc iron using the Mendelev EAM potential, but with another model setup.

The results of the performed calculations are in good agreement with the results from the literature, derivations of up to 10 % occur at large positive angles. In the graph the breakdown of Schmid's law in bcc iron is clearly visible: For the region $\chi < 0°$ the critical stresses obtained by the lattice static calculations are smaller than the stresses expected from the Schmid's law, while the critical stresses in the region $\chi > 0°$ are larger. Following [6], the CRSS is a function of the MRSSP orientation χ

$$\text{CRSS}(\chi) = \frac{\tau^*_{\text{cr}}}{\cos\chi + a_1 \cos(\chi + \pi/3)} . \tag{8.13}$$

a_1 and τ^*_{cr} are determined by fitting the atomistically determined data to Eq. (8.13) by using the least square fitting procedure. The results for all flow rule parameters are given in Tab. 8.1.

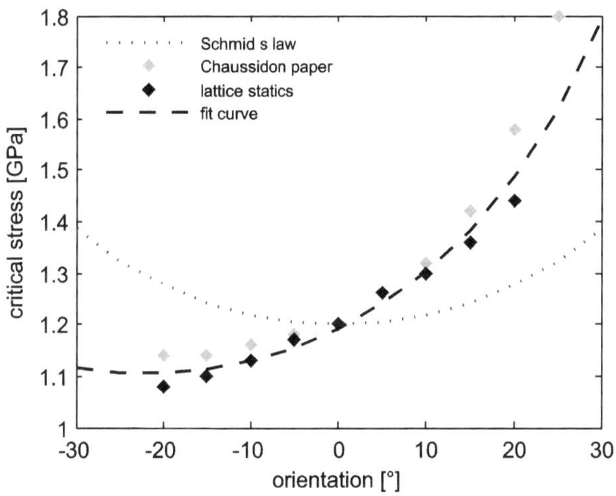

Fig. 8.3: Dependence of the critical stress on the orientation of the MRSSP applying shear load parallel to the slip direction in the MRSSP.

8.3.2 Influence of shear stresses perpendicular to the glide direction

For the identification of the influence of shear stresses perpendicular to the glide direction, tests with combined tension/compression loading perpendicular to the glide direction with the following stress tensor

$$\boldsymbol{\sigma} = \begin{bmatrix} -\sigma & 0 & 0 \\ 0 & \sigma & \tau \\ 0 & \tau & 0 \end{bmatrix} \tag{8.14}$$

are performed. To find the relationship between the shear stresses perpendicular to the slip direction and the critical stress, studies on three different MRSSP orientations (−10°, 0°, 10°) are performed. The results are shown in Fig. 8.4.

With the help of these results, the parameters a_2 and a_3 of the flow rule are fitted using the least square method by taking only the linear part into account. The fitting function of the CRSS [6] reads

$$\mathrm{CRSS}(\chi) = \frac{\tau_{\mathrm{cr}}^{*} - \tau\,[a_2 \sin 2\chi + a_3 \cos(2\chi + \pi/6)]}{\cos\chi + a_1 \cos(\chi + \pi/3)}\,. \tag{8.15}$$

Based on the approach of Groeger and Vitek [4] it is seen that the shear stresses perpendicular to the glide direction influence the dislocation core in bcc iron and thus increase or reduce the resistance to glide.

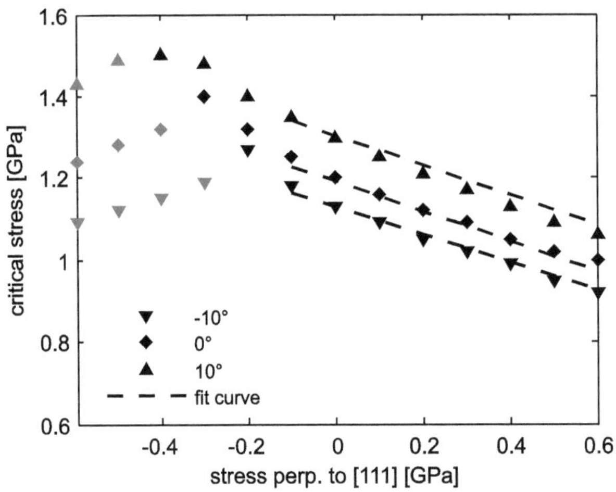

Fig. 8.4: Dependence of the critical stress on the shear stress perpendicular to the glide direction $\chi = -10°, 0°$, and $10°$.

8.3.3 Influence of tension and compression perpendicular to the glide direction

The remaining question is whether pure tension respectively compression perpendicular to the glide direction also have an influence on the dislocation core and thus hinder or support dislocation motion. For this purpose further lattice statics calculations are performed. Under the assumption of $\chi = 0°$, the MRSSP is the $(\bar{1}01)$ plane and the x-axis pointing in the $[12\bar{1}]$ direction (cf. Fig. 8.2). To identify the contribution of tension/compression load and distinguish between the participation of the shear stresses perpendicular to the glide direction and the tension/compression load to the dislocation core changes, the studies are based on the stress tensor

$$\boldsymbol{\sigma} = \begin{bmatrix} -0.2\text{G Pa} & 0 & 0 \\ 0 & 0.2\,\text{GPa} & \tau \\ 0 & \tau & 0 \end{bmatrix} + \begin{bmatrix} \Sigma_{11} & 0 & 0 \\ 0 & \Sigma_{22} & 0 \\ 0 & 0 & 0 \end{bmatrix}, \tag{8.16}$$

The first part of the overall stress tensor is equivalent to the stress tensor, which is used to study the influence of the shear stresses perpendicular to the glide direction (Eq. (8.14)). By including the fixed orthogonal shear component $\sigma = 0.2$ GPa, the influence of tension and compression in $[12\bar{1}]$ and $[\bar{1}01]$ direction could be determined by adding Σ_{11} respectivley Σ_{22} to the stress tensor. The determination of the critical stress τ according to tension and compression is shown in Fig. 8.5.

The results show that tension in $[\bar{1}01]$ direction assists the dislocation motion, while compression in this direction retards the dislocation motion. An opposing effect can be seen in the case of tension and compression loading in $[12\bar{1}]$ direction. This particular effect on the dislocation motion is correlated to elastic changes in the dislo-

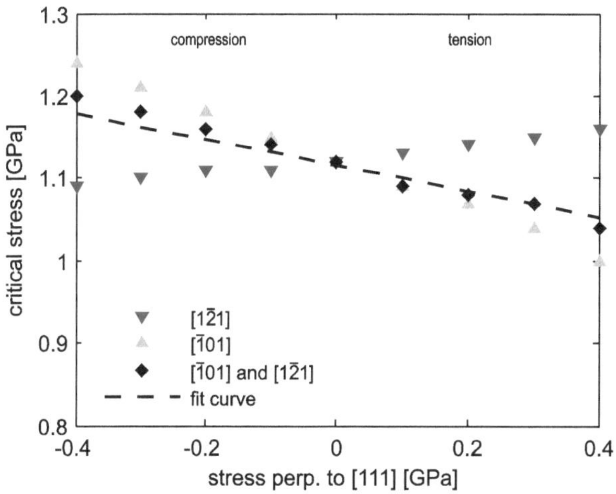

Fig. 8.5: Dependence of the critical stress on tension and compression stresses parallel and perpendicular to the glide plane normal. MRSSP $(\bar{1}01)$, $\chi = 0°$.

cation due to the stresses that do not exert any force on the dislocation. To investigate the elastic distortion of the dislocation core due to tension and compression in $[\bar{1}01]$ and $[12\bar{1}]$ direction, a stress tensor with $\tau = 0$ is applied. The distorted dislocation cores are shown for the case of tension and compression loading in $[12\bar{1}]$ direction with $\Sigma_{11} = \pm 1$ GPa. The reference core structure is the configuration that results from a stress tensor with $\tau = 0$ and $\sigma = 0.2$ GPa. Since shear stresses perpendicular to the glide direction have an influence on the dislocation core, the reference core structure is already slightly distorted. The dislocation core is stretched in the $(\bar{1}01)$ plane, while it is compressed in the $(1\bar{1}0)$ and $(0\bar{1}1)$ planes, which results in an easier motion of the dislocation in the $(\bar{1}01)$ plane. In the case of additional applied compression in the $[12\bar{1}]$ direction, the same effect on the core can be observed: the core is further stretched in the $(\bar{1}01)$ plane, while it is compressed in the $(1\bar{1}0)$ and $(0\bar{1}1)$ planes. This core distortion correlates with the observation of an easier motion of the dislocation in case of compression. The contrary effect on the core configuration can be observed in the case of applying tension in the $[12\bar{1}]$ direction. The core is compressed in the $(\bar{1}01)$ plane, while it is stretched in the $(1\bar{1}0)$ and $(0\bar{1}1)$ planes. This elastic distortion of the core results in a retardation of the dislocation motion. The core changes due to tension and compression are shown qualitatively in Fig. 8.6.

To incorporate the effect of tension and compression perpendicular to the glide direction, we suggest to extend the yield rule proposed by Groeger and Vitek [4] by three additional terms. The first one takes into consideration the tension and compression in Σ_{11} direction and the second one considers the load in Σ_{22} direction. A third term is added to make the yield law independent of hydrostatic stresses, since it has been ob-

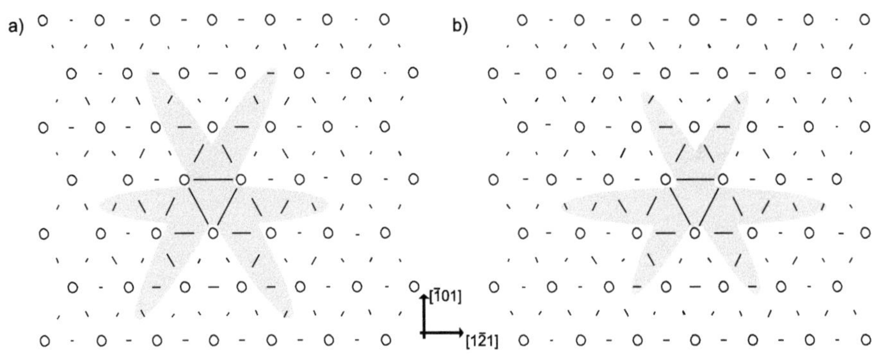

Fig. 8.6: Structure of the $\frac{1}{2}[111]$ dislocation core, which is elastically distorted due to (a) tension and (b) compression in $\langle 12\bar{1}\rangle$ direction.

served that hydrostic stresses do not have any influence on the dislocation motion [9]. This leads to the extended flow rule

$$\begin{aligned}\tau^\alpha = &\ \boldsymbol{\sigma} : \boldsymbol{m}^\alpha \otimes \boldsymbol{n}^\alpha \\ &+ a_1 \boldsymbol{\sigma} : \boldsymbol{m}^\alpha \otimes \boldsymbol{n}_1^\alpha \\ &+ a_2 \boldsymbol{\sigma} : (\boldsymbol{n}^\alpha \times \boldsymbol{m}^\alpha) \otimes \boldsymbol{n}^\alpha \\ &+ a_3 \boldsymbol{\sigma} : (\boldsymbol{n}_1^\alpha \times \boldsymbol{m}^\alpha) \otimes \boldsymbol{n}_1^\alpha \\ &+ a_4 \boldsymbol{\sigma} : \boldsymbol{n}^\alpha \otimes \boldsymbol{n}^\alpha \\ &+ a_5 \boldsymbol{\sigma} : (\boldsymbol{n}^\alpha \times \boldsymbol{m}^\alpha) \otimes (\boldsymbol{n}^\alpha \times \boldsymbol{m}^\alpha) \\ &+ a_6 \boldsymbol{\sigma} : \boldsymbol{m}^\alpha \otimes \boldsymbol{m}^\alpha \leq \tau_{\text{cr}}\end{aligned} \tag{8.17}$$

The parameters a_4 and a_5 are gained by fitting the curve of the combined loading in $[12\bar{1}]$ and $[\bar{1}01]$ direction using the least square fitting method. For this purpose the applied stress tensor (Eq. (8.16)) is transformed to the global coordinated system $x = [100]$, $y = [010]$ and $z = [001]$ and entered into Eq. (8.17) under the assumption of $a_6 = 0$. The transformed yield rule is solved for CRSS

$$\text{CRSS} = \frac{\tau_{\text{cr}}^* - \left[\sigma\left(\frac{3}{\sqrt{12}}a_3 + a_4 - a_5\right) + \Sigma_{11}\left(-\frac{3}{2\sqrt{12}}a_3 + a_5\right) + \Sigma_{22}\left(\frac{3}{2\sqrt{12}}a_3 + a_4\right)\right]}{1 + \frac{1}{2}a_1} \tag{8.18}$$

with $\sigma = 0.2$ GPa. The parameter a_6 is determined by theoretical calculations, so that the hydrostatic stress does not influence the resolved shear stress τ. The determined parameters a_1 to a_6 and τ_{cr}^* are given in Tab. 8.1.

Tab. 8.1: Parameters for the yield rule determined by atomistic studies.

a_1	a_2	a_3	a_4	a_5	a_6	τ^*_{cr}	
0.61	0.23	0.55	0.11	0.09	−0.2	1.56	GPa

8.4 FEM study of a bcc iron single crystal

The atomistically informed flow rule is incorporated into a crystal plasticity constitutive model and applied for numerical experiments on a bcc iron single crystal. With the help of the numerical calculations the orientation dependence of the plastic behavior of bcc iron single crystals as well as the qualitative strain rate and temperature dependence are investigated. To identify the model parameters of the crystal plasticity model, an optimization algorithm is used, which is based on the efficient global optimization [18]. The objective function (OF), which needs to be minimized equals the mean squared error of the experimental and numerical obtained stress-strain curves for the crystal orientations [001] and [011].

$$\mathrm{OF} = \frac{1}{N}\sum_{n=1}^{N}\left(\sigma_{100}^{\mathrm{sim}}(\epsilon_n) - \sigma_{100}^{\mathrm{exp}}(\epsilon_n)\right)^2 + \left(\sigma_{100}^{\mathrm{sim}}(\epsilon_n) - \sigma_{100}^{\mathrm{exp}}(\epsilon_n)\right)^2 . \tag{8.19}$$

The elastic constants of bcc iron are set to $C_{11} = 236$ GPa, $C_{12} = 134$ GPa and $C_{44} = 119$ GPa [19] and p, q, G_0, and s_{t0} are fitted using the experimental results of [5] and Eq. (8.8). All determined model parameters are listed in Tab. 8.2.

Tab. 8.2: Material parameters for crystal plasticity model.

Activation free energy	G_0	0.72 eV
Activation profile parameter	p	1
Activation profile parameter	q	2
Thermal resistence at 0 K	s_{t0}	395.5 MPa
Initial hardening rate	h_0	471.2 MPa
Saturation value of athermal slip	s_s	193.3 MPa
Athermal slip resistance	s_{a0}	29.8 MPa
Latent hardening	$q^{\alpha\beta}$	1.4
Hardening exponent	r	1.16
Reference strain rate	$\dot{\gamma}_0$	$2.44 \cdot 10^7\ \mathrm{s}^{-1}$

The experimental setup and results of Keh [19, 20] are used as reference to specify the performance of the atomistically informed constitutive model. To accomplish this, numerical tension tests on differently orientated single crystal specimens of the dimensions 25.4 mm × 6.35 mm × 1.27 mm are performed. The dimensions are only given for completeness, since no length scale is present in our model they do not influence the results. The numerical specimen is discretized into $20 \times 5 \times 1$ hexagonal elements with linear shape functions. Displacement boundary conditions are applied in the following way: All lateral surfaces are free surfaces, the lower surface is fixed in all directions while the upper surface is forced to move in tensile direction and constrained in the other two directions. These boundary conditions are intended to reflect the uniaxial tensile test, i.e. the specimen is forced to stay in the tensile direction, such that the plastic deformation causes a rotation of the slip planes. A strain rate of $3.3 \cdot 10^{-4}\,\mathrm{s}^{-1}$ is applied. The numerical simulations show a strong orientation dependence of the plastic behavior in bcc iron that is consistent with experimental observations. Furthermore, the material response in $[\bar{1}11]$ direction that has not been used for the fitting procedure is described with the same precision as that in the other directions. This serves as an independent test for our model (Fig. 8.7).

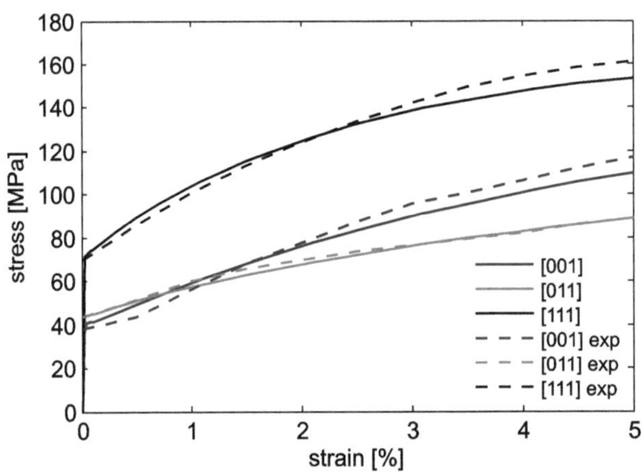

Fig. 8.7: Orientation dependency of bcc iron single crystals in tension. Comparison of numerical and experimental stress-strain curves.

By using the developed atomistically informed yield rule within the crystal plasticity model, the right choice of the parameters a_1–a_6 enable the exclusive use of $\{110\}\langle 111\rangle$ glide systems, which is in agreement to the experimental results [8, 9].

With the help of the determined model parameters the temperature and strain rate dependence of bcc iron single crystals are calculated. Via the thermal resistance s_t^α the material model becomes strain rate and temperature dependent. Fig. 8.8 shows the $[001]$ oriented crystal at four different temperatures; with decreasing temperature the initial yield stress increases. Whereas the work hardening rate is not strongly influenced by temperature. The stress-strain curves shown in Fig. 8.9 illustrate the strain rate dependence of bcc iron. The initial yield stress increases with increasing strain rate, the work hardening rate is again not strongly influenced. Such behavior is consistent with the temperature and strain rate dependence of bcc metals in general and independent studies of bcc iron [2].

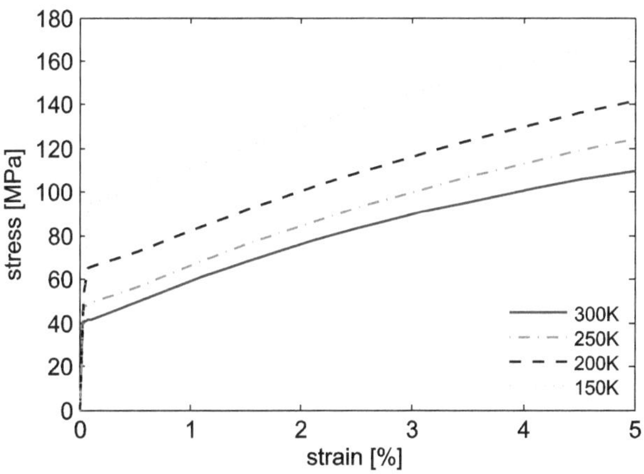

Fig. 8.8: Temperature dependence of simulated stress-strain curves of a [001] bcc iron single crystal in tension, temperature given in legend.

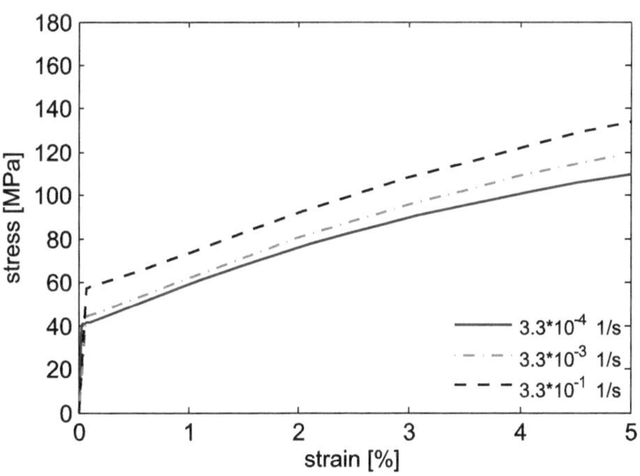

Fig. 8.9: Strain rate dependency of simulated stress-strain curves of a [001] bcc iron single crystal in tension at 300 K, loading rates given in legend.

8.5 Sensitivity analysis of the flow rule parameters

From FEM single crystal studies on the yield law parameters a_1–a_6 it is found that theses parameters have a strong influence on the hardening behavior of bcc iron. They directly influence the yield stress and the stress-strain behavior of the bcc crystal. The change of the parameter a_1 will mostly affect the tension compression asymmetry: the higher the parameter, the more pronounced the tension compression asymmetry. The parameters a_2–a_6 strongly affect the relative curve progression of the differently oriented single crystals. A high value of the parameters a_2 and a_4 influences the crossing behavior of the [001] and [011] stress-strain curves in a way that they cross each other at a higher strain level. While low values of a_3 and a_5 induce a crossing at a lower strain.

8.6 Summary

The developed constitutive model for bcc iron connects the crystal plasticity approach to atomistic studies. Via the atomistically informed formulation of a flow rule the breakdown of Schmid's law and the dependence of the plastic flow on non-glide components of the stress tensor are introduced to the crystal plasticity model. Atomistic studies reveal that not only shear stresses parallel and perpendicular to the glide direction, but also tensile and compressive stresses normal to the glide plane influence the structure of dislocation cores and hence their resistance to glide. The numerical simulations of tensile tests of a body-centred cubic (bcc) single crystal demonstrate that the atomistically informed constitutive model is in good agreement with recent experimental results showing that at room temperature glide of screw dislocations in bcc iron takes place on $\{110\}\langle 111\rangle$ glide systems [8, 9]. By exclusively using these glide systems and the extended yield rule it is possible to accurately model the plastic behavior of a bcc iron single crystal.

Acknowledgment: The authors gratefully thank Christoph Begau for support with the optimization procedure and ThyssenKrupp AG, Bayer MaterialScience AG, Salzgitter Mannesmann Forschung GmbH, Robert Bosch GmbH, Benteler Stahl/Rohr GmbH, Bayer Technology Services GmbH and the state of North-Rhine Westphalia as well as the European Commission in the framework of the European Regional Development Fund (ERDF) for financial support.

References

[1] Hollang L, Brunner D, Seeger A. Work hardening and flow stress of ultrapure molybdenum single crystals. Mater Sci Eng, A. 2001;319-321:223–236.

[2] Takeuchi T, Honda R, Iwayama K, Taoka T. Tensile deformation of iron single crystals having the [100] and [110] axes between −70 °C and 250 °C. Jpn J Appl Phys. 1967;6:1282–1291.

[3] Christian JW. Some surprising features of the plastic deformation of body-centered cubic metals and alloys. Metall Mater Trans A. 1983;14:1237–1256.

[4] Groeger R, Bailey AG, Vitek V. Plastic deformation of molybdenum and tungsten: I. Atomistic studies of the core structure and glide of 1/2⟨111⟩ screw dislocations at 0 K. Acta Mater. 2008; 56:5401–5411.

[5] Brunner D, Diehl J. Temperature and strain-rate dependence of the tensile flow stress of high-purity α-Iron below 250K. I. Stress/temperature regime III. Phys Stat Sol(a). 1991;124:455–464.

[6] Groeger R, Racherla V, Bassani JL, Vitek V. Plastic deformation of molybdenum and tungsten: II. Yield criterion for bcc metals involving the non-Schmid behavior of dislocations. Acta Mater. 2008;56:5412–5425.

[7] Groeger R, Vitek V. Plastic deformation of molybdenum and tungsten: III. Incorporation of the effects of temperature and strain rate. Acta Mater. 2008;56:5426–5439.

[8] Caillard D. Kinetics of dislocations in pure Fe Part I: In situ straining experiments at room temperature. Acta Mater. 2010;58:3493–3503.

[9] Caillard D. Kinetics of dislocations in pure Fe Part II: In situ straining experiments at low temperature. Acta Mater. 2010;58:3504–3515.

[10] Asaro R. Crystal plasticity. J Appl Mech. 1983;50:921–934.

[11] Kocks UF, Argon AS, Ashby MF. Thermodynamics and kinetics of slip. Pergamon Press; 1975.

[12] Asaro R, Needleman A. Overview no. 42 texture development and strain hardening in rate dependent polycrystals. Acta Metall. 1985;33:923–953.

[13] Brown SB, Kim KH, Anand L. An internal variable constitutive model for hot working of metals. Int J Plast. 1989;5:95–130.

[14] Mendelev MI, Han S, Srolovitz DJ, Ackland GJ, Sun DY, Asta M. Development of new interatomic potentials appropriate for crystalline and liquid iron. Philos Mag. 2003;83:3977–3994.

[15] Bitzek E, Koskinen P, Gähler F, Moseler M, Gumbsch P. Structural relaxation made simple. Phys Rev Lett. 2006;97:170–201.

[16] Vitek V, Perrin RC, Bowen DK. Core structure of 1/2(111) screw dislocations in bcc crystals. Philos Mag. 1970;21(173).

[17] Chaussidon J, Fivel M, Rodney D. The glide of screw dislocations in bcc Fe: Atomistic static and dynamic simulations. Acta Mater. 2006;54:3407–3416.

[18] Jones DR, Schonlau M, Welch WJ. Efficient global optimization of expensive black-box functions. J Global Optim. 1998;13:455–492.

[19] Yalcinkaya T, Brekelmans WAM, Geers MGB. Deformation patterning driven by rate dependent non-convex strain gradient plasticity. J Mech Phys Solids. 2011;59(1):1–17.

[20] Keh AS. Work hardening and deformation sub-structure in Iron single crystal deformed in tension at 298 °C. Philos Mag. 1964;12:9–30.

J. J. Möller, A. Prakash, and E. Bitzek*

9 FE2AT – finite element informed atomistic simulations

Abstract: Atomistic simulations play an important role in advancing our understanding of the mechanical properties of materials. Currently, most atomistic simulations are performed using relatively simple geometries under homogeneous loading conditions, and a significant part of the computer time is spent calculating the elastic response of the material, while the focus of the studies lies usually on the mechanisms of plastic deformation and failure. Here we present a simple but versatile approach called *FE2AT* to use finite element calculations to provide appropriate initial and boundary conditions for atomistic simulations. *FE2AT* allows to forgo the simulation of large parts of the elastic loading process, even in the case of complex sample geometries and loading conditions. *FE2AT* is open source and can be used in combination with different atomistic simulation codes and methods. Its application is demonstrated using the bending of a nano-beam and the determination of the displacement field around a crack tip as examples.

Keywords: finite element to atomistic coupling, finite element (FEM) simulations, atomistic simulations, displacement mapping, nanobeam bending, fracture

9.1 Introduction

Atomistic simulations play an important role in advancing our understanding of the mechanical behavior of materials and its relationship to the material's microstructure [1]. Calculations using density functional theory (DFT) have provided significant insight into the properties of individual defects, like the core structure of dislocations [2] or the critical stress intensity necessary for crack propagation [3], as well as on inherent material properties like stacking fault energies [4] or the theoretical strength [5]. Simulations using simplified – but still material specific – models of the atomic interactions, like semi-empirical environment-dependent potentials (e.g., EAM [6] or Finnis-Sinclair-Potentials [7]), many-body potentials (e.g. Stillinger-Weber [8]), or bond-order potentials (BOP, e.g. [9]) can be used to simulate many millions of atoms for time-scales of up to nanoseconds. They are particularly useful when it comes to studying the direct interaction between defects. Examples include the study of dislocation nucleation at interfaces [10], dislocation-grain boundary (GB) interactions like pinning, absorption or transmission [11–13], the interactions between cracks and dislocations or GBs [14], and the self-organization of shear transformation zones into shear bands [15]. The ability to simulate millions of atoms, furthermore,

DOI: 10.1088/0965-0393/21/5/055011, Modelling and Simulation in Materials Science and Engineering, Volume 21, Number 5.

allows to study the fundamental processes during the deformation of nano-scale structures like thin films, nano-pillars or nano-wires [16], and the one-to-one comparison with in-situ experiments [17].

The term atomistic simulation is not limited to the classical molecular dynamics (MD), molecular statics (MS) or Monte-Carlo (MC) type of simulations. It also includes, among others, methods to calculate activation energies like the nudged elastic band (NEB) method [18], accelerated MD methods [19], or newly developed methods like diffusive molecular dynamics (DMD) [20].

Usually, atomistic simulations of deformation and failure are performed using relatively simple loading conditions like uniaxial tension or compression, simple shear deformation or indentation. Compared to simulations using these kind of loading conditions, significantly fewer atomistic studies on bending or torsion can be found in the literature, e.g. [21–23]. In typical experiments, the loading situation is usually more complex, and one needs to consider the effects due to e.g., misalignments, machine stiffnesses or substrate compliance [24–26]. In addition, in many materials internal stresses are present, which are often not taken into account in atomistic simulations, as they can model only a small part of a large microstructure.

In many atomistic studies, the focus lies on the deformation and failure mechanisms rather than the study of elastic properties. Nonetheless, significant amount of computation time is usually spent on simulating the elastic response of a material or structure before plastic instability (or brittle failure) occurs. In particular, quasistatic simulations, in which a sample is iteratively loaded followed by an energy minimization to determine the critical load for the occurrence of an instability, require long simulation times. For samples with simple geometries and loading conditions, this problem can in principle be circumvented by displacing the atoms according to the linear elastic solution prior to the start of a molecular statics simulation. However, often no simple analytical expressions for the anisotropic elastic displacement field exist for more complex sample shapes and loading conditions.

The size limitation of atomistic simulations has spurred the development of several types of multiscale models (see e.g. [27–30] for a review of multiscale modeling). Hierarchical multiscale models like FE^2 [31] or HMM [32] obtain the material response at each integration point of a coarse scale FE model by solving locally the adequate microscale model. Partitioned-domain multiscale models on the other hand, describe the inhomogeneous response of a material in the vicinity of the region of interest (typically around defects) with a fine scale computation (e.g. atomistic simulation), whilst the material response far away is obtained by a relatively inexpensive continuum scale simulation, generally using a finite element (FE) model [27, 33–43].

To date, most of the methods which couple atomistics with continuum methods are restricted to simulations in two dimensions (2D) and only make use of simple atomic interaction models [28]. Partitioned-domain methods are well-suited for static simulations, where they are mostly used to avoid artifacts due to limited box sizes, rather than to apply complex loading conditions. Dynamic methods need to additionally address artifacts like wave-reflections at the interface between the two domains and ghost entropy [30]. To our knowledge, partitioned-domain methods have, furthermore, not yet been used in combination with accelerated MD methods or DMD.

In this work, we present a simple, yet elegant approach to use finite element calculations to provide appropriate initial and boundary conditions for atomistic simulations. The role of the relatively inexpensive FE computation – similar to partitioned-domain methods – is to solve for the (anisotropic) elastic response of a continuum sample subjected to (complex) loading conditions. However, contrary to partitioned-domain methods, our *finite element to atomistic* (*FE2AT*) scheme does not employ an embedding continuum. Instead a one-to-many (one finite element to many atoms) mapping of atomic positions is defined between the atomistic and finite element configurations. The elastic displacements obtained from the FE computation are then mapped onto the atomistic configuration, by interpolating between the nodal displacements of the corresponding finite element. Subsequent optimization of this configuration by energy minimization under the corresponding boundary conditions ensures that deviations from the linear elastic solution in regions around defects (e.g. surfaces) are taken into account. The nicety of this method is that the FE computation does not need to perfectly represent the atomistic situation after a certain deformation; it only needs to provide starting atomic configurations which are close to the actual minimum energy structure. The so generated *finite element informed* atomistic configuration can then be used for subsequent (quasi-) static or MD simulation as well as for other simulation methods like MC, accelerated MD or DMD.

As mentioned before, the *FE2AT* scheme does not employ an embedding continuum. Contrary to partitioned-domain methods, we start a-priori with two equivalent configurations, one defined by the FE model and the other by the atomistic sample. The FE mesh used is fine enough to capture the continuum scale solution, but much coarser than the resolution used in the atomistic configuration. The FE solution is merely used to displace the initial atomistic configuration. Fundamentally, the *FE2AT* framework leads to a one-way coupling of the different length scales, i.e. FE to atomistics. It thus conceptually corresponds to a unidirectional hierarchical multiscale model, where information is just passed from the macro to the micro scale. This approach results, nonetheless, in a simple, robust, versatile and *easy-to-use* scheme to model complex loading situations in atomistic simulations, all while significantly reducing the time spent on calculating the elastic response of the atomistic sample. However, as is generally the case in simulations without embedding continuum, artifacts due to boundary conditions have to be minimized by choosing sufficiently large simulation boxes.

Technically, converting finite element regions to atomistic regions is a part of most concurrent multiscale simulation programs. However, the framework as described in this paper has hitherto not been studied in the literature. By focusing on the simple task to provide only initial and boundary conditions, *FE2AT* can be used in combination with most atomistic simulation methods. The required additional FE computations can be performed with any commercial or free finite element method (FEM) program. FE calculations are already widely used in materials science and are part of most materials science and mechanical engineering curricula, making the *FE2AT*-method directly applicable to many scientists. *FE2AT* is freely distributed as open-source program [44]. It can be used as a preparatory stage for both concurrent multi-scale simulations and fully atomistic simulations, in the latter case helping scientists who do not have access to adaptive methods.

The paper is organized as follows. In Sec. 9.2, we first present the details of the method, followed by the algorithmic formulation. The accuracy of the *FE2AT* approach is then demonstrated on two examples in Sec. 9.3: bending of a nano-sized beam with anisotropic elastic material properties, and determination of the displacement field around a crack. The subsequent discussion in Sec. 9.4 highlights in particular the significant reduction of computation time by *FE2AT*.

9.2 Methodology of *FE2AT*

To obtain the initial and boundary conditions for an atomistic simulation of a sample under complex loading conditions from FE computations, one needs to follow three basic steps:

1. create a FE representation of the atomistic sample under consideration.
2. perform an anisotropic elastic FE calculation to obtain the equilibrium configuration under the desired loading conditions.
3. obtain the atomic positions corresponding to the elastically deformed FE configuration.

Well established procedures already exist for the first two steps. An elegant way to address the first point is to already generate the atomistic sample from a FE mesh. We recently developed a tool called nanoSCULPT [45] to create atomistic samples from arbitrarily shaped 3D structures. FE meshes of the atomic sample under consideration can also be generated by common mesh generators, e.g. [46], using the mathematical description of the sample shape which was used to create the atomic sample in the first place (typically combinations of simple geometrical objects like cylinders, spheres, cones, ...) or by determining the convex hull [47] of the sample. The anisotropic elastic calculation can then be performed with any commercial or free FE program.

The gist of the *FE2AT* method is to determine the new atomic positions from the deformed FE configuration. To this end, *FE2AT* employs a two step strategy. First, each atom in the atomistic configuration is mapped to an element in the finite element mesh of the sample. This atom-localization in the FE mesh is necessary to exploit the *partition of unity* and *compact support* properties of the FE [48–50]. In the second step, the displacement of each atom is obtained by interpolating the nodal displacements from the corresponding finite element calculation. For this purpose, we make use of the *natural coordinate space* of the FE. In what follows, this two step strategy of localizing each atom in the FE mesh and interpolating the nodal displacements is explained in more detail, and the typical workflow in the *FE2AT* approach is presented.

9.2.1 Atom-localization in a finite element mesh

The first task is the identification of an element containing the coordinates of a given atom. Mathematically, an atom is expressed by a point A with coordinates (X_A, Y_A, Z_A) and the finite element is a polyhedron with N nodes, where $N = 4$ for tetrahedra and $N = 8$ for hexahedra. To ascertain whether the atom under consideration is located in a certain FE or not, a point-in-polyhedron [47] test can be used. However, if the task is to test a high number of elements for a large number of atoms, this method becomes computationally expensive. The elements would be typically tested in the order provided by the meshing tool and *not* necessarily according to their spacial position. In the worst case, the atom would be contained in the last element, requiring a point-in-polyhedron test of every single element in the FE mesh.

Here, we use a three-dimensional search tree, a so-called octree [51], instead, to rank the elements according to their position in space. For given (X_A, Y_A, Z_A), the octree provides a most likely finite element $E(A)$ which is the first to be tested in the subsequent point-in-polyhedron test. This scheme follows the corresponding two-dimensional problem of a quaternary search tree (quadtree) [52]. It is based on the principle of recursive spatial decomposition. First, a cuboid is defined in such way that it surrounds the entire FE mesh. This cuboid is then subdivided into eight sections (octants) which are further subdivided recursively until *only one* element remains in an octant. This leads to a unique relation between octants and elements when the octree is finally built. To decide whether an element lies in an octant or not, its center of mass is calculated. If the center of mass is in between the coordinates of an octant's edges, the element is defined to be *in* this octant. Empty suboctants are not further subdivided. In other words, a link between an octant and a corresponding suboctant is only established if there is at least one element in the suboctant. Octants containing exactly *one* element's center of mass are called *leaves* [47]. If the number of leaves equals the number of elements, the octree is completely built and provides fast access to every finite element by means of its coordinates.

If a leaf is identified, the corresponding element is simultaneously stored in a separate array. This newly structured list of elements is a one-dimensional representation of adjacency in the FE mesh and will be useful if the point-in-polyhedron test with $E(A)$ was not successful.

To find $E(A)$ for a given atom A, the search is started at the octree's root, i.e., the cuboid circumscribing the entire FE mesh. This cuboid is henceforth called the parent octant P. Then, the suboctant S containing A is found by recursive comparison of the atom's coordinates (X_A, Y_A, Z_A) with the center (x_0^P, y_0^P, z_0^P) of the parent octant P. This means, only three inequalities ($X_A > x_0^P$, $Y_A > y_0^P$, $Z_A > z_0^P$) have to be assessed to find the correct suboctant. The matching suboctant becomes the parent octant ($S \rightarrow P$) and is evaluated accordingly. If a suboctant is found to be a leaf, it points to the element $E(A)$ which is most likely to contain the atom. Now, a point-in-polyhedron test is performed which leads either to a positive or a negative match. In case of negative match, elements that are stored before and after $E(A)$ in the restructured element list are evaluated further.

The localization of the appropriate element in the FE mesh for a given atom is by far the most computationally expensive part of the *FE2AT*-program. By using the octree-algorithm instead of a serial point-in-polyhedron test, the time to localize an atom in one of 12 000 elements is reduced from 37 ms to 1 ms[1]. This comparison, of course, depends strongly on the number and type of the elements (handling of tetrahedra is inherently more difficult for the octree). If the point-in-polyhedron test shows that the atom does not lie within the element $E(A)$ suggested by the octree, the neighboring elements need to be tested. Using the octree-algorithm to sort the elements according to their adjacency in the FE mesh proved here to be more efficient than constructing a list of nearest neighbors of each element (factor of four reduction in the time to locate an atom in the neighboring elements). This restructured list of elements can now be tested either in a linear fashion, or alternately, by evaluating elements that are stored before and after $E(A)$ in the array. Fig. 9.1 shows a direct comparison of both of these approaches using 18 107 tetrahedral elements and approximately 2.2 million atoms. By using the alternating search pattern, clearly fewer point-in-polyhedron tests need to be performed. The described optimized method to use octrees for localizing atoms in a FE might also help to improve the efficiency of coupled atomistic-continuum methods.

9.2.2 Interpolation of nodal displacements

Having localized each atom to an element in the FE mesh, we can interpolate the FE nodal displacements onto the atoms by making use of the so called *shape functions*, which provide for the continuity of a field variable inside an element. Note that only key equations required for completeness of this paper are provided in this section. For further information, the reader is referred to standard text books on the finite element method, e.g. [48–50].

Generally, in any finite element, the shape function N_i for node i, is defined so as to satisfy *partition of unity*, i.e.

$$\sum_{i=1}^{N_{\text{nodes}}} N_i = 1 \,. \tag{9.1}$$

1 Indicated times are for a serial code on a AMD Opteron 6134 processor with 2.3 GHz.

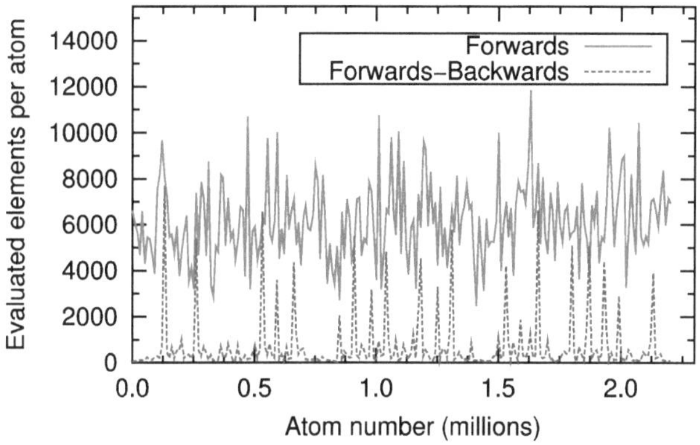

Fig. 9.1: Comparison of evaluated elements per atom using a forwards (blue, solid line) and a forwards-backwards search (red, dashed line) through the restructured element list.

Furthermore, N_i is defined to be zero in any element not touching node i. This property of shape functions, also called *compact support*, limits the region of interpolation to that of an element domain [53]. Most finite element softwares use the isoparametric formulation, permitting the use of arbitrarily shaped elements like non-rectangular hexahedra. As a result, the shape functions are defined in the *intrinsic* or *natural* coordinate system of the element.

$$N_i = \hat{N}_i(\boldsymbol{\xi}), \tag{9.2}$$

where

$$\boldsymbol{\xi} = (\xi, \eta, \kappa) \tag{9.3}$$

is a vector defining the natural coordinate system of the finite element in a three dimensional cartesian frame. The isoparametric formulation does lead to a slightly expensive computational scheme. The resulting framework is, nevertheless, more enterprising due to its generality.

To proceed further, we first need to determine the coordinates of each atom in the natural system of the finite element. Using the isoparametric formulation, it follows that

$$\boldsymbol{x} = \sum_{\alpha}^{N_{\text{nodes}}} N^{\alpha}(\boldsymbol{\xi}) \cdot \boldsymbol{X}^{\alpha}, \tag{9.4}$$

where $\boldsymbol{x}$ is the vector of coordinates of the atom under consideration, and $\boldsymbol{X}^{\alpha}$ defines the coordinates of node α in the global cartesian coordinate system; N^{α} denotes the shape function of node α defined in the natural coordinate system of the element. The summation in the above equation is performed only over the nodes of the element enclosing the atom (note that the element was identified in the previous step). The definition of the shape functions depends on the type of element used. For eight node hexahedral (brick) elements, the shape functions are defined by

$$N^{\alpha}(\boldsymbol{\xi}) = \frac{1}{8}(1 + \xi\,\xi^{\alpha})(1 + \eta\,\eta^{\alpha})(1 + \kappa\,\kappa^{\alpha}), \quad \text{where } \boldsymbol{\xi} = (-1, 1), \tag{9.5}$$

and for the four node tetrahedral elements as

$$\left.\begin{aligned} N^1(\boldsymbol{\xi}) &= \xi, \\ N^2(\boldsymbol{\xi}) &= \eta, \\ N^3(\boldsymbol{\xi}) &= \kappa, \\ N^4(\boldsymbol{\xi}) &= 1 - \xi - \eta - \kappa, \end{aligned}\right\} \quad \text{where } \boldsymbol{\xi} = (0, 1)\,. \tag{9.6}$$

The superscripts in Eq. (9.6) denote the node number α under consideration. It is evident that Eq. (9.4) results in a linear system of equations for the four node tetrahedron, and a non-linear system of equations for the eight node hexahedron.

The displacements of the atom under consideration, as a function of the nodal displacements, can now be determined by

$$\boldsymbol{u} = \sum_{\alpha}^{N_{\text{nodes}}} N^{\alpha}(\boldsymbol{\xi}) \cdot \boldsymbol{d}^{\alpha}\,, \tag{9.7}$$

where $\boldsymbol{u}$ defines the displacement of the atom, and $\boldsymbol{d}^{\alpha}$ denotes the displacements of the node α. The summation in the above equation is again performed over the nodes of the element enclosing the atom.

9.2.3 The *FE2AT* approach

The typical workflow to perform atomistic simulations which make use of FE calculations to generate appropriate initial and boundary conditions can be summarized as follows:

- *Construction of an atomistic sample*: This is also the first step of any atomistic simulation and remains unchanged for the *FE2AT* approach. The geometrically constructed sample should be optimized by the usual energy minimization techniques to start with an equilibrium structure.
- *FE mesh generation*: The generation of the finite element mesh is the first step for each FE computation. The usual tools can be used to generate the equivalent geometry to the (optimized) atomistic configuration. It is recommended that the FE mesh is slightly larger than the atomistic configuration. This ensures that even the boundary atoms in the atomistic configuration fall well within the domain of the finite elements, thus fulfilling the requirements for interpolation.
- *FE computation*: Standard FE codes can be used to calculate the equilibrium shape of the sample under the desired loading conditions according to linear elasticity. Required inputs are the generated mesh, the boundary conditions, and the anisotropic elastic constants of the material (e.g., of the used potential) to be studied in the atomistic simulations. The FE calculations should be performed using loads and displacements where no plastic deformation is expected.
- *Running* FE2AT *to generate the deformed atomistic configuration*: *FE2AT* takes as input the (optimized) atomistic configuration, the initial FE mesh and the result of

the FE computation. The output of *FE2AT* is an atomistic configuration in which all atoms of the provided configuration are displaced according to the result of the FE computation. Internally this is achieved by first mapping each atom of the atomistic input configuration to an element in the undeformed FE mesh. The domain of a finite element is, in general, much larger than the typical interatomic distances. As a result, the created mapping array is likely to be a *many-to-one* mapping array. The coordinates of each atom are then expressed in the *natural coordinate* system of the FE using Eq. (9.4). Using these, a vector of numerical values of element shape functions for the atom is generated. The displacements of each node are then extracted from the deformed FE mesh. The displacements of each atom are then determined by interpolating the nodal displacements using Eq. (9.7).

- *Geometry optimization of the deformed atomistic configuration*: The geometry of the atomic structure obtained from *FE2AT* needs to be optimized by standard energy minimization techniques, since effects from nonlinear elasticity or surface relaxations are not taken into account in the FE computation. Care has to be taken that the imposed boundary conditions match the ones in the FE computation. Typically these are imposed by applying forces to selected atoms or by fixing the positions of certain atoms. In case the atomic sample deforms plastically or fractures during the energy minimization, the FE computation needs to be performed at lower loads.
- *Perform the desired atomistic simulations*: The relaxed atomistic configuration can now be used as starting configuration for the simulations of interest. In case quasi-static loading is to be simulated, the atomic displacements between the initial, relaxed but undeformed and the relaxed *FE2AT*-deformed configuration can be determined and further loading can be incrementally applied to the sample by correspondingly scaling the displacement field. For MD-simulations at finite temperatures, the relaxed sample would need to be homogeneously expanded according to the thermal expansion and then equilibrated at the desired temperature, before dynamically loading the sample by applying appropriate boundary conditions. For NEB calculations, the chain of replicas can be generated by interpolating between two configurations generated by the *FE2AT*-approach.

The *FE2AT*-approach only provides initial configurations according to the (anisotropic) linear elastic material response to some given loading conditions. Boundary conditions in accordance to the elastic deformation can be imposed by fixing atomic positions. Unlike in FE-atomistic coupling methods [27, 33–43], these boundary conditions will, however, not be able to adapt to any non-homogeneous deformation taking place during the atomistic simulation. This effect of boundary conditions can, nevertheless, be minimized by using sufficiently large samples, as is indeed the case in the second example presented in this paper. It should be, furthermore, noted that the *FE2AT*-approach should only be used for situations where the material can be expected to deform purely elastic. Metallic glasses or nanocrystalline materials, for example, can show irreversible deformation on the atomic scale before any macroscopic deviation from elasticity is observed [13, 15], and *FE2AT* may not be suitable in such cases.

By focusing only on the task to provide initial configurations for atomistic simulations, *FE2AT* can be easily used in combination with different atomistic simulation codes and methods, including classical and accelerated MD methods, NEB and DMD. It is furthermore independent of the method used to calculate the interatomic interactions (e.g., semi-empirical potentials, BOP, tight-binding or DFT).

9.3 Application examples

In what follows, two examples for the application of *FE2AT* to typical problems are presented. The first example demonstrates the usefulness of the *FE2AT*-approach on a simple problem, the bending of a nano-sized beam with anisotropic elastic material properties. This example clearly highlights the potential of *FE2AT* to significantly reduce the total computing time required to study nanoscale plasticity by performing quasi-static simulations. The atomistic study of cracks in elastically isotropic and anisotropic materials is used as the second example. Here, *FE2AT* provides a simple way to generate appropriate initial and boundary conditions from FE calculations.

The FE computations were performed using the commercial FE software ABAQUS®, and the MD software package IMD (ITAP molecular dynamics, [54, 55]) was used for the atomistic simulations. Atomistic configurations were visualized with AtomEye [56].

9.3.1 Bending of a nano-beam

The mechanical properties of nano-scale structures have recently attracted a lot of attention [16]. Experiments as well as simulations have shown that the yield stress of nanowires increases drastically with decreasing size, and that the deformation mechanisms [17] as well as the elastic constants [16, 21, 23] change with decreasing diameter. Simulations on nanowires are for the most part performed in tension [17, 21], simulations of nanowires under bending loads were only recently reported [21, 22]. The bending of a beam is, furthermore, a classic engineering problem, and therefore a good first test of our method.

To study the atomic scale deformation mechanism during bending, one can perform MD simulations e.g., using a cantilever geometry and applying the load by an indenter, similar to the situation in in-situ nano mechanical testing [57]. However, due to the short time scales available in MD simulations an the subsequently high strain rates of the order of $10^8\ s^{-1}$, the elastic deformation of the beam is characterized by the propagation of elastic waves and the shape of the beam as function of indenter displacement is different from the one at more realistic strain rates. An alternative approach is to perform quasi-static simulations, in which the indenter is displaced by a small increment, the entire structure is relaxed by energy minimization, and the process is repeated with the resulting structure. This approach also allows to determine critical stresses, e.g., for the nucleation of a dislocation. This iterative procedure, however, requires many computations in which the beam just reacts elastically to the applied load – which is usually not of interest whilst studying small scale plasticity.

In the following section, we show how the *FE2AT*-approach can be used to reduce the computational effort to study the bending of a Molybdenum nano-beam. Besides serving to validate our method, the focus of this section lies on assessing the overall performance of *FE2AT*.

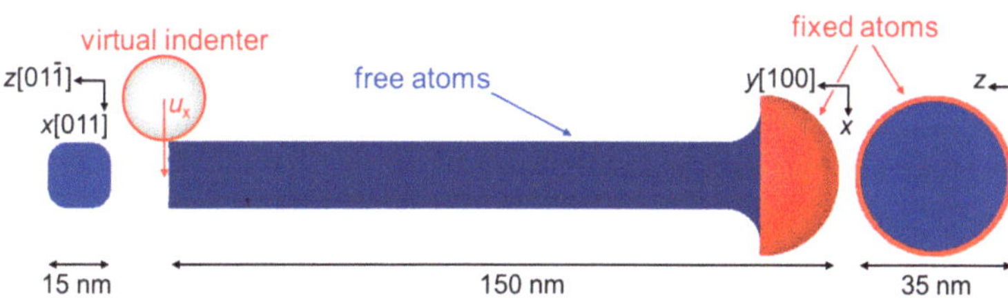

Fig. 9.2: Schematic illustration of the simulation set-up for the nano-beam bending.

9.3.1.1 Simulation set-up

The simulation set-up for the nano-beam is shown in Fig. 9.2. It consists of of three regions: the beam itself, a transition region from the beam to the grip, and the clamped support. The beam has a cross-section of $15 \times 15\,\text{nm}^2$ and is 125 nm long. The edges were rounded similar to the experimentally observed fibers [58]. To minimize unrealistic stress concentrations at atomically sharp corners, a smooth transition with a filleted corner (7.5 nm radius) from the beam to the clamp was used. The clamp is a half sphere consisting of free inner atoms and an outer shell region of 1 nm thickness where the atoms are fixed. The sample consists of approximately 2.4 million atoms. Bending loads are applied to the beam by x-displacement of a spherical indenter, placed exactly at the center of the beam's free end (see Fig. 9.2). The indenter was modeled by a repulsive potential according to [59] with the indenter hardness $d = 0.08\,\text{nm}$ and a radius of 16 nm.

The interaction of the molybdenum atoms was modeled with the Finnis-Sinclair potential [7, 60]. The elastic constants of this potential, which were also used in the FE calculation, are given in Tab. 9.1. Prior to a structure optimization using the energy minimization algorithm FIRE [61] the sample was annealed for 60 ps at about half of the melting temperature $T_\text{m}/2 \approx 1500\,\text{K}$.

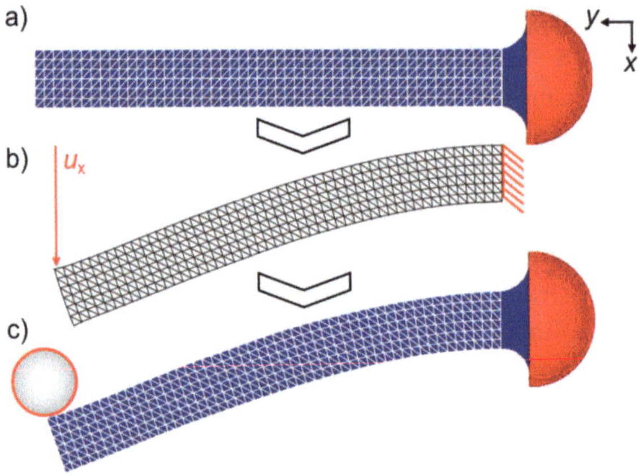

Fig. 9.3: Schematic illustration of the application of *FE2AT* in order to provide an elastically bent configuration as input for further simulations; (a) meshing the undeformed relaxed configuration; (b) finite element simulation of the beam bending by displacing the free end of the beam by $u_x = 32.5$ nm; (c) mapping of the nodal displacements onto atoms via *FE2AT*, placement of the virtual indenter, and relaxation of the configuration.

The relaxed atomic structure and a corresponding FE mesh are then used as as input reference configuration for *FE2AT*. As shown in Fig. 9.3, it is sufficient to only model that part of the sample where most of the elastic deformation is expected, by finite elements. The structure optimization using the atomic positions provided by *FE2AT* as starting configuration ensures that the *entire* atomistic sample is in equilibrium. The 3D FE mesh is used to calculate the equilibrium shape of the beam for a fixed displacement of the free end of the beam by u_x using anisotropic linear elasticity with the elastic constants of Tab. 9.1. As an example, the deformed mesh for $u_x = 32.5$ nm is shown in Fig. 9.3 (b). With this mesh, *FE2AT* is used to generate the atomic configuration corresponding to the elastic deformation. Fig. 9.3 (c) shows the mesh superposed to the resulting atomistic configuration after structure optimization with FIRE, where the displacement has been imposed by the spherical indenter.

Further loading of the sample was performed by appropriately scaling the atomic displacements. This approach takes into account effects due to surfaces and nonlinear elasticity as well as the response of the clamp. This requires of course that no plastic deformation has taken place. The atomic displacements $u_\mathrm{A}(x, y, z)$ are computed as the difference between the relaxed initial configuration the relaxed configuration provided by *FE2AT*. To determine the x-displacement necessary for onset of plastic deformation, atoms were incrementally displaced according to a displacement field $D_\mathrm{i} \cdot u_\mathrm{A}(x, y, z)$ and subsequently relaxed by FIRE. The factor D_i was chosen such that the deformation step corresponds to a shift of the indenter position by $i \cdot \delta u_x$, with $\delta u_x = 0.5$ nm.

Tab. 9.1: Elastic constants of the used potentials.

Potential	c_{11} in GPa	c_{12} in GPa	c_{44} in GPa
Molybdenum [7, 60]	464.7	161.5	108.9
Tungsten [7, 60]	522.5	204.4	160.6
α-Iron [62]	243.4	145.0	115.9

As reference, a MD simulation in the NVT-ensemble at $T = 0$ K and an indenter velocity of $v_x = 0.05$ nm/ps, as well as a quasi-static simulation with indenter displacement of $\delta u_x = 0.1$ nm between relaxations with FIRE were performed.

9.3.1.2 Comparison of *FE2AT* with quasistatic and dynamic simulations

FE2AT was used to generate atomistic structures according to indenter displacements of $u_x = 0.1$, 1, 32.5, and 35 nm. The shape of the relaxed *FE2AT*-structure at $u_x = 1$ nm is shown in Fig. 9.4. The atomic positions of the relaxed *FE2AT*-structure are nearly identical to the ones obtained iteratively after ten displacement increments of $\delta u_x = 0.1$ nm and FIRE-relaxations in the quasistatic approach. The average position difference between an atom in the relaxed *FE2AT*-structure and in the quasistatically displaced structure is less than $3 \cdot 10^{-3}$ nm.

The shape of the dynamically deformed beam at the identical indenter position is also shown in Fig. 9.4. For the given length of the beam $L_y = 125$ nm, a parallel displacement of one end of the beam with $v_x = 0.05$ nm/ps would correspond to a (for atomistic simulations) comparably low shear rate of about $4 \cdot 10^7$ 1/s. However, compared to the velocity of shear waves in Mo $c_s \approx 3.6$ nm/ps, the indenter velocity is relatively fast. Within the 20 ps of simulated time to reach $u_x = 1$ nm, information about the initial displacement can only travel about half the way through the beam. The shape of the beam in the dynamic simulation therefore differs from the equilibrium shape at this indenter displacement, as shown in Fig. 9.4. In order to reach equilibrium shape, the sound waves would need to travel multiple times through the beam, which would require significantly lower indenter velocities and subsequently longer simulated times.

In the quasi-static simulation of beam bending, an indenter displacement of $\delta u_x = 0.1$ nm resulted in 269 617 calls to the force calculation routine to reach a force norm of 10^{-7} eV/nm, whereas for the same indenter displacement the structure generated by *FE2AT* required only 105 721 force-calls in FIRE to reach the same force norm. In the quasi-static simulation procedure, the computational costs scale with the number of iterative displacements, whereas it stays roughly constant in the *FE2AT*-approach. For example, for an indenter displacement of $u_x = 1.0$ nm, 1 721 523 force-calls in FIRE were needed in total in the quasi-static approach, compared to 100 972 with *FE2AT*.

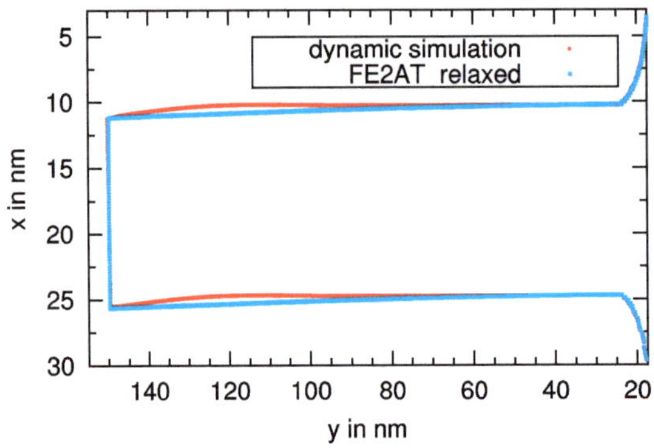

Fig. 9.4: Outline of the beam (at an indenter displacement of $u_x = 1.0$ nm) according to *FE2AT* after relaxation (blue) and in a MD simulation with an indenter velocity $v_x = 0.05$ nm/ps.

9.3.1.3 Plastic deformation of the beam

For an indenter displacement of $u_x = 35$ nm the relaxation of the configuration provided by *FE2AT* showed a plastic instability. Therefore, a relaxed *FE2AT* configuration at $u_x = 32.5$ nm was used as starting point for a quasi-static approach, using however the scaled displacement field determined from that configuration. With this approach, dislocation emission occurred at an indenter displacement of $u_x = 35.5$ nm, close to (but not in) the transition region. The dislocation was identified to be an $a_0/2[\bar{1}11](101)$ edge dislocation. Fig. 9.5 displays the corresponding configuration.

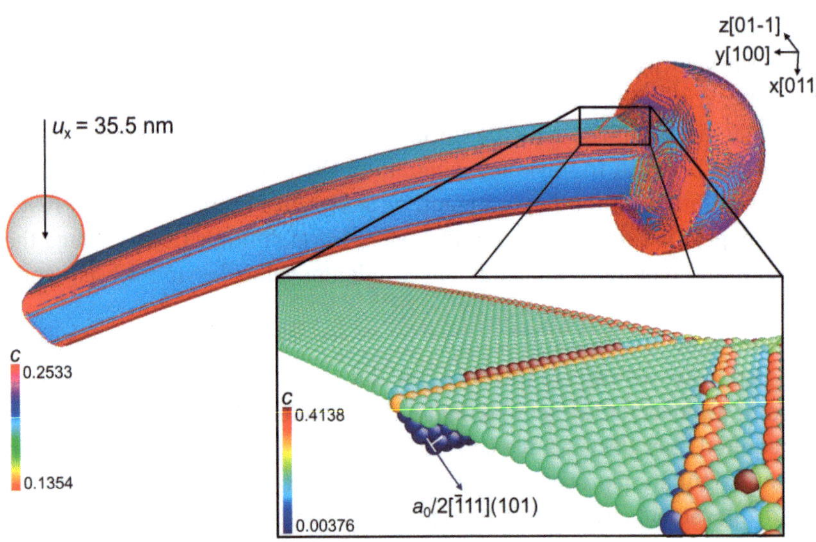

Fig. 9.5: Illustration of the atomistic configuration at an indenter displacement of $u_x = 35.5$ nm, after the first dislocation is nucleated. The atoms are colored according to their centrosymmetry parameter [63].

To estimate the critical resolved shear stress required for the nucleation of a dislocation, the usual virial expression for the atomic stress tensor can be used [64]. The required atomic volume is determined on a per-atom basis by a Voronoi construction [47]. The averaging over the atomic stresses was carried out in the region where the dislocation was emitted at $u_x = 35.0\,\text{nm}$, but in the relaxed configuration of the previous loading step ($u_x = 35.0\,\text{nm}$), i.e. when no dislocation was emitted. The critical resolved shear stress was estimated to $\tau_{\text{crss}} \approx 8\,\text{GPa}$. A more detailed analysis of the plastic deformation of Mo nano-beams under bending loads and its size and orientation dependence will be published elsewhere.

9.3.2 Fracture

Our second example for the application of *FE2AT* is the atomistic study of fracture. Atomistic simulations of cracks have played an important part in advancing our understanding of dislocation-crack interactions [14], grain boundary fracture [65], the lattice-trapping effect [66] and the emission of dislocations from crack tips [14, 67]. For these kind of studies, careful control of the loading of the crack is required. Usually, the fracture toughness K_{Ic} is determined by applying the linear elastic displacement field around a crack tip in a semi-infinite anisotropic body and relaxing the sample, see e.g. [66]. For isotropic materials, the displacement field around the tip of a crack under mode I-loading can be found in textbooks [68]. In the case of anisotropic materials, the displacement field can be calculated by numerically finding the roots of a fourth-order polynomial equation [69]. For more general problems, the situation is more complex.

Here, we use *FE2AT* to provide the elastic displacements of atoms around a crack tip. The method is applied to a crack in an isotropic tungsten single crystal and compared to the theoretical solution [69], but can in principle be applied to more complex situations like grain boundary fracture in anisotropic materials.

9.3.2.1 Simulation set-up

The general simulation set-up for a quantitative study of fracture toughness of cracks under mode I-loading in shown in Fig. 9.6. Periodic boundary conditions (PBC) are applied along the crack front direction, and the atoms within a cylinder are displaced according to the corresponding displacement field around the central crack tip. Atoms in an outer layer are held fixed. Structure optimization by energy minimization is used to determine the equilibrium configuration at a certain applied stress intensity facto K_{I}. Similar to the previous example, the atomic displacements between two relaxed configurations at different K_{I} can be used to further load the sample by adding the scaled displacements to a relaxed configuration. This incremental displacement field then corresponds to an additional K_I increase by ΔK_I. The critical stress intensity factor K_{Ic} is determined as the K_I where the crack starts to propagate during relaxation.

In case of a homogeneous, elastic isotropic or anisotropic material, the displacement field around a crack tip in a semi-infinite anisotropic body is readily available from the literature and its determination is presented in the next paragraph. The general approach to determine the atomic displacements around a crack tip in an elastically homogeneous or inhomogeneous material under complex loading conditions using *FE2AT* is discussed in the following paragraph.

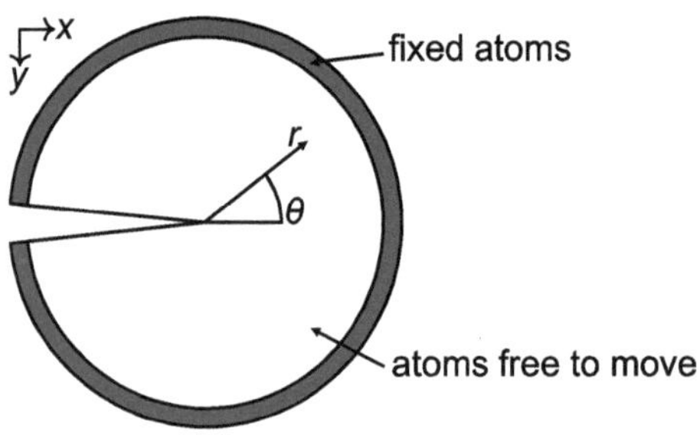

Fig. 9.6: Schematic illustration of the simulation set-up for cracks in single crystals: all atoms in the simulation box are displaced from their initial position according to the anisotropic linear-elastic solution of a crack under plane strain and mode I loading; the inner atoms are free to move while the boundary atoms are fixed on their positions.

Theoretical crack-tip displacement field in a single crystal. The displacement displacement field $u_{x,y}$ around the crack tip of a mode I crack in a semi-infinite anisotropic body is according to anisotropic linear elastic fracture mechanics a function of the stress intensity factor K_I [69, 70]:

$$u_x(r,\theta) = \frac{K_I\sqrt{2r}}{\sqrt{\pi}}\,\mathrm{Re}\left\{\frac{1}{s_1 - s_2}\left[s_1 p_2\sqrt{\cos\theta + s_2\sin\theta} - s_2 p_1\sqrt{\cos\theta + s_1\sin\theta}\right]\right\} \quad (9.8)$$

$$u_y(r,\theta) = \frac{K_I\sqrt{2r}}{\sqrt{\pi}}\,\mathrm{Re}\left\{\frac{1}{s_1 - s_2}\left[s_1 q_2\sqrt{\cos\theta + s_2\sin\theta} - s_2 q_1\sqrt{\cos\theta + s_1\sin\theta}\right]\right\} \quad (9.9)$$

where r is the distance between crack-tip and the present position and θ, is the angle between the crack plane and the direction of r, see Fig. 9.6. The complex parameters p_1, p_2, q_1, and q_2 are given by Eq. (4.13) in [69] while s_1 and s_2 are either complex or purely imaginary [71] and solve Eq. (4.8) in [69].

According to Irwin [72], the stress intensity factor can be expressed in terms of the energy release rate G_I, i.e., the change in elastic strain energy per unit area of crack advance, by:

$$K_I = \sqrt{G_I B^{-1}}\,, \quad (9.10)$$

where B describes the elastic response of the material. For an anisotropic material under plane strain conditions, B is defined, as follows [69]:

$$B = \sqrt{\frac{b_{11}b_{22}}{2}\left(\sqrt{\frac{b_{22}}{b_{11}}} + \frac{2b_{12} + b_{66}}{2b_{11}}\right)} \tag{9.11}$$

with b_{ij} being the plane strain moduli obtained from the anisotropic elastic compliances s_{ij} for the orientation under consideration [70, 73, 74]:

$$\begin{aligned} b_{11} &= \frac{s_{11}s_{33} - s_{13}^2}{s_{33}}, & b_{22} &= \frac{s_{22}s_{33} - s_{23}^2}{s_{33}}, \\ b_{12} &= \frac{s_{12}s_{33} - s_{13}s_{23}}{s_{33}}, & b_{66} &= \frac{s_{66}s_{33} - s_{26}^2}{s_{33}}. \end{aligned} \tag{9.12}$$

Crack-tip displacements by the *FE2AT*-approach. The *FE2AT*-approach uses FE calculations to provide the atomic coordinates of a sample under the desired loading conditions. As long as the FE set-up represents the elastic properties of the atomistic sample and the desired loading conditions, the details of the FE set-up are irrelevant to the *FE2AT*-approach. In particular, the atomistic sample can represent only a small region of the FE geometry. For the fracture problem, we choose the atomic sample to correspond to a small cylindrical region around the crack tip in the well defined compact tension (CT) specimen geometry according to the ASTM Standard E399-90 [75]. The general approach is, however, independent of the geometry and loading condition. It must be noted, nevertheless, that particularly in the case of interfacial cracks between two elastically dissimilar bodies, the usual LEFM-concepts for homogeneous materials do not hold, and the stress intensity factor becomes complex-valued.

Fig. 9.7 shows the schematic illustration of the two-dimensional simulation set-up. The load P is applied symmetrically on a CT specimen with an effective crack length $a = W/2$. The set-up is equal to a common single crystalline CT specimen using two-dimensional square elements [46]. The region selected for application of *FE2AT*, i.e., the region that will be represented atomistically, has a radius of 15 nm around the crack-tip and consists of approximately 100 000 atoms.

For single crystalline CT specimens, the applied stress intensity factor K_{I} is related to the load P according to the ASTM Standard E399-90 [75]:

$$K_{\mathrm{I}} = \frac{P}{T\sqrt{W}} f\left(\frac{a}{W}\right) \tag{9.13}$$

with a, P, and W as introduced in Fig. 9.7 and $T = 0.1$ nm being the plane strain thickness. For $a/W = 0.5$, the dimensionless weight function is given as $f(0.5) = 9.66$ [75]. In turn, the relationship presented in Eq. (9.13) allows the specification of a target stress intensity factor K_{It} by means of the load P_{app} that should be applied:

$$P_{\mathrm{app}} = \frac{K_{\mathrm{It}} T \sqrt{W}}{9.66} \tag{9.14}$$

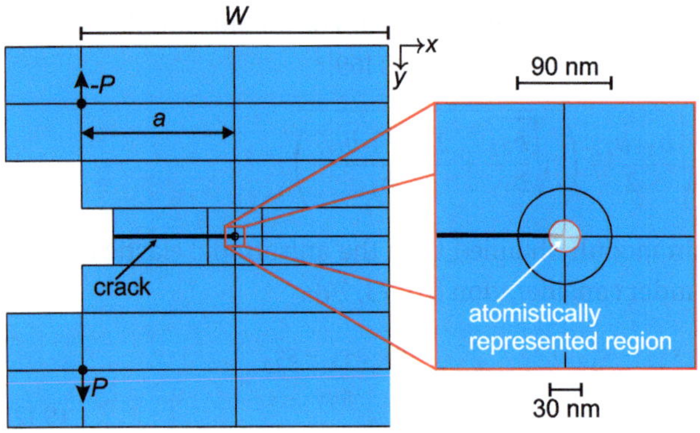

Fig. 9.7: Schematic illustration of the FE simulation set-up for a CT specimen. Only a circular region around the crack tip is used to determine the atomic displacements in the *FE2AT*-approach.

This purely geometrically derived target value for the stress intensity factor has to be equal to the actual energetic state of stress intensity around the crack-tip:

$$K_{\mathrm{I}}^{\mathrm{J}} = \sqrt{JB^{-1}} \tag{9.15}$$

with B from Eq. (9.11) and J being the J-Integral according to Rice [76]. The validity of this relation, $K_{\mathrm{I}}^{\mathrm{J}} = K_{\mathrm{It}}$, was tested by evaluating the J-Integral in the sample under a given load P. The elastic properties of α-iron and tungsten were modeled according to the potentials used in the atomistic simulations, see Tab. 9.1. The J-Integral was calculated by evaluating 50 contour integrals around the crack-tip using the standard output variables provided by ABAQUS®. The results of the J-integral evaluation differ in less than 0.5 % from the geometrically based target values K_{It} for the single crystals and for a grain boundary in the elastically isotropic tungsten.

9.3.2.2 Comparison of displacement fields: *FE2AT* vs. theoretical solution

In order to represent a crack in a semi-infinite anisotropic body, the size of the FE specimen needs to be sufficiently large compared to the region that is to be modeled atomistically. To evaluate the influence of the specimen size W (see Fig. 9.8) on the atomic displacement field, W was systematically varied between 500 and 8000 nm while keeping the radius of the atomistically described cylinder fixed to $r = 15$ nm. The resulting nodal displacements were mapped onto atoms using *FE2AT* and compared to the analytical solution, Eqs. (9.8) and (9.9), which was also used to determine the atomic positions.

The crack-tip opening displacements were evaluated at $x = -15$ nm apart from the crack-tip, i.e., at the maximum opening of the crack in the atomistic configuration. This value is compared to the target value of the analytical solution. The difference in percent is denoted by ΔCTOD. Fig. 9.8 and Tab. 9.2 summarize the size-effect on the differences between the theoretical displacement field and the displacements generated by the *FE2AT*-approach. The differences decrease with increasing box size. For a FE specimen size of $W > 4000$ nm, the difference between the *FE2AT*-approach and the theoretical solution are smaller than the displacements corresponding to typical loading increments, e.g. $\Delta K_{\mathrm{I}} = 0.01\,\mathrm{MPa}\sqrt{\mathrm{m}}$, [65, 67].

For $W = 4000$ nm, the CTOD was also evaluated using anisotropic-elastic constants, i.e., of α-iron, see Tab. 9.1. In this case, the difference to the analytical solution was somewhat larger than in the isotropic case (ΔCTOD = 0.62 %), but still smaller than half of the displacement caused by an incremental increase of ΔK_{I}.

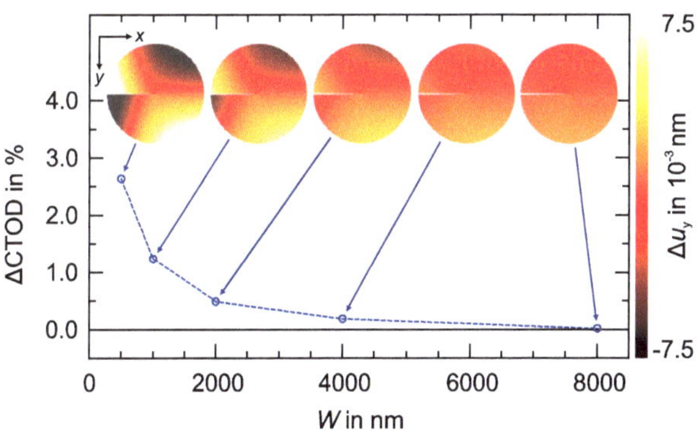

Fig. 9.8: The effect of FE specimen size W, see Fig. 9.7, on the crack-tip displacement fields compared to the analytical solution for tungsten single crystals; the diagram shows the differences in crack-tip opening displacement (ΔCTOD) in dependence of the box size W; inset figures show the differences in displacement field Δu_y where dark and bright colors represent high negative and high positive mismatch respectively.

Tab. 9.2: Summary of the effect of the FE simulation box size and the number of elements, n_{elements}, on the differences to the analytical solution for a tungsten single crystal in the (010)[001] crack system; the analytical reference solution is CTOD = $861 \cdot 10^{-3}$ nm.

W in nm	n_{elements}	CTOD in 10^{-3} nm	ΔCTOD in %
500	21 653	887	2.64
1000	59 896	873	1.23
1000	83 119	873	–
2000	253 539	866	0.49
4000	1 022 101	862	–
4000	15 505	862	0.19
8000	4 102 717	861	–
8000	16 033	861	0.02

In all cases, the size of the square elements increased radially symmetric from the crack-tip (approximately $0.1 \times 0.1\,\text{nm}^2$) to the outline of the circular region around the crack-tip (approximately $2.5 \times 2.5\,\text{nm}^2$), see Fig. 9.7. The remaining specimen was meshed with elements of $50 \times 50\,\text{nm}^2$ size. It was verified that the simulation result did not change if the mesh size was reduced by a factor of $1/5$, see also Tab. 9.2 for $W = 4000$ nm and $W = 8000$ nm.

9.4 Discussion

Both application examples show that the *FE2AT*-approach of FE informed atomistic calculations leads to the correct atomistic configurations, both when compared to an established alternative simulation method (quasistatic loading in the case of the beam) or a theoretical solution (crack tip displacement field). In the *FE2AT*-approach, the FE calculations do not need to perfectly represent the corresponding atomistic situation, as was shown by the omission of the clamp in the finite element mesh of the nano-beam example. As long as the regions undergoing large elastic deformations are included, *FE2AT* should be able to provide atomistic starting configurations for more complex sample geometries and loading conditions, which are close to the actual minimum energy structure.

The example of the nano-beam bending clearly highlights the significant reduction in computational costs by the *FE2AT*-approach. Using the iterative quasistatic loading to reach an indenter displacement $u_x = 35.5$ nm would have required about 96 days on 72 cores (Xeon 5650 processor, 2.66 GHz[2]). Using *FE2AT*, as described in Sec. 9.3.1.1, the time needed for mapping was 24 minutes (12 853 tetrahedral elements, single-core, AMD Opteron 6134, 2.3 GHz), 16 hours for the initial relaxation and 4 hours per relaxation of a scaled configuration (both on 72 cores, Xeon 5650 processor, 2.66 GHz). Creating the FE mesh for this simple geometry and performing the anisotropic elastic FE computation required less than an hour.

Already from this simple example, further possible applications for the *FE2AT*-approach become obvious: by comparing the stress state in the FE simulation with the atomistic stresses, finite size effects on the elastic properties can be directly evaluated, and the yield-criteria of constitutive material models could be directly tested.

The general applicability of *FE2AT* to fracture problems was demonstrated in the second example. The theoretical solution for the crack tip displacement field in [69] is in principle only valid for orthotropic bodies where the planes of elastic symmetry are coincident with the coordinate planes. *FE2AT* provides here a more general approach.

---

2 See the webpage of the LiMa Cluster (Regionales Rechenzentrum Erlangen) for more detailed information: http://www.rrze.uni-erlangen.de/dienste/arbeiten-rechnen/hpc/systeme/lima-cluster.shtml

In particular, *FE2AT* can also be used to generate realistic initial and boundary conditions for cracks in elastically inhomogeneous materials, including cracks located in the interface between two elastically dissimilar, anisotropic materials, see e.g. [77–81]. In this case, prescribing well defined loading conditions is more complicated as even under an external mode I loading, the crack experiences locally a mixed-mode loading [80]. *FE2AT* could in this case be used to model the situation in typical experimental set-ups for interfacial fracture, e.g. [82, 83].

9.5 Summary

A simple methodology is introduced to generate initial and boundary conditions for atomistic calculations from anisotropic elastic finite element calculations. These can be used to study deformation processes in complex geometries under realistic loading conditions. The presented *FE2AT*-approach is independent of the interatomic interaction model and can easily be used in combination with different atomistic simulation codes and methods. Application of *FE2AT* to nano beam bending and fracture show that it reproduces the correct atomic displacements, all while leading to a significant reduction in computer time necessary to simulate the process of elastic deformation.

Acknowledgment: Financial support from the German Science Foundation (DFG) in the context of its priority program "Life$^{\infty}$ – infinite lifetime of high-performance materials subjected to cyclic loading" (SPP 1466) under the contract number DFG/HO 2187/6-1 is gratefully acknowledged. We are grateful to F. Iqbal for helpful discussions.

References

[1] Li J, Ngan A, Gumbsch P. Atomistic modeling of mechanical behavior. Ann Math. 2003;51(19): 5711–5742.
[2] Woodward C, Rao S. Flexible ab initio boundary conditions: Simulating isolated dislocations in bcc Mo and Ta. Phys Rev Lett. 2002;88(21).
[3] Pérez R, Gumbsch P. Directional anisotropy in the cleavage fracture of silicon. Phys Rev Lett. 2000;84(23):5347–5350.
[4] Brandl C, Derlet PM, Van Swygenhoven H. General-stacking-fault energies in highly strained metallic environments: Ab initio calculations. Phys Rev B. 2007;76(5).
[5] Ogata S, Li J, Yip S. Ideal pure shear strength of aluminum and copper. 2002;298(5594):807–811.
[6] Daw MS, Baskes MI. Embedded-Atom Method: Derivation and application to impurities, surfaces, and other defects in metals. Phys Rev B. 1984;29(12):6443–6453.
[7] Finnis MW, Sinclair JE. A simple empirical n-body potential for transition metals. Phil Mag A. 1984;50:45–55.
[8] Stillinger FH, Weber TA. Computer simulation of local order in condensed phases of silicon. Phys Rev B. 1985;31(8):5262–5271.

[9] Tersoff J. New empirical model for the structural properties of silicon. Phys Rev Lett. 1986; 56(6):632–635.
[10] Derlet PM, Gumbsch P, Hoagland R, Li J, McDowell DL, Van Swygenhoven H, et al. Atomistic simulations of dislocations in confined volumes. MRS Bulletin. 2009;34(3):184–189.
[11] Bitzek E, Brandl C, Derlet PM, Van Swygenhoven H. Dislocation cross-slip in nanocrystalline fcc metals. Phys Rev Lett. 2008;100(23).
[12] Brandl C, Bitzek E, Derlet PM, Van Swygenhoven H. Slip transfer through a general high angle grain boundary in nanocrystalline aluminum. Appl Phys Lett. 2007;91(11).
[13] Bitzek E, Derlet PM, Anderson PM, Van Swygenhoven H. The stress-strain response of nanocrystalline metals: A statistical analysis of atomistic simulations. Ann Math. 2008;56(17): 4846–4857.
[14] Bitzek E, Gumbsch P. Mechanisms of dislocation multiplication at crack tips. Ann Math. 2013; 61:1394 – 1403.
[15] Rodney D, Tanguy A, Vandembroucq D. Modeling the mechanics of amorphous solids at different length scale and time scale. 2011;19(8).
[16] Greer JR, De Hosson JTM. Plasticity in small-sized metallic systems: Intrinsic versus extrinsic size effect. 2011;56(6, SI):654–724.
[17] Sedlmayr A, Bitzek E, Gianola DS, Richter G, Moenig R, Kraft O. Existence of two twinning-mediated plastic deformation modes in Au nanowhiskers. 2012;60(9):3985–3993.
[18] Henkelman G, Uberuaga B, Jonsson H. A climbing image nudged elastic band method for finding saddle points and minimum energy paths. J Comp Phys. 2000;113(22):9901–9904.
[19] Voter A, Montalenti F, Germann T. Extending the time scale in atomistic simulation of materials. Ann Rev Mater Res. 2002;32:321–346.
[20] Li J, Sarkar S, Cox WT, Lenosky TJ, Bitzek E, Wang Y. Diffusive molecular dynamics and its application to nanoindentation and sintering. Phys Rev B. 2011;84(5).
[21] McDowell MT, Leach AM, Gall K. Bending and tensile deformation of metallic nanowires. 2008; 16(4).
[22] Yun G, Park HS. Surface stress effects on the bending properties of fcc metal nanowires. Phys Rev B. 2009;79(19).
[23] Gao Y, Fu Y, Sun W, Sun Y, Wang H, Wang F, et al. Investigation on the mechanical behavior of fivefold twinned silver nanowires. 2012;55:322–328.
[24] Kiener D, Motz C, Dehm G. Micro-compression testing: A critical discussion of experimental constraints. 2009;505(1-2):79–87.
[25] Kiener D, Grosinger W, Dehm G. On the importance of sample compliance in uniaxial microtesting. 2009;60(3):148–151.
[26] Shade PA, Wheeler R, Choi YS, Uchic MD, Dimiduk DM, Fraser HL. A combined experimental and simulation study to examine lateral constraint effects on microcompression of single-slip oriented single crystals. 2009;57(15):4580–4587.
[27] Curtin W, Miller R. Atomistic/continuum coupling in computational materials science. 2003; 11(3):R33–R68.
[28] Miller RE, Tadmor EB. Hybrid continuum mechanics and atomistic methods for simulating materials deformation and failure. MRS bulletin. 2007;32(11):920–926.
[29] Miller RE, Tadmor EB. A unified framework and performance benchmark of fourteen multiscale atomistic/continuum coupling methods. 2009;17:053001.
[30] Tadmor EB, Miller RE. Modeling Materials: Continuum, Atomistic and Multiscale Techniques. Cambridge University Press; 2011.
[31] Feyel F. A multilevel finite element method (FE2) to describe the response of highly non-linear structures using generalized continua. Computer Methods in Applied Mechanics and Engineering. 2003;192:3233–3244.

[32] Weinan E, Engquist B, Li X, Ren W, Vanden-Eijnden E. Heterogeneous multiscale methods: a review. Commun Comput Phys. 2007;2(3):367–450.
[33] Kohlhoff S, Gumbsch P, Fischmeister HF. Crack propagation in b.c.c. crystals studied with a combined finite-element and atomistic model. Phil Mag A. 1991;64:851–878.
[34] Tadmor EB, Ortiz M, Phillips R. Quasicontinuum analysis of defects in solids. Phil Mag A. 1996; 73:1529–1563.
[35] Shenoy V, Miller R, Tadmor E, Phillips R, Ortiz M. Quasicontinuum models of interfacial structure and deformation. Phys Rev Lett. 1998;80(4):742–745.
[36] Knap J, Ortiz M. An analysis of the quasicontinuum method. J Mech Phys Sol. 2001;49:1899–1923.
[37] Shilkrot LE, Miller RE, Curtin WA. Multiscale plasticity modeling: Coupled atomistic and discrete dislocation mechanics. J Mech Phys Sol. 2004;52:755–787.
[38] Badia S, Bochev P, Lehoucq R, Parks ML, Fish J, Nuggehally M, et al. A force-based blending model for atomistic-to-continuum coupling. Int J Multiscale Com. 2007;5:387–406.
[39] Luan BQ, Hyun S, Molinari JF, Bernstein N, Robbins MO. Multiscale modeling of two-dimensional contacts. Phys Rev E. 2007;74:046710.
[40] Parks ML, Bochev PB, Lehoucq RB. Connecting atomistic-to-continuum coupling and domain decomposition. Multiscale Model Sim. 2008;7:362–380.
[41] Eidel B, Stukowski A. A variational formulation of the quasicontinuum method based on energy sampling of clusters. J Mech Phys Sol. 2009;57:87–108.
[42] Gracie R, Belytschko T. An adaptive concurrent multiscale method for the dynamic simulation of dislocations. International Journal of Numerical Methods in Engineering. 2011;86:575–597.
[43] Xiao SP, Belytschko T. A bridging domain method for coupling continua with molecular dynamics. Computer Methods in Applied Mechanics and Engineering. 2004;193:1645–1669.
[44] Möller J, Prakash A, Bitzek E. Fe2at; 2012. Available from: http://www.gmp.ww.uni-erlangen.de/FE2AT.php.
[45] Prakash A, Hummel M, Schmauder S, Bitzek E. NanoSCULPT: A methodology for generating complex realistic configurations for atomistic simulations. MethodsX. 2016;3:219–230.
[46] Systemes D. ABAQUS Documentation 6.12-1; 2012.
[47] de Berg M, Cheong O, van Kreveld M, Overmars M. Computational Geometry. 3rd ed. Berlin: Springer; 2008.
[48] Zienkewicz O, Taylor R. The Finite Element Method. London: McGraw–Hill; 1989.
[49] Cook DR, Malkus DS, Plesha ME. Concepts and applications of finite element analysis. 3rd ed. John Wiley & Sons; 1989.
[50] Hughes TJR. The finite element method. Linear static and dynamic finite element analysis. 2nd ed. Dover Publications Inc.; 2000.
[51] Meagher D. Geometric modeling using octree encoding. Comput Vision Graph. 1982;19:129–147.
[52] Krause R, Rank E. A fast algorithm for point-location in a finite element mesh. Computing. 1996;57:49–62.
[53] Tadmor EB, Miller RE, Elliott RS. Continuum Mechanics and Thermodynamics. Cambridge University Press; 2011.
[54] Stadler J, Mikulla R, Trebin HR. Imd: A software package for molecular dynamics studies on parallel computers. Int J Mod Phys C. 1997;8:1131–1140.
[55] Bitzek E, Gähler F, Hahn J, Kohler C, Krdzalic G, Roth J, et al. Recent developments in IMD: Interactions of covalent and metallic systems. In: Krause E, Jäger W, editors, High Performance Computing in Science and Engineering 2000. Heidelberg: Springer; 2001. pp. 37–47.
[56] Li J. Atomeye: an efficient atomistic configuration viewer. Modelling Simul Mater Sci Eng. 2003; 11:173–177.

[57] Wurster S, Motz C, Jenko M, Pippan R. Micrometer-sized specimen preparation based on ion slicing technique. Adv Eng Mater. 2010;12:61–64.

[58] Bei H, Shim S, George E, Miller M, Herbert E, Pharr G. Compressive strengths of molybdenum alloy micro-pillars prepared using a new technique. Scripta Mater. 2007;57:397–400.

[59] Van Vliet KJ, Li J, Zhu T, Yip S, Suresh S. Quantifying the early stages of plasticity through nanoscale experiments and simulations. Phys Rev B. 2003;67:104105.

[60] Ackland GJ, Thetford R. An improved n-body semi-empirical model for body-centred cubic transition metals. Phil Mag A. 1987;56:15–30.

[61] Bitzek E, Koskinen P, Gähler F, Moseler M, Gumbsch P. Structural relaxation made simple. Phys Rev Lett. 2006;97:170201.

[62] Mendelev M, Han S, Srolovitz D, Ackland G, Sun D, Asta M. Development of new interatomic potentials appropriate for crystalline and liquid iron. Philos Mag. 2003;83:3977–3994.

[63] Kelchner CL, Plimpton SJ, Hamilton JC. Dislocation nucleation and defect structure during surface indentation. Phys Rev B. 1998;58:11085–11088.

[64] Allen MP, Tildesley DJ. Computer simulation of liquids. Oxford, UK: Clarendon Press; 1996.

[65] Möller J, Bitzek E. Fracture toughness and bond trapping in grain boundary cracks; unpublished.

[66] Perez R, Gumbsch P. Directional anisotropy in the cleavage fracture of silicon. Phys Rev Lett. 2000;84(23):5347–5350.

[67] Gordon PA, Neeraj T, Luton MJ, Farkas D. Crack-tip deformation mechanisms in α-fe and binary fe alloys: An atomistic study on single crystals. Metall Mater Trans A. 2007;38:2191–2202.

[68] Lawn B. Fracture of Brittle Solids. 2nd ed. Cambridge: University Press; 1993.

[69] Sih GC, Liebowitz H. Fracture, An Advanced Treatise, ch. 2. New York: Academic Press; 1968. pp. 67–190.

[70] Sih GC, Paris PC, Irwin GR. On cracks in rectlinearly anisotropic bodies. Int J Fract. 1965;1(3): 189–203.

[71] Lekhnitskii SG. Theory of Elasticity of an Anisotropic Elastic Body. San Francisco: Holdenday; 1963.

[72] Irwin G. Analysis of stresses and strains near the end of a crack traversing a plate. J Appl Mech. 1957;24:361–364.

[73] Liebermann DS, Zirinsky S. A simplified calculation for the elastic constants of arbitrarily oriented single crystals. Acta Cryst. 1956;9:431–436.

[74] Sadd MH. Elasticity – Theory, Applications, and Numerics. Burlington: Academic Press; 2009.

[75] ASTM International. ASTM Standard E 399-90; 1997.

[76] Rice JR. A path independent integral and the approximate analysis of strain concentration by notches an cracks. J Appl Mech. 1968;35:379–386.

[77] Gotoh H. Some problems of bonded anisotropic plates with cracks along the bond. Int J Fract. 1967;3:253–265.

[78] Suo Z. Singularities, interfaces and cracks in dissimilar ansisotropic media. Proc Roy Soc A. 1990;427:331–358.

[79] Wu KC. Stress intensity factor and energy release rate for interfacial cracks between dissimilar anisotropic materials. J Appl Mech. 1990;57:882–886.

[80] Gao H, Abbudi M, Barnett DM. Interfacial crack-tip field in anisotropic elastic solids. J Mech Phys Solids. 1992;40:393–416.

[81] Hwu C. Explicit solutions for collinear interface crack problems. Int J Solids Struct. 1993;30: 301–312.

[82] Charalambides P, Lund J, Evans A, McMeeking R. A test specimen for determining the fracture resistarim of bimaterial interfaces. J Appl Meteorol. 1989;56(1):77–82.

[83] Suo Z, Hutchinson J. Sandwich test specimens for measuring interface crack toughness. 1989; 107:135–143.

Ž. Božić*, S. Schmauder, M. Mlikota, and M. Hummel

10 Multiscale fatigue crack growth modelling for welded stiffened panels

Abstract: The influence of welding residual stresses in stiffened panels on effective stress intensity factor values and fatigue crack growth rate is studied in this paper. Interpretation of relevant effects on different length scales such as dislocation appearance and microstructural crack nucleation and propagation are taken into account using Molecular Dynamics (MD) simulations as well as a Tanaka-Mura approach for the analysis of the problem. Mode I stress intensity factors (SIF), K_I, were calculated by the finite element method (FEM) using shell elements and the crack tip displacement extrapolation technique. The total SIF value, K_{tot}, is derived by a part due to the applied load K_{appl}, and by a part due to welding residual stresses, K_{res}. Fatigue crack propagation simulations based on power law models showed that high tensile residual stresses in the vicinity of a stiffener significantly increase the crack growth rate, which is in good agreement with experimental results.

Keywords: dislocation, microstructurally small cracks, fatigue crack growth rate, residual stress

10.1 Introduction

Structural health monitoring and damage detection in aircraft, ship, offshore and other structures are highly important for their fitness for service assessment. Under cyclic loading fatigue cracks may initiate at sites of stress concentration and further propagate, which can eventually result in unstable fracture and structural failure. In aircraft, ships and other thin-walled structures stiffened panels are widely used owing to their light weight and high strength and stiffness. Welded stiffened panels are mostly implemented in the deck and side structure of a ship. The crack growth rate in welded stiffened panels can be significantly affected by the residual stresses which are introduced by the welding process. The high heat input from the welding process causes tensile residual stresses in the vicinity of a stiffener. These tensile stresses are equilibrated by compressive stresses in the region between the stiffeners. Residual stresses should be taken into account for a proper fatigue life assessment of welded stiffened panels under cyclic tension loading.

From a physics point of view the fatigue phenomenon involves multiple length scales due to the presence of microcracks or inclusions that are small compared to the large size of structural components. Therefore, it is necessary to consider the fatigue process at all scales. A scale dependent physics-based model is required for accurate simulation and understanding of material behavior in various operational environments to assess fatigue life of a structure [1–5].

The process of fatigue failure of mechanical components may be divided into the following stages: (1) crack nucleation; (2) small crack growth; (3) long crack growth; and (4) occurrence of final failure. In engineering applications, the first two stages are usually termed as the "crack initiation or small crack formation period" while long crack growth is termed as the "crack propagation period".

In pure metals and some alloys without pores or inclusions, irreversible dislocations glide under cyclic loading. This leads to the development of persistent slip bands, extrusions and intrusions in surface grains that are optimally oriented for slip. Dislocation development can be simulated by using the molecular dynamics (MD) simulation code IMD [6]. To analyze dislocation development, atomistic scale simulation methods are implemented, [7–10]. With continued strain cycling, a fatigue crack can be nucleated at an extrusion or intrusion within a persistent slip band [11–18]. Non-metallic inclusions, which are present in commercial materials as a result of the production process, can also act as potential sites for fatigue crack nucleation. In the high cycle regime fatigue cracks initiate from inclusions and defects on the surface of a specimen or component. For very high cycle fatigue, fatigue cracks initiate from defects located under the surface of the specimen [19, 20]. Micro-crack nucleation can be analyzed by using the Tanaka-Mura model or some of its modifications [11, 21–24].

Fatigue crack growth prediction models based on fracture mechanics have been developed to support the damage tolerance concepts in metallic structures [25]. A well-known method for predicting fatigue crack propagation under constant stress range is a power law described by Paris and Erdogan [26]. Dexter et al. [27] and Mahmoud and Dexter [28] analysed the growth of long fatigue cracks in stiffened panels and simulated the crack propagation in box girders with welded stiffeners. They conducted cyclic tension fatigue tests on approximately half-scale welded stiffened panels to study propagation of large cracks as they interact with the stiffeners. Measured welding residual stresses were introduced in the finite element model and crack propagation life was simulated. Sumi et al. [29] studied the fatigue growth of long cracks in stiffened panels of a ship deck structure under cyclic tension loading. For that purpose fatigue tests were carried out on welded stiffened panel specimens damaged with a single crack or an array of collinear cracks.

In order to analyze the total fatigue life of a structural component or a test specimen, from crack initiation through cyclic slip mechanism up to long crack propagation and final failure, a multiscale approach is needed. Fig. 10.1 shows a schematic description of bridging between the three considered scales: Nano, Micro, and Macro. The relevant material's property parameter from atomistic scale needed for the micromechanics modeling is the critical resolved shear stress (CRSS). The CRSS is inferred from MD simulation and it is the input parameter for a modified Tanaka-Mura model. The modified Tanaka-Mura micromechanics model provides information on the number of loading cycles to initiate a small crack and its size. This information is further an in-

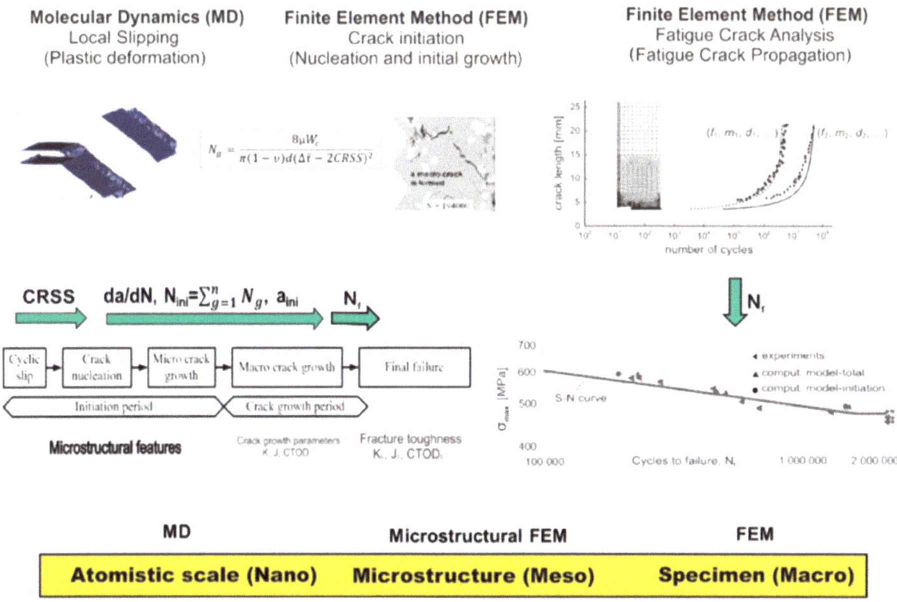

Fig. 10.1: A schematic description of bridging between the three considered scales.

put for the macro scale fatigue crack growth model based on power law equations, by means of which then a total fatigue life up to fracture as a final event can be assessed. Based on the presented procedure the fatigue behavior of a material can be simulated and new materials can be modeled and analyzed. In this way Wöhler curves can be obtained for new materials without experiments.

This paper presents a study of the influence of welding residual stresses in stiffened panels on effective stress intensity factor values and fatigue crack growth rate. A total SIF value, K_{tot}, was obtained by a linear superposition of the SIF values due to the applied load, K_{appl}, and due to weld residual stresses, K_{res}. The effective SIF value, K_{eff}, was considered as a crack growth driving force in a power law model [30, 31]. Mode I SIF values, K_I, were calculated by the FE software package ANSYS using shell elements and the crack tip displacement extrapolation method in an automatic post processing procedure [32]. Simulated fatigue crack propagation life was compared with the experimental results as obtained by Sumi et al. [29]. The molecular dynamics (MD) simulation was implemented to analyze dislocation development in an iron cuboid model with a triangular notch tip. Numerical simulations of the fatigue crack initiation and growth for martensitic steel, based on modified Tanaka-Mura, were carried out.

10.2 Molecular dynamics (MD) simulation of dislocation development in iron

10.2.1 Methods and model

Taking a close look on dislocation development leads to the necessity of atomistic scale simulation methods. Therefore, we used for the present work the molecular dynamics (MD) simulation code IMD [6]. It was developed at the Institute of Theoretical and Applied Physics (ITAP) belonging to the University of Stuttgart. In MD the atoms are seen as mass m points at the position $\vec{r}$ for which the elementary Newton's equations of motion:

$$F(\vec{r}, t) = m\frac{\partial^2 \vec{r}}{\partial^2 t} \tag{10.1}$$

are solved in every time step. The force $F(\vec{r}, t)$ is given by the derivative of the interatomic embedded atom method (EAM) [7] pair potential $U(\vec{r}, t)$ (10.2):

$$F(\vec{r}, t) = -\nabla U(\vec{r}, t)\,. \tag{10.2}$$

The system we investigated contains about half a million iron atoms (see Fig. 10.2). They form a cuboid of the size $286 \times 143 \times 143$ Å^3 where a notch (dimensions 15×90 Å^2) with a triangular notch tip was inserted along the (110) plane.

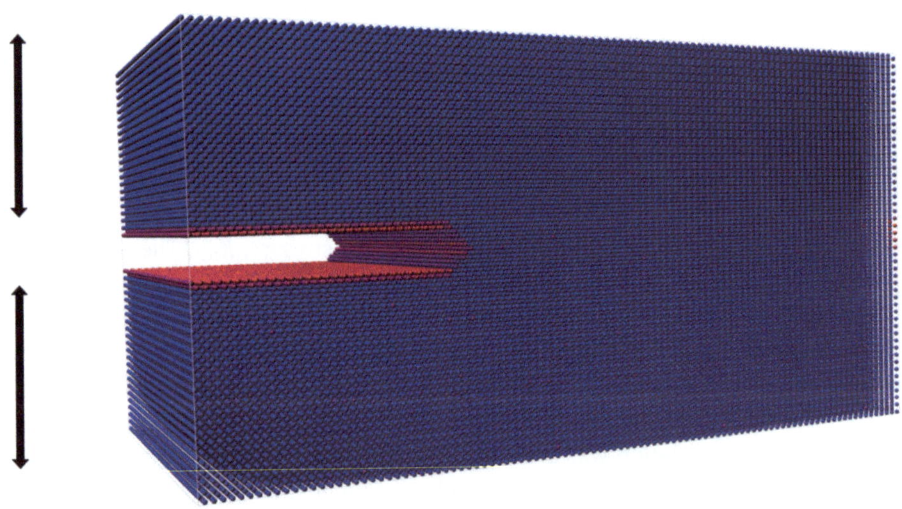

Fig. 10.2: Bcc iron cuboid 286 × 143 × 143 Å^3 with a 15 × 90 Å^2 notch on a (110) plane. The 486 000 atoms are color coded via von Mises stress (red ≙ high stress, blue ≙ low stress). Image by MegaMol™ [8]. Full periodic boundary conditions are applied. The loading direction is marked by the arrows.

Cyclic deformation of the simulation box was applied in the [001]-direction. Therefore, the z-component of the simulation cell was elongated with a constant rate of $5 \cdot 10^{-7}$ at each time step. The value $5 \cdot 10^{-7}$ represents a factor by which z-components of the system are multiplied at each time step, therefore, it is without units. The strain rate could be calculated as $2.5 \cdot 10^{8}\ \mathrm{s}^{-1}$. Such high strain rates are typical for MD simulations, but very high for experimentalists. After reaching a strain of seven percent we applied pressure at the same rate until we reach seven percent of strain in compression. Seven percent tend to be at the upper end of the elastic elongation regime. In order to reduce computation time and still can observe changes in the structure already in the very first cycles one has to apply these high values of maximum strain. Still, this value should be realistic. For that reason the authors used an input strain value of 7 percent. There have been other simulations with lower maximum strain, where the formation of dislocations occurs after a higher number of cycles. The loading and unloading was repeated continuously. The temperature was chosen to be room temperature (300 K). The time step was fixed to 2 fs. Periodic boundary conditions were used in every direction.

10.2.2 Results and discussion

During the continuous cyclic change from elongation to compression different stages of the system appeared (see Fig. 10.3).

- Stage I: Configuration under no pressure.
- Stage II: Initiation of reversible local restructuring under tensile loading.
- Stage III: Formation of one continuous plane with face centered cubic (fcc) structure.
- Stage IV: Compression leads to a resolution of the deformation introduced in the previous steps into the structure and to bending of the middle of the notch surfaces towards each other up to a minimum distance of 6.8 Å.
- Stage V: During the fourth loading cycle dislocations are initiated. The Dislocation Extraction Algorithm (DXA) [9] detects "defect surfaces". "The defect surface consists of those parts of the interface mesh, which have not been swept by elastic Burgers circuits." [10]
- Stage VI: Dislocations still remain in the structure even though no pressure remains in the system.

The MD simulations presented here, illustrate the development of the system from reversible plastic deformation to irreversible plastic deformation including dislocation nucleation, propagation, back-propagation and multiplication. The quite high stress level of the system in the regime of 5.5 GPa during dislocation nucleation is attributed to the fact that the underlying model is assumed to be an internally defect free single crystal, except for the external notch, and to a minor degree to the high strain rate

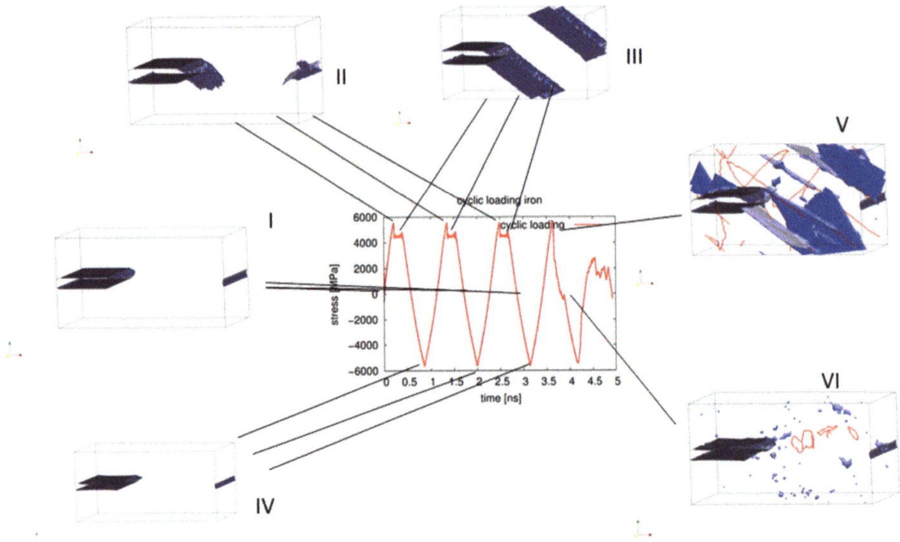

Fig. 10.3: Stress [MPa] in *z*-direction in terms of the time [ns] during cyclic loading of a nanostructure of a notched iron cuboid. System configurations at different times are depicted: blue are according to DXA [9] "defect surfaces", red represent dislocations. The view is from lower left.

which is applied in this simulation. Strain rate effects in MD simulations have been previously discussed in literature, e.g. [33] and will thus not be taken into further account here. The stress to propagate an artificially introduced dislocation in an MD model is frequently obtained via pure shear simulations in literature [34] and at the IMWF institute as demonstrated, e.g. in Kohler et al. [35] and in Molnar et al. [36] where typical values of 84 MPa were obtained for 0 K in close agreement to values of 81 MPa for 10 K obtained by Kumar et al. [37] with the Fe-Fe potential presented by Bonny et al. in [7].

In the here presented simulations the intention was to force dislocation nucleation and to study such naturally formed dislocations as a realistic basis for dislocation movement analyses rather than to use artificially implanted dislocations in the model. In the first two peaks a reversible bcc–fcc Bain transition from α- to γ-Fe takes place (Stage III, Fig. 10.3). The oscillation of the stress-time curve (Stage III, Fig. 10.3) is explained by the formation of stacking faults inside the transition fcc phase. In contrast to Farkas et al. [38], where it was believed that the emission of Shockley partial dislocations is relevant for the stacking fault formation, no dislocations are observed in our simulation at this early stage.

Dislocation nucleation takes place as the irreversible deformation begins. The small spikes in the stress-time curve are identified as single dislocation movement. The height of the spike, which is related to the CRSS to move a dislocation, is calculated to a value of 293 MPa, taking into account that this value is higher than the CRSS due to the angle of the dislocations glide plane with respect to the loading axis and the, thus, involved Schmid factor which amounts to 0.40 in the present case: The CRSS value obtained from the present simulation is calculated according to Schmid's law to be $\tau = 293\,\text{MPa}\cos 26.5° \cos 63.5° = 117\,\text{MPa}$ which is in very good agreement to pure shear simulations mentioned above [33, 36] and to an even better degree with an experimental value of 108 MPa which has been previously used for microscale modelling of fatigue by Jezernik et al. [24] for the present material.

10.3 Microstructural crack nucleation and propagation

To solve problems of crack nucleation the Tanaka-Mura model [11, 21–24] is frequently used. In the two articles [21, 22] Tanaka and Mura proposed dislocation models for treating fatigue crack nucleation at slip bands. Tanaka and Mura envisioned that fatigue crack nucleation occurs by the accumulation of dislocation dipoles in a single grain during strain cycling. In the theory of fatigue crack nucleation in slip bands, the forward and reverse plastic flows within slip bands are caused by dislocations with different signs moving on two closely located layers. It is assumed that their movements are irreversible. Based on the Tanaka-Mura model, the monotonic build-up of dislocation dipoles is systematically derived from the theory of continuously distributed dislocations. The number of stress cycles up to the nucleation of a crack about one grain diameter in length is reached when the self-strain energy of the accumulated dislocation dipoles reaches a critical value. The number of stress cycles N_g required for crack nucleation in a single grain can be determined as follows:

$$N_g = \frac{8GW_c}{\pi(1-\nu)d(\Delta\bar{\tau} - 2\text{CRSS})^2} \tag{10.3}$$

Eq. (10.3) presumes that cracks form along slip bands within grains, depending on slip band length d and average shear stress range $\Delta\bar{\tau}$ on the slip band. Slip band length represents distance along the slip band between grain boundaries in a single grain. Other material constants (shear modulus G, specific fracture energy per unit area W_c, Poisson's ratio ν and frictional stress of dislocations on the slip plane i.e. critical resolved shear stress (CRSS) can be found in the specialized literature [23] or calculated by means of MD (CRSS). According to Nakai [39] the initiation conditions of small fatigue cracks still have not been clarified enough, since no method for successive, direct and quantitative observation of the process had been devised. In the presented model cracks nucleate sequentially, but on the segmental level. In each simulating iteration just one segment of a particular grain is broken. It means that in one iteration a segment belonging to one grain brakes, while in the following iteration a segment of some other grain can break. In the presented case, the number of segments in each individual grain is equal to four. The total number of stress cycles N_{ini} needed for the initiation of a small crack is calculated by summing the cycles spent to nucleate all cracks, including those coalesced to form the final small crack.

Jezernik et al. [24] used the Tanaka-Mura model to numerically simulate the small crack formation process. Three improvements were added to this model: (a) multiple slip bands inside each crystal grain as potential sites for crack nucleation, (b) crack coalescence between two grains, and (c) segmented crack generation inside one grain. A numerical model was directed at simulating fatigue properties of thermally cut steel. The authors took into account accompanying residual stresses in order to simulate the properties of the thermally cut edge as faithfully as possible. As residual stresses are not taken into account in the original Tanaka-Mura model according to Eq. (10.3), in the present study they are imposed as additional external loading. The superpositioning principle in linear elastic micromechanics analysis has been applied by taking residual stresses as part of the total load into account which leads to the shear stress distribution shown in Fig. 10.4. Therefore, the residual stresses are implicitly evaluated in the Tanaka-Mura equation through the average shear stress range on the slip band $\Delta\bar{\tau}$.

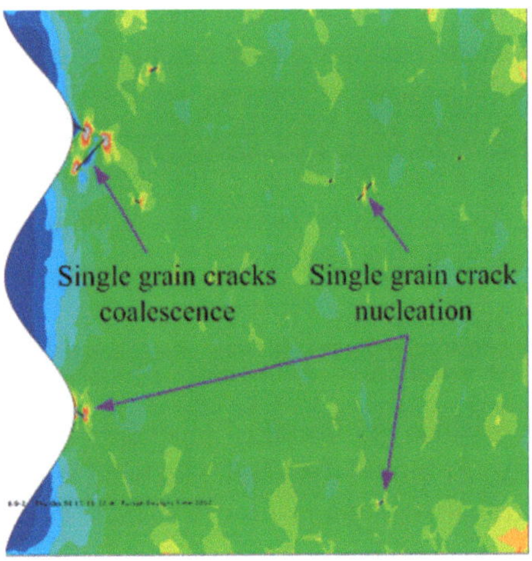

Fig. 10.4: Micro-crack nucleation and subsequent coalescence [24].

Fig. 10.4 shows the shear stress distribution and nucleated cracks for a typical high cycle fatigue regime load level (450 MPa). In the beginning, cracks tended to occur scattered in the model and form in larger grains that are favorably oriented and show higher shear stresses. But after a while, existing single grain cracks started coalescing, causing local stress concentrations and amplifying the likelihood of new cracks forming near already coalesced cracks. When calculating cycles required for crack initiation, no cycles were attributed to crack coalescences (it is simulated as being instantaneous). The total number of cycles of crack initiation equals the sum of cycles needed for each micro-crack to nucleate.

When the crack depth is less than a critical value, the crack growth behavior has been found to be highly dependent upon the microstructure, [40–42]. With increasing length, the growing cracks leave the originally 45°-oriented slip planes and tend to propagate perpendicular to the external stress axis. The change of the crack plane from the active slip plane to a non-crystallographic plane perpendicular to the stress axis is called the transition from Stage I (crystallographic propagation) to Stage II (non-crystallographic propagation) or transition from crack initiation to crack propagation. In Stage II of fatigue-crack propagation, only one crack usually propagates while most of other cracks usually stop within Stage I, [15]. Stage II of fatigue-crack propagation is simulated by using the power law crack propagation models based on the linear elastic fracture mechanics (LEFM).

10.4 Modeling and simulation of crack propagation in welded stiffened panels

It is well-known that the residual stress in a welded stiffened panel is tensile along a welded stiffener and compressive in between the stiffeners. Residual stresses may significantly influence the stress intensity factor (SIF) values and fatigue crack growth rate. A total SIF value, K_{tot}, is contributed by the part due to the applied load, K_{appl}, and by the part due to weld residual stresses, K_{res}, as given by Eq. (10.4):

$$K_{\text{tot}} = K_{\text{appl}} + K_{\text{res}} \,. \tag{10.4}$$

The so-called residual stress intensity factor, K_{res}, is required in the prediction of fatigue crack growth rates. The considered analysis method is based on the superposition rule of LEFM. The finite element method has been widely employed for calculating SIFs. In the FE software package ANSYS [32] the command INISTATE is used for defining the initial stress conditions. For evaluating K_{res}, it is important to input correct initial stress conditions to numerical models in order to characterize residual stresses [28, 43].

The effective SIF range ΔK_{eff} was considered in crack growth models in order to take crack closure effects on fatigue crack growth rate into account. The Elber [30] and the Donahue [31] crack growth models are employed to simulate fatigue lifetime for welded stiffened panel specimens. In the Donahue model the effective SIF range values are calculated based on the applied load, without taking welding residual stresses into account. The Elber model takes into account both the applied load and welding residual stresses, and the effective SIF range values are calculated based on the effective SIF ratio R_{eff}, which depends on the $K_{\text{tot,max}}$ and $K_{\text{tot,min}}$ values. It is important to determine the total SIF values K_{tot} accurately, to model effects of residual stresses on crack propagation rate.

10.4.1 Specimen's geometry and loading conditions

Fatigue tests with constant stress range and frequency were carried out on a stiffened panel specimen with a central crack [29]. The specimen geometry is shown in Fig. 10.5. The material properties and chemical composition of the used mild steel for welding are given in Tab. 10.1 [44].

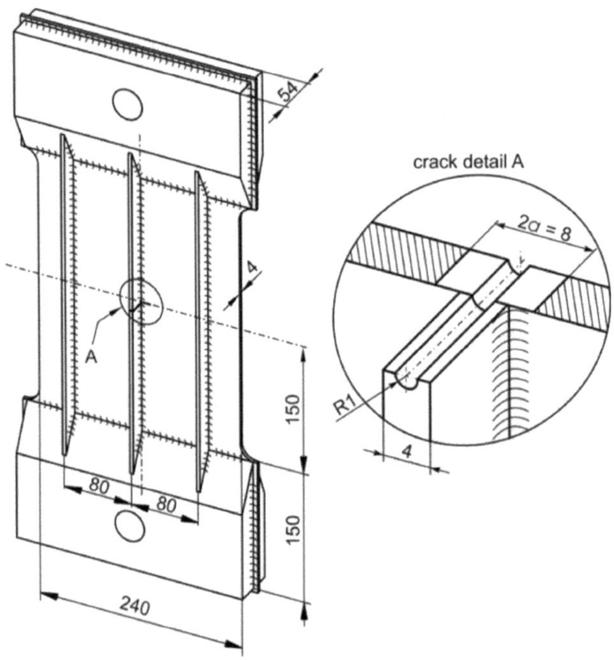

Fig. 10.5: Stiffened panel specimen.

Tab. 10.1: Material properties.

Mechanical properties		
E	Young's modulus	206 000 MPa
v	Poisson's coefficient	0.3
σ_0	Yield strength	235 MPa
Chemical composition (%)		
C_{max}		0.18
Si_{max}		0.35
Mn_{min}		0.70
P_{max}		0.035
S_{max}		0.035

Tab. 10.2 shows the fatigue test conditions applied in the experiment. The cross sectional area of the intact section, and the average stress range away from the notch, are denoted as, A_0 and $\Delta\sigma$, respectively. The force range, and the stress ratio are denoted by $\Delta F = F_{\max} - F_{\min}$, and $R = F_{\min}/F_{\max}$, respectively. The applied stress range was $\Delta\sigma = 80$ MPa. The initial notch length was $2a = 8$ mm and the loading frequency was 3 Hz.

Tab. 10.2: Fatigue test conditions.

A_0 [mm^2]	ΔF [N]	$\Delta\sigma$ [MPa]	R
1200	96 000	80	0.0204

In the experiment the crack lengths were measured by using so-called crack gauges. The crack gauges are bonded to a specimen's surface in front of a crack tip in order to measure length of a growing crack with respect to applied number of loading cycles. Different from usual strain gauges, the grid of crack gauges is cut along with crack development, resulting in resistance change.

10.4.2 Modeling of welding residual stresses in a stiffened panel by using FEM

The residual stress distribution implemented in present simulations follows Faulkner's model [45] in which the tensile regions around the stiffeners are modeled as rectangular shapes with a base width proportional to the plate thickness. For ship structures the rectangular width typically ranges from 3.5 to 4 times the plate thickness. Dexter et al. [27] performed fatigue tests on half scale welded stiffened panel specimens which model a part of ship deck structure. The authors measured welding residual stresses in stiffened panel specimens and observed that the measured stresses vary mostly between a rectangular and a triangular shape [27, 28]. Subsequent crack growth simulations showed that the residual stress distribution can be well described by a rectangular shape, where the maximum tensile stress equals to yield stress of the considered material and compressive stresses between the stiffeners provide the equilibrium of internal forces [28].

In this study the distribution of welding residual stresses in the stiffened panel specimen is taken into account in a similar manner as in the model developed by Mahmoud and Dexter [28]. Residual stress distribution depicted in Fig. 10.6 was utilized for the considered specimen. This model presents the tensile regions around the stiffeners as rectangular shapes with a base width equal to 10 mm and with a stress level equal to yield strength $\sigma_0 = 235$ MPa. Estimated base width is in good agreement with Dexter's model [28], where approximately $\frac{1}{4}$ of the span between the two stiffeners is exposed to tension. Compared with Faulkner's model the base width used in this study is slightly shorter. The compressive residual stresses between the stiffeners and on the stiffeners were applied in the model to satisfy equilibrium.

To evaluate the SIF value contributed by the residual stresses, K_{res}, it is important to input correct initial stress conditions in the numerical model. Experimental measurements [27, 28] showed that the profiles of welding residual stresses are almost identical along the axis parallel to the weld line. Correspondingly, in the FE model elements with the same horizontal coordinate should have the same initial stress condition. For that purpose regular 1 mm size FE mesh was used, so that elements can be selected in columns and associated with the corresponding initial stress level as given in Fig. 10.6. Applying initial stresses for the used 8 node shell elements is based on the element integration point. These initial stresses are equilibrated in the first analysis step. Due to the symmetry of specimen's geometry and loading conditions it was sufficient to model only one quarter of the specimen. In ANSYS software package the command INISTATE was implemented to define assumed initial stress conditions [32]. Fig. 10.7 shows the σ_y component of the welding residual stresses in the stiffened panel specimen obtained for the implemented stress distribution as given in Fig. 10.6 (the σ_y stress component acts parallel to the loading axis).

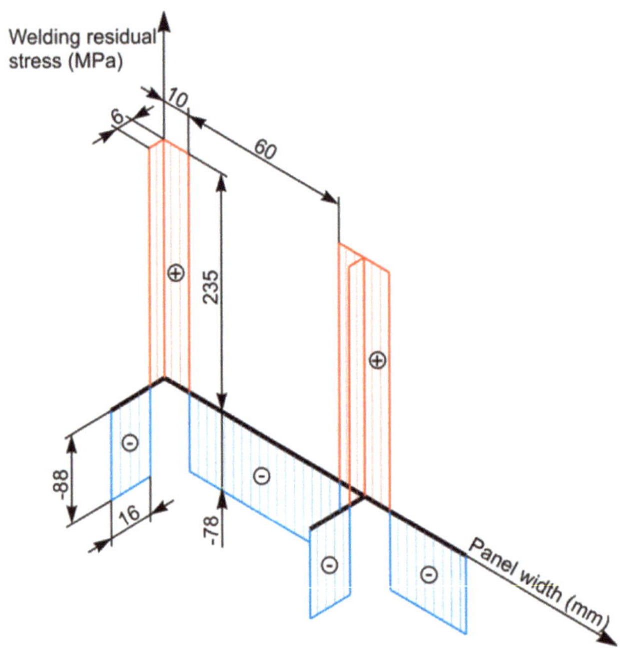

Fig. 10.6: Welding residual stress distribution.

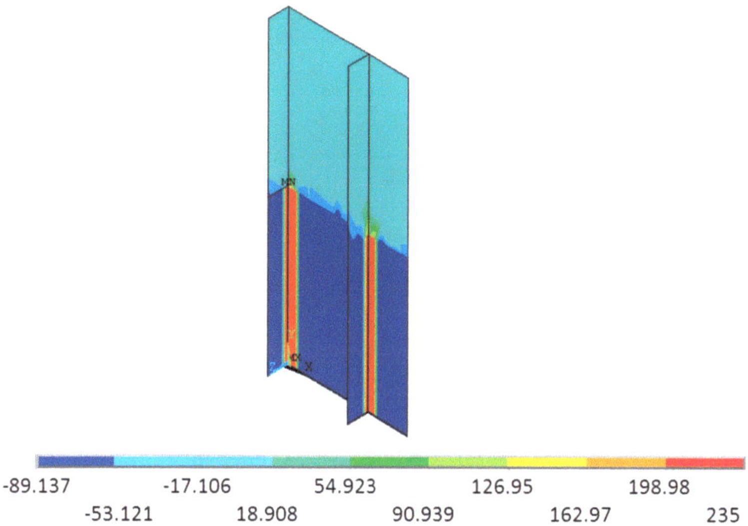

Fig. 10.7: Welding residual stresses [MPa] in the stiffened panel specimen.

10.4.3 Stress intensity factors and fatigue crack growth rate

For evaluating SIFs by FEM, the crack tip displacements extrapolation method was implemented [32]. The SIFs were determined in a linear elastic FE analysis. In the FEM modelling the crack tip region was meshed by singular elements. The procedure for the calculation of SIFs is based on the application of well-known "quarter-point" elements introduced by Barsoum [46] and by Henshell and Shaw [47].

The near crack tip displacements and stresses of LEFM are usually related to the three fundamental deformation modes of fracture where mode I is the opening mode, mode II is the shearing mode, and mode III is the tearing mode [25, 32]. Opening displacement v in the vicinity of the crack tip, as depicted in Fig. 10.8, is given by Eq. (10.5):

$$v = \frac{K_{\mathrm{I}}}{4G}\sqrt{\frac{r}{2\pi}}\left[(2\kappa+1)\sin\frac{\theta}{2} - \sin\frac{3\theta}{2}\right] + \frac{K_{\mathrm{II}}}{4G}\sqrt{\frac{r}{2\pi}}\left[(2\kappa-3)\cos\frac{\theta}{2} + \cos\frac{3\theta}{2}\right], \quad (10.5)$$

where K_{I} and K_{II} are the stress intensity factors associated with modes I and II. G is the shear modulus; v is Poissons's ratio, v is displacement in loading direction; r and Θ are local polar coordinates. The conversion factor κ for plain stress conditions is given by Eq. (10.6).

$$\kappa = (3 - v)/(1 + v) \quad (10.6)$$

In Eq. (10.5) the higher order terms are neglected and the equation is therefore only valid in the vicinity of the crack tip. It should be noted that stress distribution is singular for $r = 0$. When Eq. (10.5) is applied to the half crack configuration illustrated in

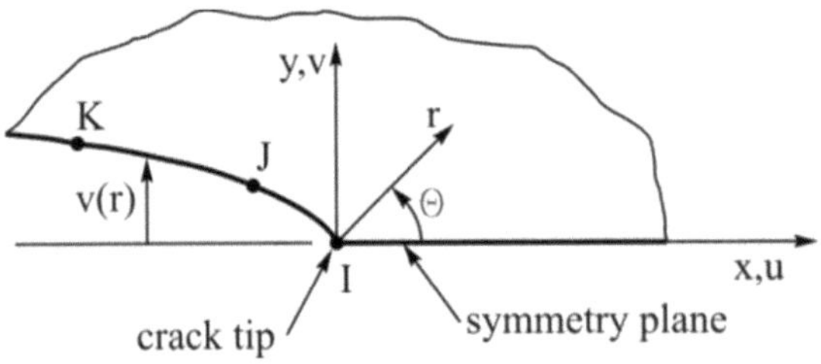

Fig. 10.8: Crack opening profile for a half crack model.

Fig. 10.8, the displacements v, across the faces of the crack are given by Eq. (10.7).

$$v = \frac{K_I}{2G}\sqrt{\frac{r}{2\pi}}(1+\kappa) \tag{10.7}$$

In the next step is described how K is obtained by using finite element results and the theoretical equations given above. Attention is restricted to calculating the K_I factor for mode I displacements. The displacement v on the crack face of the half crack configuration depicted in Fig. 10.8, can be approximated by

$$v/\sqrt{r} = A + Br\,, \tag{10.8}$$

where A and B are constants determined from a linear curve fit of nodal displacements. Once A and B are determined, the limit $r \to 0$ is taken.

$$\lim_{r\to 0}\,(v/\sqrt{r}) = A \tag{10.9}$$

By combining Eqs. (10.5) and (10.9) for the half crack model, SIF is obtained as

$$K_I = \frac{2G\sqrt{2\pi}A}{(1+\kappa)}\,. \tag{10.10}$$

Based on the above described technique, in the general post processing procedure, the KCALC command was used to calculate SIFs. The Mode I SIF values, K_I, are determined for a stiffened panel specimen for a loading stress range $\Delta\sigma = 80$ MPa, assuming the presence of residual stresses as described above. SIF values with respect to half crack length a are given in Fig. 10.9. K_{appl} represents the SIF values due to the applied stress range only, without residual stresses. K_{tot} represents the SIF values for the case when the residual stresses are taken into account along with the external loading stress range. It can be seen that residual stresses significantly increase K_{tot} values for shorter crack lengths, where tensile residual stresses prevail. Between the stiffeners residual stresses reduce the K_{tot} values.

A well-known method for predicting fatigue crack propagation under constant stress range is the power law (10.11) introduced by Paris and Erdogan [26]. In Eq. (10.11) da/dN and ΔK represent the crack growth rate and the stress intensity factor range, respectively. C and m are material constants which are determined experimentally.

$$\frac{da}{dN} = C(\Delta K)^m \tag{10.11}$$

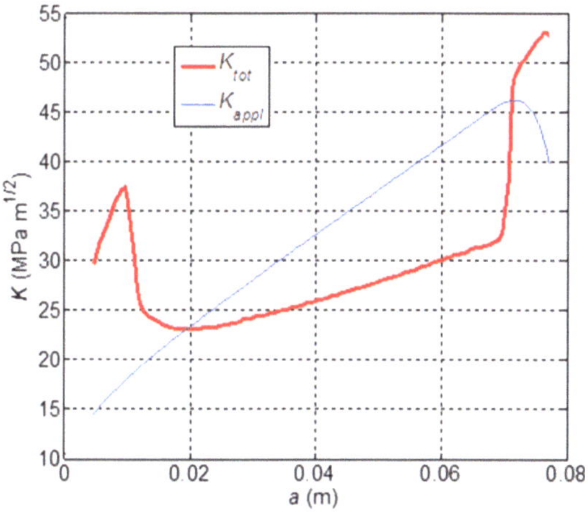

Fig. 10.9: K_{tot} and K_{appl} values.

Elber [30] and Donahue et al. [31] further developed Paris' law assuming an effective stress intensity factor range, ΔK_{eff}, as the crack growth driving force parameter. Donahue et al. [31] defined the effective stress intensity factor range as $\Delta K_{\text{eff}} = K_{\max} - K_{\text{th}}$, where $K_{\max}$ is the maximum SIF value in a loading cycle and K_{th} is a SIF threshold value below which no crack propagation occurs. Božić et al. [48, 49] used this model to analyse fatigue crack propagation in plates damaged with a single and multiple cracks, respectively. Assuming the stress ratio $R = 0$ as applied in the experiment, the threshold SIF value for the used mild steel was taken as $K_{\text{th}} = 6.8$ MPa, [50]. The Donahue model predicted well fatigue lifetime and crack growth rate for centrally cracked un-stiffened plate specimens without welds and with a constant stress ratio $R = 0$, [48, 49]. However, this model does not take into account variable stress ratio R, which occurs in welded specimens due to residual stresses and can significantly influence the fatigue crack growth rate.

Elber [30] observed that crack closure decreases the fatigue crack growth rate by reducing the effective SIF range. He proposed an equation for the effective stress intensity factor range which takes the load ratio R into account:

$$\Delta K_{\text{eff}} = (0.5 + 0.4R)\Delta K_{\text{appl}} \,. \tag{10.12}$$

In the present study the nominal ratio R in Eq. (10.12) was replaced by the effective SIF ratio R_{eff}, in order to take into account the influence of welding residual stresses on fatigue crack propagation rate. The effective SIF ratio R_{eff} is defined as:

$$R_{\text{eff}} = \frac{K_{\text{tot,min}}}{K_{\text{tot,max}}} = \frac{K_{\text{appl,min}} + K_{\text{res}}}{K_{\text{appl,max}} + K_{\text{res}}} \tag{10.13}$$

This superposition method based on the principle of LEFM was originally proposed by Glinka [51], and was recently implemented in the finite element modeling of fatigue crack growth rate in welded butt joints by Servetti and Zhang [52]. As the crack propagates under cyclic loading, the effective SIF ratio R_{eff} changes due to the presence of residual stresses.

The number of constant amplitude loading cycles due to which a crack propagates from its initial crack length, a_0, to a final crack length, a_{fin}, can be determined by the integration of Eq. (10.11), which becomes:

$$N = \int_{a_0}^{a_{\text{fin}}} \frac{da}{C[\Delta K_{\text{eff}}]^m} . \tag{10.14}$$

The integration of Eq. (10.14) can be performed numerically. Fatigue life was simulated for the specimen by integrating Eq. (10.14), where the material constants C and m were as given in Tab. 10.3. The exponent m for the considered material was determined in a previous study by means of crack growth rate diagrams based on a-N data obtained for centrally cracked plate specimens [45]. The C constants from Tab. 10.3 were estimated in a way to provide a good agreement of simulated fatigue lifetime with experimentally obtained a-N curve.

Tab. 10.3: Material's constants*.

Model	C	m
Donahue	$6.50 \cdot 10^{-11}$	2.75
Elber	$1.67 \cdot 10^{-10}$	2.75

* The units for ΔK and $\Delta a/\Delta N$ are [MPa · m$^{1/2}$] and [m], respectively.

The Donahue model was implemented to the welded stiffened panel specimen, considering the effective SIF range values due to applied load only, $\Delta K_{\text{eff}} = K_{\text{appl}} - K_{\text{th}}$, without taking into account residual stresses. The Elber model takes into account the influence of residual stresses on fatigue crack growth rate by using the effective SIF range defined by Eq. (10.12), and the effective SIF ratio R_{eff} given by Eq. (10.13). For the two cases considered, the fatigue crack propagation life was obtained as shown in Fig. 10.10 (a) and (b). The presented measured crack lengths a, are to be considered as averaged half crack lengths. These values are obtained by averaging measured crack lengths of the two propagating crack tips with respect to applied number of cycles N. The averaged half crack lengths are used in order to be comparable with the simulation results for semi crack lengths obtained by the FE model where only one quarter of the specimen was modeled.

The FE analysis for the stiffened panel specimens showed that high tensile residual stresses in the vicinity of a stiffener significantly increase K_{res} and K_{tot}, as shown in Fig. 10.9. Correspondingly, the simulated crack growth rate was higher in this region, which is in good agreement with experimental results, as can be seen in Fig. 10.10 (b). Compressive weld residual stresses decreased the total SIF value K_{tot}. The Donahue model, which does not take account of welding residual stresses, could not simulate high crack growth rates in the vicinity of the stiffener, as can be seen in Fig. 10.10 (a). Fatigue crack growth simulation based on the Elber model, which takes into account the welding residual stresses, provides thus better agreement with experimental results in terms of crack growth rate and total number of cycles. In conclusion, residual stresses in welded stiffened panels should be taken into account for a proper evaluation of SIFs and fatigue crack growth rates.

In further work one should do microstructural analyses in experiment and simulation (micro and nano) in order to foster the multiscale procedure and use the method to predict materials with improved fatigue properties in the future.

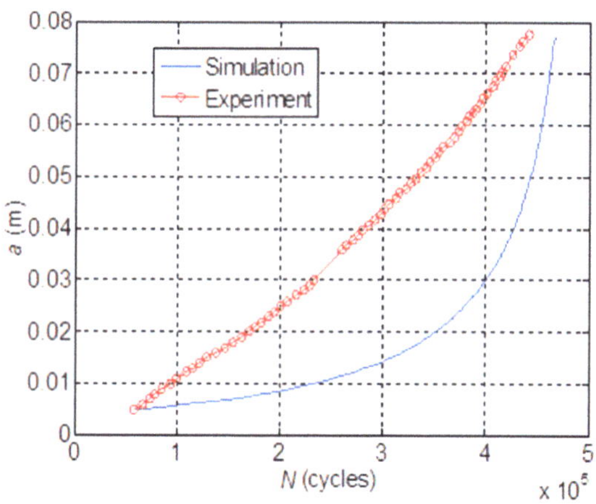

(a)

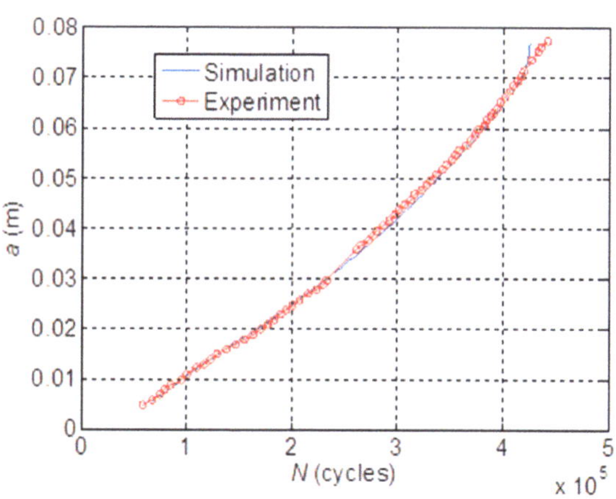

(b)

Fig. 10.10: Fatigue crack growth life for the applied stress range $\Delta\sigma_0 = 80$ MPa: (a) without residual stresses, (b) including residual stresses.

10.5 Conclusions

Simulation of cyclic loading to model dislocation nucleation as an initial step in fatigue initiation is possible in MD. Already after the very few cycles, essential changes in the system behavior were observed under respective loading conditions. Contrary to the first cycles, where reversible changes were dominant, not dissolving restructuring occurs in the sense of dislocations and remaining lattice defects or in other words, plasticity.

Numerical simulations of the fatigue crack initiation and growth of martensitic steel, based on a modified Tanaka-Mura model, was presented. A simulation model related to the micro-crack nucleation along slip bands was presented. Results obtained by using the proposed simulation model were compared to high cyclic fatigue tests and showed reasonably good agreement.

Crack propagation simulation based on numerical integration of a power law equation, taking account of welding residual stresses, was implemented to welded stiffened panel specimens. The FE analysis of the stiffened panel specimens showed that high tensile residual stresses in the vicinity of a stiffener significantly increase K_{res} and K_{tot}. The simulated crack growth rate was higher in this region, which is in good agreement with experimental results. Compressive welding residual stresses decreased the total SIF value K_{tot}, and the crack growth rate between the two stiffeners. Residual stresses should thus be taken into account for a proper evaluation of SIFs and fatigue crack growth rates in welded stiffened panels.

Acknowledgment: This work was supported by the Deutsche Forschungsgemeinschaft (DFG) under Grant No. Schm 746/132-1 and as part of the Collaborative Research Centre SFB 716 at the University of Stuttgart, and by the Croatian Science Foundation Grant No. 120-0362321-2198. The support is gratefully acknowledged.

Nomenclature

a	half crack length
a_0	initial crack length
a_{fin}	final crack length
C	material constant of the Paris equation
CRSS	critical resolved shear stress
d	slip band length
da/dN	crack growth rate
E	Young's modulus
$F(\vec{r}, t)$	interatomic force
F_{max}	maximum applied force
F_{min}	minimum applied force
G	shear modulus
K	stress intensity factor (SIF)

K_{appl}	stress intensity factor due to the applied load
K_{res}	stress intensity factor due to weld residual stresses
K_{th}	stress intensity factor threshold
K_{tot}	total stress intensity factor
m	atomic mass
m	material constant of the Paris equation
N	number of stress cycles for the fatigue crack propagation
N_f	number of stress cycles for fatigue failure
N_g	number of stress cycles required for crack nucleation in a single grain
N_{ini}	number of stress cycles needed for the initiation of a smal crack
R	stress ratio
R_{eff}	effective stress intensity factor ratio
$U(\vec{r}, t)$	interatomic embedded atom method (EAM) pair potential
W_c	specific fracture energy per unit area
ΔF	applied force range
ΔK	stress intensity factor range
ΔK_{eff}	effective stress intensity factor range
$\Delta\sigma$	average applied stress range
$\Delta\bar{\tau}$	average shear stress range on the slip band
σ_0	Yield stress

References

[1] Luo C and Chattopadhyay A (2011). Prediction of fatigue crack initial stage based on a multiscale damage criterion, *International Journal of Fatigue*, **33**, 403–413.

[2] Curtin W A, Deshpande V S, Needleman A, Van der Giessen E and Wallin M (2010). Hybrid discrete dislocation models for fatigue crack growth, *International Journal of Fatigue*, **32**, 1511–1520.

[3] White P (2012). Molecular dynamic modelling of fatigue crack growth in aluminium using LEFM boundary conditions, *International Journal of Fatigue*, **44**, 141–150.

[4] Horstemeyer M F, Farkas D, Kim S, Tang T and Potirniche G (2010). Nanostructurally small cracks (NSC): A review on atomistic modeling of fatigue, *International Journal of Fatigue*, **32**, 1473–1502.

[5] Božić Ž, Schmauder S, Mlikota M and Hummel M (2013). Fatigue Crack Growth Modelling in Welded Stiffened Panels under Cyclic Tension, *13th International Conference on Fracture*, Beijing, China.

[6] Stadler J, Mikulla R and Trebin H R (1997). IMD: A software package for molecular dynamics studies on parallel computers, *Int. J. Mod. Phys.*, **8**, 1131.

[7] Bonny G, Pasianot R C, Castin N and Malerba L (2009). Ternary Fe-Cu-Ni many-body potential to model reactor pressure vessel steels: First validation by simulated thermal annealing, *Phil. Mag.*, **89**, 3531–3546.

[8] Grottel S, Reina G, Dachsbacher C and Ertl T (2010). Coherent Culling and Shading for Large Molecular Dynamics Visualization, *Computer Graphics Forum Proc. of EUROVIS 2010*, **29**(3), 953–962.

[9] Stukowski A, Bulatov V V and Arsenlis A (2012). Automated identification and indexing of dislocations in crystal interfaces, *Modelling Simul. Mater. Sci. Eng.*, **20**, 085007.

[10] Stukowski A (2010). DXA user manual Version 1.3.4 http://dxa.ovito.org/README.txt

[11] Glodez S, Jezernik N, Kramberger J and Lassen T (2010). Numerical modelling of fatigue crack initiation of martensitic steel, *Advances in Engineering Software*, **41**(5), 823–829.
[12] Wood W A (1956). *Fatigue in aircraft structures*, New York: Academic Press.
[13] Fine M E and Ritchie R O (1978). Fatigue-crack initiation and near-threshold crack growth, *Fatigue and microstructure. Metals Park (OH): ASM* ed M Meshii, 245–278.
[14] Laird C (1978). Mechanisms and theories of fatigue, *Fatigue and microstructure. Metals Park (OH): ASM,* ed M Meshii, 149–203.
[15] Klesnil M and Lukas P (1980). *Fatigue of metallic materials*, New York: Elsevier, 57–80.
[16] Mughrabi H (1988). *Rev. Phys. Appl.*, **23**, 367–379.
[17] Mughrabi H (2002). *Fatigue: David L. Davidson symposium, Warrendale (PA): TMS* ed K S Chan, P K Liaw, R S Bellows, T Zogas and W O Soboyejo, 3–15.
[18] Davidson D L and Chan KS (1989). Crystallography of fatigue crack initiation in Astrology at ambient temperature, *Acta Metall.*, **37**(4), 1089–1097.
[19] Wang Q Y, Bathias C, Kawagoishi N and Chen Q (2002). Effect of inclusion on subsurface crack initiation and gigacycle fatigue strength, *International Journal of Fatigue*, **24**(12), 1269–1274.
[20] Murakami Y, Nomoto T and Ueda T (2000). On the mechanism of fatigue failure in the super-long life regime ($N > 10^7$ cycles). Part 1: influence of hydrogen trapped by inclusions, *Fatigue Fract. Engng. Mater. Struct.*, **23**(11), 893–902.
[21] Tanaka K and Mura T (1981). A dislocation model for fatigue crack initiation, *J. Appl. Mech.*, **48**, 97–103.
[22] Tanaka K and Mura T (1982). A theory of fatigue crack initiation at inclusions', *Metallurgical Transactions A*, **13**(1), 117–123.
[23] Brückner-Foit A and Huang X (2006). Numerical simulation of micro-crack initiation of martensitic steel under fatigue loading, *International Journal of Fatigue*, **28**(9), 963–971.
[24] Jezernik N, Kramberger J, Lassen T and Glodez S (2010). Numerical modelling of fatigue crack initiation and growth of martensitic steels, *Fatigue & Fracture of Engineering Materials & Structures*, **33**, 714–723.
[25] Broek D (1989). *The Practical Use of Fracture Mechanics*, Kluwer Academic Publishers, Dordrecht, The Netherlands.
[26] Paris P and Erdogan F (1963). A critical analysis of crack propagation laws, *Journal of Basic Engineering*, **85**, 528–534.
[27] Dexter R J, Pilarski P J and Mahmoud H N (2003). Analysis of crack propagation in welded stiffened panels, *International Journal of Fatigue*, **25**, 1169–1174.
[28] Mahmoud H N and Dexter R J (2005). Propagation rate of large cracks in stiffened panels under tension loading, *Marine Structures*, **18**, 265–288.
[29] Sumi Y, Božić Ž, Iyama H and Kawamura Y (1996). Multiple Fatigue Cracks Propagating in a Stiffened Panel, *Journal of the Society of Naval Architects of Japan*, **179**, 407–412.
[30] Elber W (1971). The significance of fatigue crack closure, *Damage tolerance in aircraft structures. ASTM STP 486. American Society for Testing & Materials*, 230–242.
[31] Donahue R J, Clark H M, Atanmo P, Kumble R and McEvily A J (1972). Crack Opening Displacement and the Rate of Fatigue Crack Growth, *International Journal of Fracture Mechanics*, **8**, 209–219.
[32] Swanson Analysis System (2009). Inc. ANSYS User's Manual Revision 11.0.
[33] Han T, Luo Y and Wang C (2014). Effects of temperature and strain rate on the mechanical properties of hexagonal boron nitride nanosheets, *J. Phys. D: Appl. Phys.*, **47**, 025303.
[34] Tapasa K, Bacon D J and Osetsky Y N (2005). Simulation of dislocation glide in dilute Fe-Cu alloys, *Materials Science & Engineering A.*, **400-401**, 109–113.

[35] Kohler C, Kizler P, Schmauder S (2005). Atomistic simulation of precipitation hardening in α-iron: influence of precipitate shape and chemical composition, *Modelling and Simulation in Materials Science and Engineering*, **13**, 35-45.

[36] Molnar D, et al. (2014). *Unpublished research.*

[37] Naveen Kumar N, Durgaprasad P V, Dutta B K and Dey G K (2012). Modeling of radiation hardening in ferritic/martensitic steel using multi-scale approach. *Computational Materials Science*, **53**, 258-267.

[38] Latapie A and Farkas D (2003). Molecular dynamics simulations of stress-induced phase transformations and grain nucleation at crack tips in Fe, *Modelling Simul. Mater. Sci. Eng.*, **11**, 745–753.

[39] Nakai Y (2001). Evaluation of Fatigue Damage and Fatigue Crack Initiation Process by Means of Atomic-Force Microscopy, *Materials Science Research International*, **7**(2), 1–9.

[40] Zabett A and Plumtree A (1995). Microstructural Effects on the Small Fatigue Crack Behaviour of an Aluminum Alloy Plate, *Fatigue & Fracture of Engineering Materials & Structures*, **18**(7-8), 801-809.

[41] Taylor D and Knott J F (1981). Fatigue Crack Propagation Behaviour of Short Cracks; The Effect of Microstructure, *Fatigue & Fracture of Engineering Materials & Structures*, **4**(2), 147–155.

[42] Miller K J (1987). The Behaviour of Short Fatigue Cracks and Their Initiation Part II-A General Summary, *Fatigue & Fracture of Engineering Materials & Structures*, **10**(2), 93-113.

[43] Bao R, Zhang X and Yahaya N A (2010). Evaluating stress intensity factors due to weld residual stresses by the weight function and finite element methods, *Engineering Fracture Mechanics*, **77**, 2550–2566.

[44] Croatian Register of Shipping (2012). Rules for the Classification of Ships, *Part 25 - Metallic Materials*.

[45] Faulkner D (1975). A Review of Effective Plating for Use in the Analysis of Stiffened Plating in Bending and Compression. *J Ship Res*, **19**, 1–17.

[46] Barsoum, RS (1976). On the Use of Isoparametric Finite Elements in Linear Fracture Mechanics, *International Journal for Numerical Methods in Engineering*, **10**, 25–37.

[47] Henshell R D and Shaw K G (1975). Crack Tip Finite Elements are Unnecessary, *International Journal for Numerical Methods in Engineering*, **9**, 495–507.

[48] Božić Ž, Mlikota M and Schmauder S (2011). Application of the ΔK, ΔJ and ΔCTOD parameters in fatigue crack growth modelling, *Technical Gazette*, **18**(3), 459–466.

[49] Božić Ž, Schmauder S and Mlikota M (2012). Fatigue growth models for multiple long cracks in plates under cyclic tension based on ΔK_I, ΔJ-integral and $\Delta CTOD$ parameter, *Key Engineering Materials*, **488-489**, 525–528.

[50] Liu Y and Mahadevan S (2007). Threshold stress intensity factor and crack growth rate prediction under mixed-mode loading, *Engineering Fracture Mechanics*, **74**, 332–345.

[51] Glinka G (1979). Effect of residual stresses on fatigue crack growth in steel weldments under constant and variable amplitude load, *Fracture mechanics, ASTM STP 677, American Society for Testing and Materials*, 198–214.

[52] Servetti G and Zhang X (2009). Predicting fatigue crack growth rate in a welded butt joint: the role of effective R ratio in accounting for residual stress effect, *Engng Fract Mech.*, **76**, 1589–1602.

Y. Furuya, H. Noguchi, and S. Schmauder

11 Molecular dynamics study on low temperature brittleness in tungsten single crystals

Abstract: This study employed a numerical model combining molecular dynamics and micromechanics to study the low temperature fracture of tungsten. In the simulations a pre-crack was introduced on/the (110) planes and cleavage was observed along the (121) planes. Cleavage along (121) planes has also been observed in experiments. Simulations were performed with three sizes of molecular dynamic regions at 77 K, and it was found that the results were independent of the size. Brittle fracture processes were simulated at temperatures between 77 K and 225 K with the combined model. The fracture toughness obtained in the simulations showed clear temperature dependency, although the values showed poor agreement with experimental results. A brittle fracture process at 77 K was discussed considering driving forces for dislocation emissions and cleavage in an atomic scale region of the crack tip. The driving force for dislocation emissions was saturated after the first dislocation emission, whilst the driving force for cleavage gradually increased with the loading K-field. The increased driving force caused cleavage when it reached a critical value. The critical values of driving force, which were close to the theoretical strength of the materials, were not influenced by temperature. This indicates that the temperature dependency of fracture toughness is not caused by the temperature dependency of dislocation emissions, but by that of dislocation mobility.

Keywords: molecular dynamics, micromechanics, brittle fracture, fracture toughness, crack propagation, dislocation

11.1 Introduction

The brittleness and ductility of materials have been major subjects of materials science. Research into brittle and ductile characteristics has advanced greatly in recent years. One source of this progress is the Rice–Thomson formulation [28] of a dislocation emission, with its later improvement [26]. The formulation, based on the competition between dislocation emission from a crack tip and cleavage, has successfully explained the intrinsic ductility of most fee metals and cleavability in most bcc metals. However, a thermal activation process had not been considered, and the formulation was not sufficient to explain the brittle to ductile transition (BDT).

Fracture toughness in most bcc metals is influenced by temperature [10, 11]. The toughness of metals increases with temperature, and the materials never cleave above a critical temperature. Even semiconductors [14] and ionic crystals [23], which are concluded to be intrinsically brittle materials in the Rice-Thomson formulation, show BDT characteristics. What mechanisms cause the temperature dependency of toughness and the brittle to ductile transition, is still a question that has not been answered satisfactorily until now. Some groups [27, 31, 33] insist that dislocation emission is the controlling factor, whereas others [12, 18] insist on dislocation mobility. Several remarkable models have been proposed in the discussion of dislocation emissions and dislocation mobility. Zhou and Thomson [33] proposed a dislocation emission model from the ledge of a crack front, which enables dislocations to be emitted at much lower external loading than in the Rice–Thomson formulation. The dislocation emission from the ledge of a crack front provides a good explanation for the river patterns on the fracture surfaces [9], as well as the observation of ten or fewer dislocations per slip plane [19]. Hirsch [12] proposed a computer simulation method for the generation and motion of the dislocations from crack tips, where the dynamics of emitted dislocations were taken into account.

In this study molecular dynamics (MD) has been applied to investigate the process of brittle fracture and the temperature dependency of fracture toughness. Molecular dynamics is an effective tool for the analysis of a crack. The technique enables us to analyse directly the events occurring on an atomic scale, such as dislocation emissions and cleavage in the crack tip region. However, modem computers are only capable of treating nano-scale material specimens – in the order of 10^6 atoms. This problem is fatal in the simulation of a crack because periodic boundary conditions cannot be assumed in all directions. In such case it is necessary to combine molecular dynamics with continuum mechanics. Molecular dynamics should be applied only to the crack tip region and continuum mechanics should then be applied to the surrounding region.

In early research into finding a method of combining molecular dynamics with a continuum, a major area of investigation had been how to synchronise the deformation of a continuum region with that of a molecular dynamics region. Several groups [15, 16, 20, 21] proposed a method to combine molecular dynamics with a finite element method (FEM), where nodes of finite elements were synchronised with atoms of MD in a boundary region. Others [13, 30] proposed a method of correcting the boundary conditions using Green's function. However, the problem with these methods is that the emitted dislocations from a crack tip cannot pass smoothly through the boundary between the molecular dynamics and the continuum regions. Yang et al. [32] firstly proposed a method where emitted dislocations could pass through the boundary. In their method the continuum region was divided into two regions. The outer region was calculated with a finite element method, and the inner region was calculated with an elastic continuum where the movement of dislocations was analysed dynamically. Yang's method, however, has a limitation in the number of emitted dislocations, and in the validity of the method, which was not examined satisfactorily.

The authors developed a new method [7, 24] in which molecular dynamics was combined directly with linear elastic theory, that is micromechanics [2, 5, 17, 22]. A thorough examination of the validity of the method was undertaken. In the new method the dislocations emitted in the molecular dynamics region can pass through the boundary of the two regions smoothly, and are distributed at the equilibrium positions in the micromechanics region according to the elastic solution. That is to say that the dynamics of dislocations was not under consideration. The simulation was then presumed to be quasi-static. The limitation of dislocation emissions was removed by moving the molecular dynamics region with the crack propagation. Crack tip opening displacements calculated in the simulation with the method showed good agreement with an analytical solution derived by Rice [25]. The combined model is limited to two-dimensional and quasi-static simulations. It means that the ledge of a crack front and the effect of strain rate cannot be taken into account. However, the limitation does not mean that the temperature dependency of dislocation mobility is neglected, because friction forces acting on each dislocation, which reflect dislocation mobility, depend on temperature. The difference between a dynamic simulation and a quasi-static simulation is merely whether the distribution of dislocations is dynamic or in equilibrium.

In this paper brittle fracture processes at low temperature are simulated with the combined model of molecular dynamics and micromechanics in tungsten single crystals. The mechanisms of brittle fracture and the temperature dependency of fracture toughness are investigated.

11.2 A combined model of molecular dynamics with micromechanics

11.2.1 The principle of the combined model

The combined model of molecular dynamics with micromechanics is shown in Fig. 11.1. An infinite plate exhibiting a crack and dislocations is subjected to uniform tension applied at infinity. The deformation of the hatched region in Fig. 11.1 is analyzed with molecular dynamics, and that of the surrounding region is analyzed with micromechanics. A periodic boundary condition is applied to the molecular dynamics region in the direction of plate thickness, and the micromechanics region is analyzed as a plane strain problem in two dimensions.

A model in the molecular dynamics region is shown in Fig. 11.2. A free atom is defined as an atom that moves according to the molecular dynamics algorithm with thermal oscillations.

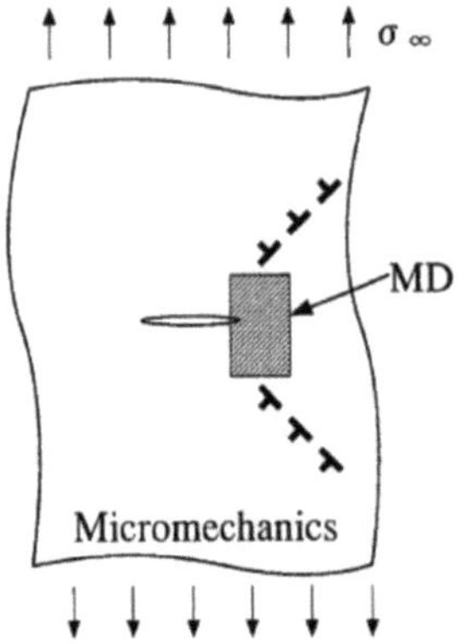

Fig. 11.1: Combined molecular dynamics and micromechanics model.

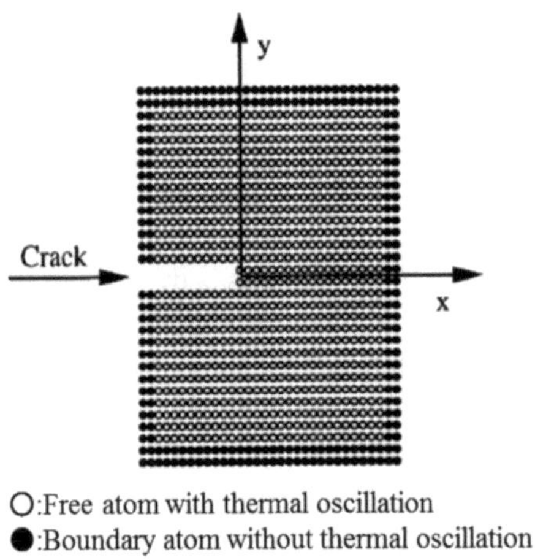

Fig. 11.2: Molecular dynamics model.

A boundary atom is defined as an atom that moves without thermal oscillation according to the displacement calculated with micromechanics, that is the boundary atom layer is a part of the micromechanics region. As shown in Fig. 11.2, the crack in the molecular dynamics region is expressed by removing two layers of atoms. A quasi-static simulation, the detail of which is explained elsewhere [7], is presumed in this model. Temperature of the molecular dynamics region is kept constant using a velocity scaling technique.

Remarkable points in this model are:

1. The boundary condition to combine two regions is flexible and both displacement and stress fields are continuous at the boundary.
2. Emitted dislocations in the molecular dynamics region pass through the boundary smoothly.
3. The molecular dynamics region moves with the crack propagation.

The details of these three points are explained in the following sections.

11.2.2 Flexible boundary conditions using body forces

Fig. 11.3 shows stress–strain curves of smooth specimens in the process of elastic deformation. A stress–strain curve calculated with molecular dynamics is shown together with a linear line used in micromechanics. As shown in Fig. 11.3, the deformation of the molecular dynamics region is intrinsically non-linear. In turn a rigid boundary condition, where only a displacement field is continuous at the boundary is not satisfactory, because a stress field is discontinuous in this case.

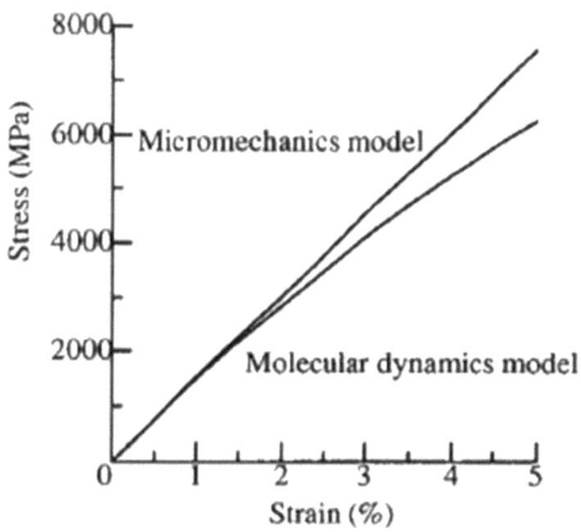

Fig. 11.3: Stress–strain curves.

In our combined model the boundary condition, which is basically rigid, is corrected with body forces [4]. Body forces distributed at the boundary influence both the stress field and the displacement field. This procedure is similar to Flex-II [30]. In Flex-II the balance of forces acting on each atom is considered at the boundary and Green's functions are used for correction. In our model the balance of stress at the boundary is considered and the body forces are used for correction.

11.2.3 Transformation from an atomistic dislocation to an elastic dislocation

Dislocations are distributed at equilibrium positions in quasi-static simulations, that is the dynamics of dislocation movements are not considered in the combined model. In the case when the equilibrium position of an emitted dislocation is in the micromechanics region, the dislocation must move across the boundary of the two regions. A method for moving an emitted dislocation from the molecular dynamics region to the micromechanics region is illustrated in Fig. 11.4. A displacement field caused by slip is applied to the boundary atom's layer. After that, the molecular dynamics region is smoothed. The re-smoothing procedure contributes to avoidance of difficulties arising from hard distortion of the molecular dynamics region after several slips have

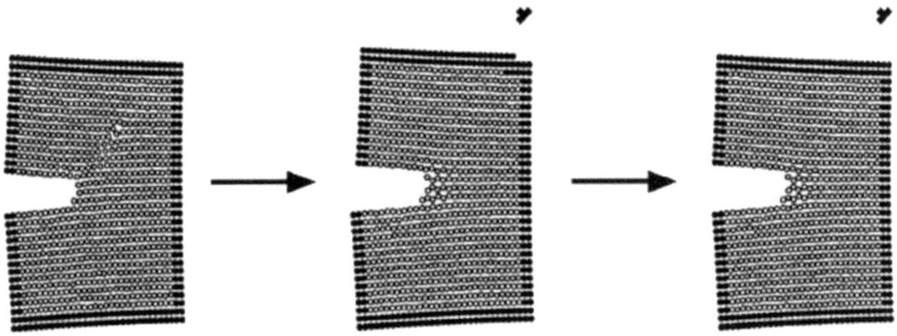

Fig. 11.4: Transformation from an atomistic dislocation to an elastic dislocation.

occurred. The dislocation from the molecular dynamics region is transformed into an elastic dislocation and distributed at the equilibrium position. The dislocations become edge type. In the case when the equilibrium position of the emitted dislocation is in the molecular dynamics region, the above procedure should not be applied.

Equilibrium positions of the dislocations are derived from a balance between driving forces and friction forces acting on each dislocation. The driving forces include loading stress, interaction with other dislocations and an attractive force from the free surface of a crack. The friction force should be Peierls stress in the case of a perfect crystal. However, a critical resolved shear stress (CRSS), obtained in experiments, may be better for simulations in comparison with experimental results. This is because practical crystals used in experiments include defects such as pre-existing dislocations and inclusions.

11.2.4 Movement of a molecular dynamics region with crack propagation

Fig. 11.5 shows the simulated result of the crack propagation process with emission of dislocations. The crack tip moves with crack propagation so quickly reaches the boundary between the molecular dynamics region and the micromechanics region. The crack propagation simulation must be stopped when the crack tip reaches the boundary. This means that the length of crack propagation that can be simulated and the number of dislocations that can be emitted, depends on the size of the molecular dynamics region. This limitation is fatal because of the limited capacity of computers.

In order to remove the limitation, the molecular dynamics region moves with crack propagation. The basic idea is illustrated in Fig. 11.6. The crack tip could be kept in the molecular dynamics region with this procedure. Therefore, the simulation is no longer limited by the size of a molecular dynamics region.

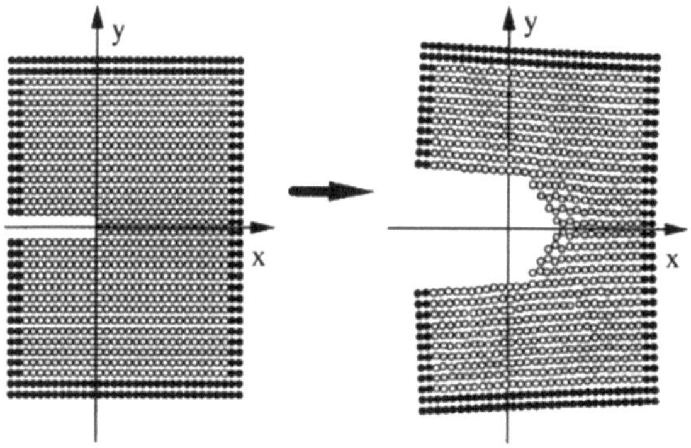

Fig. 11.5: Simulation result of crack propagation.

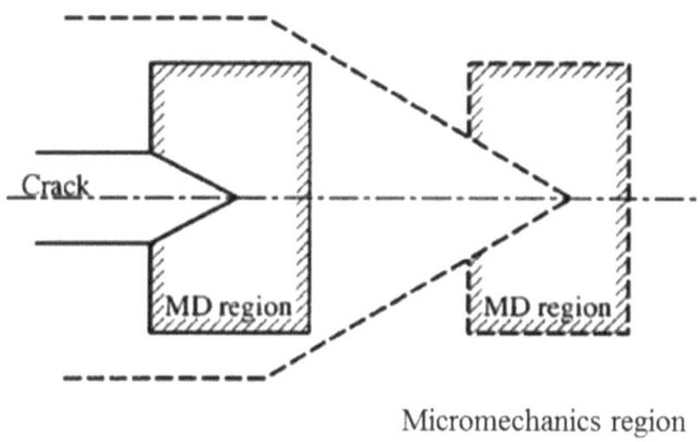

Fig. 11.6: Movement of the molecular dynamics region with crack propagation.

11.3 Simulation of a brittle fracture process in tungsten single crystals

11.3.1 Calculation conditions and additional procedures for the simulation of tungsten single crystals

Tungsten single crystals are appropriate specimens for brittle fracture simulations because of the brittleness in spite of being single crystals. An N-body potential derived by Finnis and Sinclair [6] was used in the simulations. Fig. 11.7 shows a molecular dynamics model for tungsten single crystals. Two layers of atoms were accumulated in the direction of plate thickness and a periodic boundary condition was applied. In this model the crack face was in the (110) plane and the crack front direction was ⟨110⟩. Temperature, pre-crack length, bulk moduli and CRSS for the simulations are shown in Tab. 11.1. The CRSS value, obtained by experiment [1], was used as the friction force of a dislocation. In the simulations cleavage was defined as fatal fracture because crack propagation with cleavage can not be stopped. Crack propagation with dislocation emissions can be stopped because of the shielding forces of the emitted dislocations. That is in crack propagation with cleavage, shielding forces to stop crack

Tab. 11.1: Calculation conditions and material properties.

Temperature	77 K
Pre-crack length	2 mm
Young's modulus (plane strain)	445.7 GPa
Poisson's ration (plane strain)	0.390
Shear modulus	160.6 GPa
CRSS [1]	450 MPa

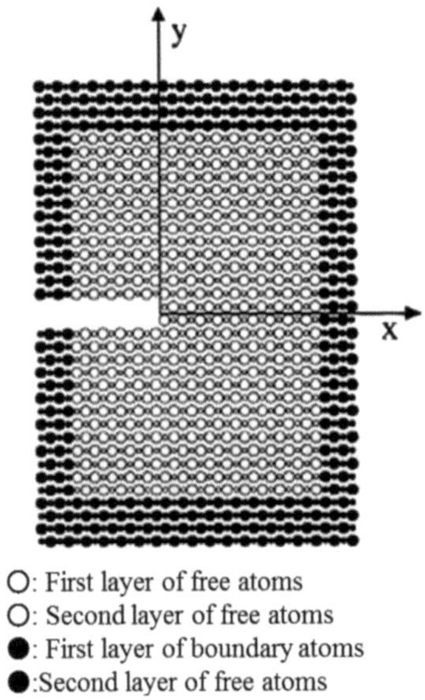

Fig. 11.7: Molecular dynamics model for tungsten single crystals.

propagation never increase but the driving forces to propagate a crack increase with the crack length.

Fig. 11.8 shows a broken specimen of a fracture toughness test at 77 K in a tungsten single crystal [29]. The specimen, which contains a (110) re-crack cleaved along (121) planes inclined with respect to the pre-crack. Other experiments [3], whose crystallographic orientations were a little different from those of Riedle's experiments, showed (100) cleavage against (110) half elliptic pre-cracks. A conclusion derived from these experiments is that (110) planes are resistant to cleavage in tungsten single crystals hence cleavage occurs along other planes against (110) pre-cracks.

In simulations in which the crystallographic orientations corresponded to those of Riedle's Experiments. (121) cleavage was expected to occur. To achieve (121) cleavage the combined model required an additional procedure. Fig. 11.9 shows crack tip shape and the position of the molecular dynamics region in the crack tip after the crack has been opened. In the case of (121) cleavage, the origins of the cleavage are not expected to be the center of the crack tip but the edges of the crack tip as indicated

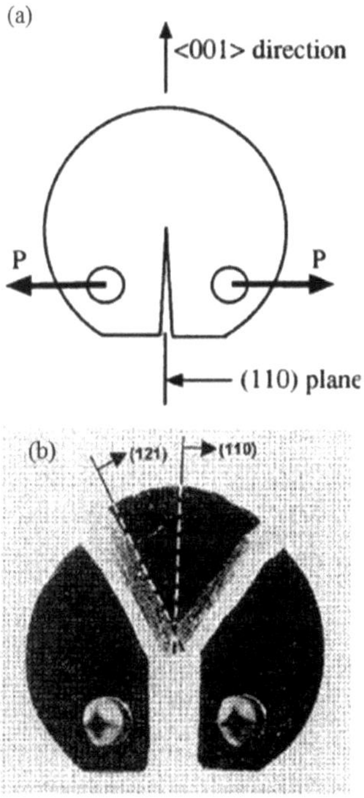

Fig. 11.8: Broken specimen of a fracture toughness test at 77 K in a tungsten single crystal [29].

in Fig. 11.9. The problem then is that origins exist outside of the molecular dynamics region. In the combined model, cleavage occurrence is dependent on the molecular dynamics calculation. In turn simulations would contain errors if the origins were not in the molecular dynamics region. Fig. 11.10 shows the result of one such simulation run without any additional correcting procedure [8]. In this case cleavage occurred from a boundary between the molecular dynamics region and the micromechanics region. The result is obviously wrong because the boundary between the two regions does not exist in real materials.

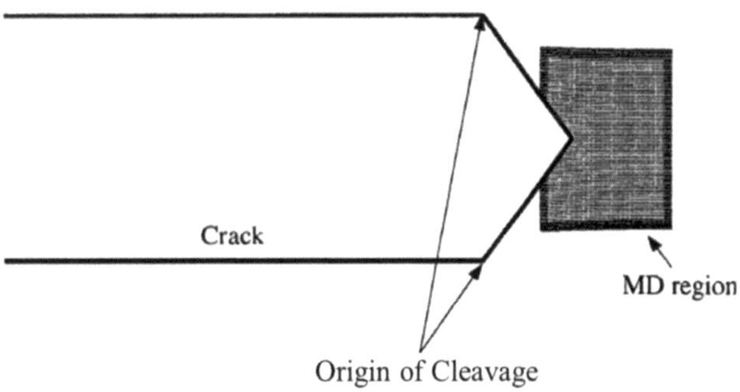

Fig. 11.9: Origin of cleavage in case of (121) cleavage.

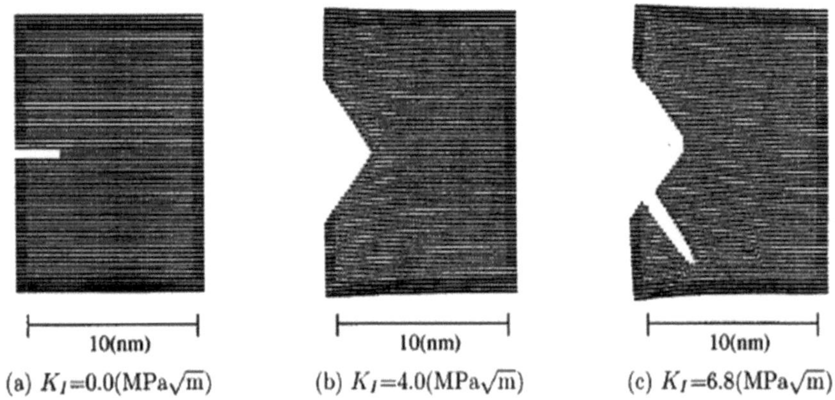

Fig. 11.10: Result [8] of a simulation for cleavage in case the crack tip is blunted. In this case friction forces acting on each dislocation were 1000 MPa.

The additional procedure for correcting this problem is illustrated in Fig. 11.11. Dislocation emissions cause crack opening and crack tip blunting. The more the dislocations are emitted, the more the crack tip is blunted. This crack tip blunting leads to the crack tip edges escaping from the molecular dynamics region. The point of the additional procedure is to keep the crack closed. In the procedure shown in Fig. 11.11, the open space created by a dislocation emission is filled with new atoms and relaxation calculations are performed. This procedure prevents the crack from opening and keeps the crack tip edges in the molecular dynamics region. This is an approximation and is accompanied by the problem that now the influence of crack tip blunting is removed. However, the procedure is hardly expected to influence the intrinsic crack behavior, and the radius of a crack tip with blunting would remain so small compared to the crack length that its effects are negligible in comparison to the brittleness of tungsten single crystals.

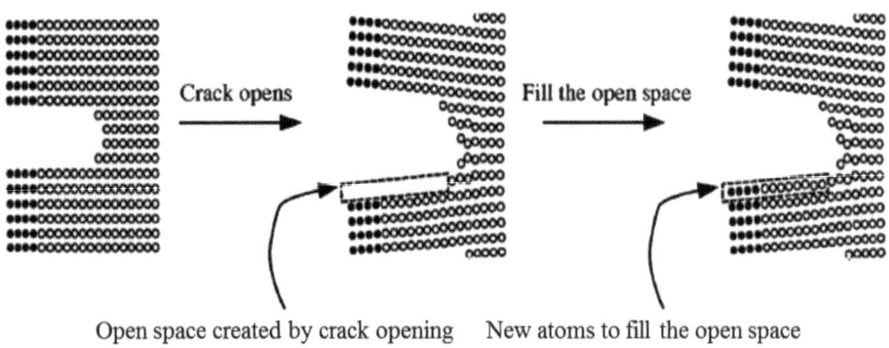

Fig. 11.11: Method to prevent a crack from opening by filling the open space with new atoms.

11.3.2 Simulation results and size dependency of the molecular dynamics region on the results

Three sizes of molecular dynamics regions (see Fig. 11.12) were used in simulations to investigate size dependency. The numbers of free atoms in each were 2304, 3600, and 10 000. Fig. 11.13 shows the result of a simulation containing 3600 free atoms. In the simulation, cleavage occurred not from the boundary between the molecular dynamics region and the micromechanics region but from the edge of the crack tip along a (121) plane. The (121) cleavage corresponds to experimental results (see Fig. 11.8). An interesting feature in this result was the presence of backward twins from the edges of the crack tip at $K = 5.0\,\mathrm{MPa}\sqrt{\mathrm{m}}$. Backward twins have also been observed in other simulations [16] for α-iron.

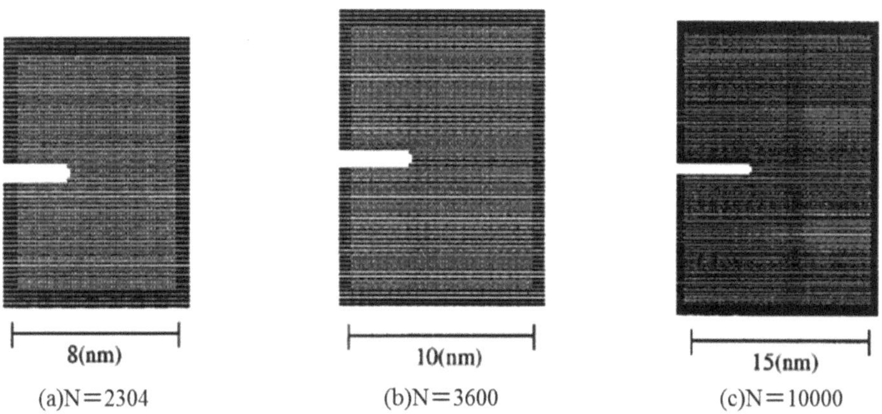

Fig. 11.12: Three sizes of molecular dynamics regions (*N*: number of free atoms)

In this simulation, 155 dislocations were emitted before cleavage occurred although the crack tip in Fig. 11.13 was not blunted due to the additional procedure applied as explained in the previous section. The distribution of dislocations is shown in Fig. 11.14 where only the upper half of the model ($y \geq 0$) is plotted, both halves of the model being considered in the simulation.

Values of fracture toughness evaluated for each size of the molecular dynamics regions are displayed in Tab. 11.2. The differences in fracture toughness were quite small. It was concluded that the simulation result was independent of the size of the molecular dynamics region.

(a) K_I=0.0(MPa $\sqrt{m}$) (b) K_I=5.0(MPa $\sqrt{m}$) (c) K_I=6.2(MPa $\sqrt{m}$)

Fig. 11.13: Result of a simulation with an N = 3600 model (N: number of free atoms).

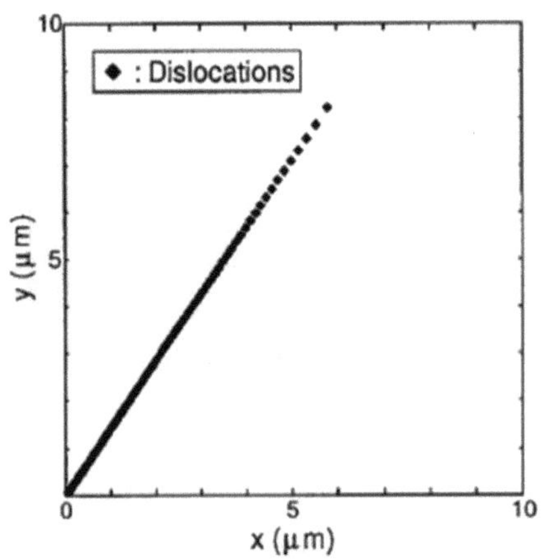

Fig. 11.14: Distribution of dislocations at the loading when the material was fractured at 77 K. Only upper side is shown and the origin means a crack tip.

Tab. 11.2: Fracture toughnesses obtained in Simulations (N: number of free atoms).

Type of model	Fracture toughness K_{IC}
N = 2 304 model	6.3 (MPa $\sqrt{m}$)
N = 3 600 model	6.2 (MPa $\sqrt{m}$)
N = 10 000 model	6.2 (MPa $\sqrt{m}$)

11.4 Investigation of brittle fracture processes and temperature dependency of fracture toughness at low temperature

11.4.1 Simulation results at low temperature

Fig. 11.15 shows experimental results [10] for fracture toughness in tungsten single crystals. The brittle to ductile transition temperature and the fracture toughness at high temperature are both influenced by strain rate. However, fracture toughness at low temperature (77–225 K) is not influenced by strain rate. In the present study strain rates cannot be taken into account because of the quasi-static simulations. At high temperatures, near the transition temperature, there are expected to be too many dislocations emitted to simulate, i.e. the more dislocations that are emitted, the greater the number of calculations to be performed with the combined model. This is because the transformation from an atomic dislocation to an elasticdislocation requires relaxation calculations of molecular dynamics with 10 000 steps or more. The experimental results and the limitations of the simulations mean that low temperatures (77–225 K) are appropriate conditions for the simulations.

The model with 2304 free atoms was used under these conditions. The temperature dependency of critical resolved shear stress (CRSS) [1] is shown in Fig. 11.16. Brittle fracture processes were simulated in the low temperature regime (77–225 K). Fig. 11.17 shows fracture toughness values obtained from the simulations together with previ-

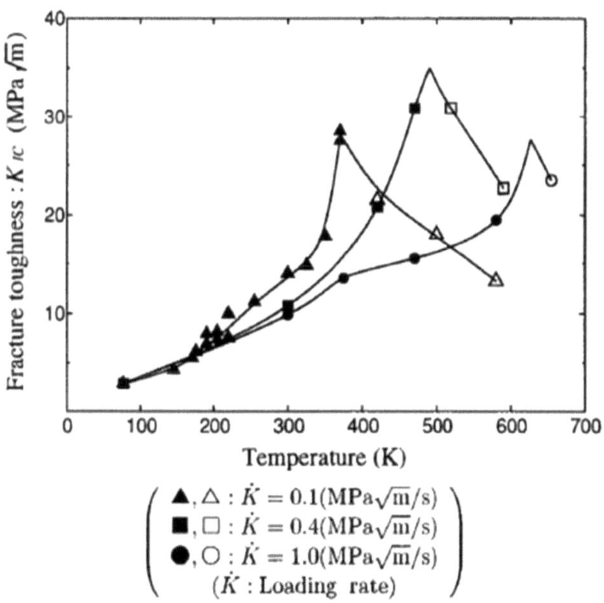

Fig. 11.15: Experimental results [10] of fracture toughnesses in tungsten single crystals. Solid marks show fracture toughnesses and open marks show stress intensities at failure in ductile manner.

ous experimental results [10]. The number of dislocations emitted and the plastic zone lengths at failure are displayed in Tab. 11.3. In the simulations, fracture toughness showed clear dependency on temperature and the tendency of the fracture toughness to increase showed good correlation to experimental data. The values of fracture toughness, however, varied from experimental results. Also, it was still unknown whether the steep increase in fracture toughness at high temperature, near the transition temperature, could be obtained in the simulations. It was good that the temperature dependency of fracture toughness was obtained at low temperature. However, the accuracy of the simulations and performance of the simulations at high temperature still remain subjects to be addressed.

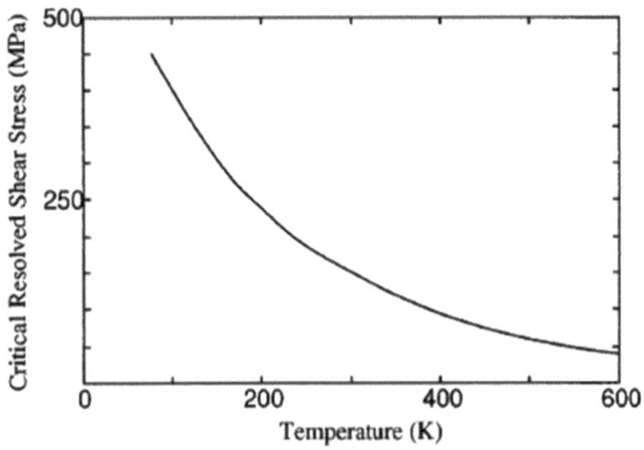

Fig. 11.16: Temperature dependency of CRSS obtained in experiments [1].

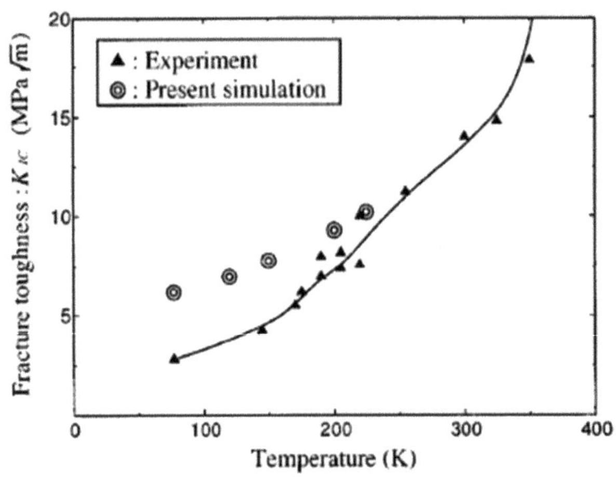

Fig. 11.17: Comparison of fracture toughnesses between simulations and experiments [10]. Loading rates in experiments: $\dot{K}_I = 0.1\,\text{MPa}\sqrt{\text{m}}\,\text{s}^{-1}$.

Tab. 11.3: Number of emitted dislocations and lengths of plastic zones at the loading when the materials were fractured in simulations.

Temperature (K)	Dislocations	Plastic zones (μm)
77	155	10.0
120	298	25.4
150	452	46.3
200	907	120.9
225	1266	208.5

11.4.2 A brittle fracture process

In this section the mechanism of brittle fracture will be discussed, based on the simulation results using the combined model. The main phenomena of the brittle fracture process are dislocation emissions and cleavage. The point of the process is to determine whether the material cleaves or emits dislocations at a crack tip during external loading. Two simple models, based on local stress analyses in the crack tip region at an atomic scale, were introduced, one model for dislocation emissions and the other for cleavage (see Fig. 11.18). The driving force τ_{local} is a resolved shear stress causing slip with dislocation emission. When the driving force τ_{local} reaches a critical value τ_c, a dislocation is emitted. The driving force σ_{local} is a normal stress that causes cleavage, i.e. when the driving force σ_{local} reaches a critical value σ_c then cleavage occurs.

The driving forces τ_{local} and σ_{local}, which were calculated elastically through continuum mechanics, consist of a loading K-field and the shielding forces of dislocations. The information about the K-field and the dislocations had already been obtained from simulations with the combined model. The problem with these models

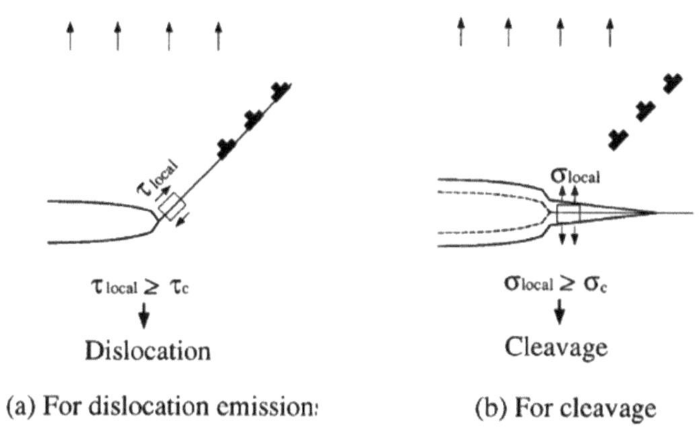

Fig. 11.18: Simple models for dislocation emissions and cleavage considering driving forces in an atomistic scale. τ_c and σ_c are critical values for dislocation emissions and cleavage, respectively.

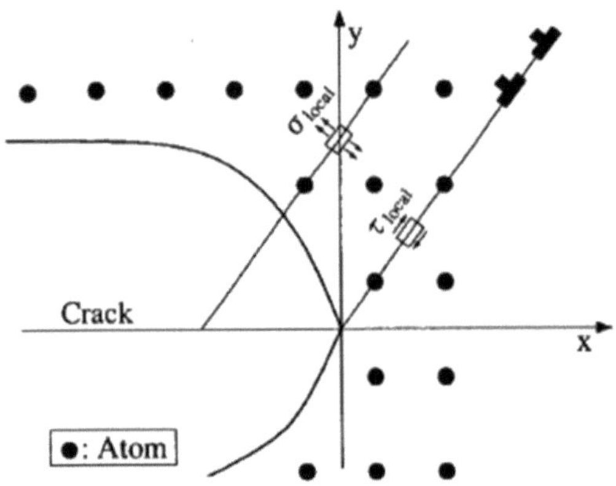

Fig. 11.19: Continuum model for τ_{local} and σ_{local} calculations, compared with atom positions. Calculation points of τ_{local} and σ_{local} were $(\frac{1}{2}a, \frac{\sqrt{2}}{2}a)$ and, $(0, \sqrt{2}a)$, respectively (a: lattice constant). A radius of the crack tip was $\rho = \sqrt{2}a$.

was where to calculate τ_{local} and σ_{local}, because the crack tip was a singular point, i.e. τ_{local} andσ_{local} depended not only on the loading but also on the calculation points. In the present study, the calculation points were one or two atoms from the crack tip. The details of calculation points are shown in Fig. 11.19.σ_{local} was calculated on a (121) plane because cleavage occurred along (121) planes in this case. The radius of the crack tip was assumed to be two atoms space ($\rho = \sqrt{2}a$). Although the models have the problem of where driving forces should be calculated from, they are useful for understanding the brittle fracture process.

Fig. 11.20 shows the driving forces τ_{local} and σ_{local} at 77 K obtained from the calculations. Young's modulus E', displayed in Tab. 11.1, is for a plane strain problem. The relationship between E' and E is $E' = E/(1 - \nu^2)$.

The driving force τ_{local} for dislocation emissions, which increased linearly with K-field in elastic deformation, was saturated after the first dislocation emission. The saturation is a result of the shielding force of the emitted dislocation. Dislocations begin to be emitted when τ_{local} exceeds a critical value τ_c ($\tau_{local} \geq \tau_c$). The driving force τ_{local} is, however, immediately relaxed because of the shielding forces of emitted dislocations. In turn dislocations continue to be emitted while τ_{local} exceeds τ_c ($\tau_{local} \geq \tau_c$), and the emissions are stopped when τ_{local} is relaxed below τ_c ($\tau_{local} < \tau_c$). The driving force σ_{local} for cleavage, which is also relaxed by the shielding forces of dislocations, gradually increased with K-field even after the first dislocation emission. In the simulation at $K_I = 6.3\,\text{MPa}\sqrt{\text{m}}$, the driving force σ_{local} reached a critical value σ_c and cleavage occurred. In summary, the brittle fracture process of the simulation would be as follows. Whilst loading K_I from zero to K_{IC}, the driving force τ_{local} for dislocation emissions reaches τ_c. After that τ_{local} is saturated by the shielding forces of emitted dislocations, while the driving force σ_{local} for cleavage continually increases with K_I. In turn cleavage, which leads to fatal fracture, occurs when σ_{local} reaches σ_c.

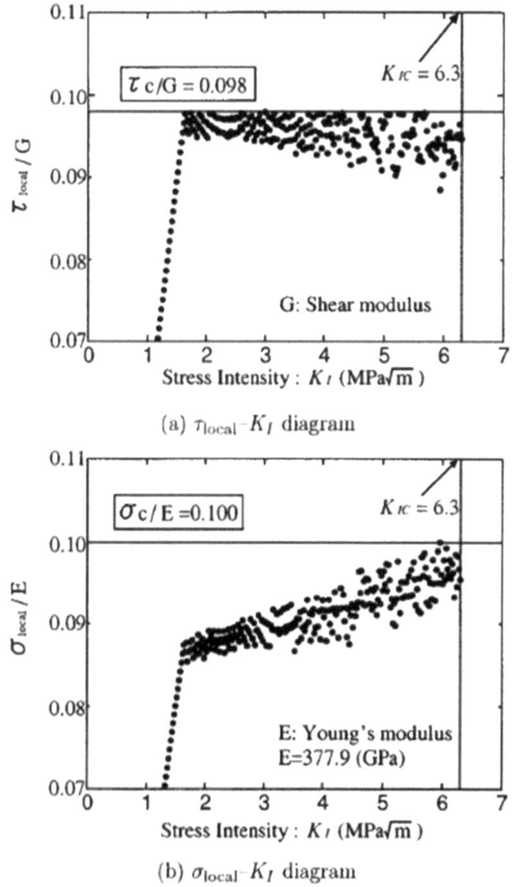

Fig. 11.20: Calculation results of τ_{local} and σ_{local}.

It is interesting to note that from the results in this section that the critical values τ_c and σ_c were quite close to the theoretical shear strength ($G/2\pi$) and the theoretical tensile strength ($E/10$) of the materials. This implies that the strength in the atomic scale region around the origin of fracture might be close to the theoretical strength of the material even if the macroscopic strength is much lower than the theoretical strength.

11.4.3 Temperature dependency of fracture toughness

Some groups insist that dislocation emission is the controlling factor in brittle to ductile transition and others insist on dislocation mobility (see Introduction). In the present study, dislocation emission was determined from the critical value τ_c, and the dislocation mobility from friction forces acting on each dislocation (CRSS). The critical values τ_c and σ_c calculated from simulation results at each temperature are displayed

Tab. 11.4: Temperature dependencies of τ_c and σ_c.

Temperature (K)	τ_c/G	σ_c/E
77	0.098	0.100
120	0.096	0.099
150	0.096	0.098
200	0.097	0.100
225	0.096	0.099

in Tab.11.4. Both values show no dependency on temperature. In comparison, the friction forces (CRSS) do vary with temperature (see Fig. 11.16). These calculation results mean that the temperature dependency of fracture toughness in the simulations is due to the temperature dependency of dislocation mobility. Temperature dependency of fracture toughness is a characteristic in the low temperature region of brittle to ductile transition. The results of the present study, therefore, support the latter insistence of dislocation mobility. The mechanism is as follows. Increasing temperature causes an increase of dislocation mobility, which causes differences in dislocation distribution, and hence leads to differences in the shielding forces of dislocations. The difference in shielding forces causes the slope of the σ_{local} to decrease. In turn the fracture limit (K_{IC}), which is the load at which σ_{local} reaches σ_c, increases with temperature.

11.5 Discussion

As mentioned above, the accuracy of simulations and performing simulations at high temperatures remain problems still unsolved by this study. The key points to solving these problems are an extension to three-dimensional simulations and an increase in calculation speed. Two-dimensional simulations lead not only to a decline of accuracy but also to principal limitations in the simulations. With a pre-crack on a (100) plane, which is a primary cleavage plane, the material fractured in a perfect brittle manner with the combined model, i.e. the material cleaved before it emitted dislocations. In turn it was observed that fracture toughness was independent of temperature. The perfect brittle fracture might be caused by an absence of ledge sites [33] on the crack front because of the two-dimensional nature of the model. The extension to three-dimensional simulations, involving ledges of a crack front, might be absolutely necessary both to improve the accuracy and to extend the applicability of the simulations. Increasing the speed of the calculations is necessary both to extend to three-dimensional simulations and to simulate at high temperatures. The speed increase may require the improvement not only of computers but also of software.

In this study the molecular dynamics simulations brought much benefit to solving the problem of brittle fracture and the temperature dependency of fracture toughness. This might be a good demonstration of the usefulness and applicability of simulations in research into material strength. The present study is successful in showing the range of possibilities of molecular dynamics simulations, while the simulations still have several deficiencies.

11.6 Conclusion

The combined model of molecular dynamics with micromechanics was applied to Simulations of brittle fracture processes in tungsten single crystals at low temperatures. The pre-cracks were introduced on (110) planes and cleavage was observed along (121) planes in the simulations. The cleavage along (121) planes had previously been observed in experiments [29]. In the simulations the material was twinned backwards from edges of the crack tip. The backward twins have also been observed in other simulations [16]. Three sizes of molecular dynamics regions were tested at 77 K, and the results of the simulations were found to be independent of the sizes used.

Brittle fracture processes were simulated at the temperatures between 77 K and 225 K. The fracture toughness values obtained in these simulations showed clear temperature dependency, but did not show good agreement with those from experiments. The main problem with the simulation was the limitation in the number of calculations that could realistically be performed which affected both accuracy and limited the simulations to low temperatures only. A brittle fracture process at 77 K was discussed by considering the driving forces for dislocation emissions and cleavage in an atomic scale region of a crack tip. It was found that the driving force for dislocation emissions was saturated after the first dislocation emission, whereas the driving force for cleavage gradually increased with loading K-field. When the driving force for cleavage reached a critical value, the material cleaved. The critical values of driving forces for both dislocation emissions and for cleavage, which were quite close to the theoretical strengths of the material, were not influenced by temperature. This means that the temperature dependency of fracture toughness is not caused by a temperature dependency of dislocation emissions but by that of dislocation mobility.

References

[1] Bucki, M., Novak, V., Savitsky, Y.M., Burkhanov, G.S. and Kirillova, V.M. (1979). Work-hardening in tungsten single crystals between 77 and 1343 K. *Strength of Metals and Alloys. Proceedings of the 5th International Conference*, xxx+760, 145–150.

[2] Chou, Y.T. (1967). Dislocation pile-ups against a locked dislocation of different Burgers vector. *Journal of Applied Physics* 38, 2080–2085.

[3] Cordwell, J.E. and Hull, D. (1972). Observation of {110} cleavage in ⟨110⟩ axis tungsten single crystals. *Philosophical Magazine*, 26(1), 215–224.

[4] Eshelby, J.D. (1957). The determination of the elastic field of an ellipsoidal inclusion, and related problems. *Proceedings, Royal Society London* A241, 376–396.

[5] Eshelby, J.D., Frank, F.C. and Nabarro, F.R.N. (1951). The equilibrium of linear arrays of dislocations, *Philosophical Magazine* 42, 351–364.
[6] Finnis, M.W. Sinclair, J.E. (1984). A simple empirical N-body potential for transition metals. *Philosophical Magazine* A50, 45–55.
[7] Furuya, Y. and Noguchi, H. (1998). A combined method of molecular dynamics with micromechanics improved by moving the molecular dynamics region successively in the simulation of elastic-plastic crack propagation. *International Journal of Fracture* 94, 17–31.
[8] Furuya, Y. and Noguchi, H. (1999). Simulation of crack propagation with (molecular dynamics+micromechanics) model. *Proceedings of 9th CIMTEC* 18, 57–64.
[9] George, A. and Michot, G. (1993). Dislocation loops at crack tips: nucleation and growth – an experimental study in silicon. *Materials Science and Engineering* A164, 118–134.
[10] Gumbsch, P., Riedle, J Hartmaier, A. and Fischmeister, H. (1998). Controlling factors for the brittle-to-ductile transition in tungsten single crystals, *Science* 282, 1293–1295.
[11] Ha, K.F., Yang, C. and Bao, J.S. (1994). Effect of dislocation density on the ductile-brittle transition in bulk Fe-3%Si single crystals. *Scripta Metallurgica et Materialia* 30(8), 1065–1070.
[12] Hirsch, P.B. and Roberts, S.G. (1991). The brittle-to-ductile transition in silicon. *Philosophical Magazine* A64, 55–80.
[13] Hirth, J.P., Hoagland, R.G. and Gehlen, P.C. (1974). The interaction between line force arrays and planar cracks. *International Journal of Solids and Structures* 10(9), 977–984.
[14] John, C. St. (1975). The brittle-to-ductile transition in precleaved silicon single crystals. *Philosophical Magazine* 32, 1193–1212.
[15] Kohlhoff, S., Gumbsch, P. and Fichmeister, H.F. (1991). Crack propagation in b.c.c. crystals studied with a combined finite-element and atomistic model. *Philosophical Magazine* 64(4), 851–878.
[16] Kohlhoff, S. and Schmauder, S. (1988). A new method for coupled elastic-atomistic modeling. (Edited by V. Vitek and D.J. Srolovitz), *Large Atomistic Simulation of Materials – Beyond Pair Potentials* – Plenum Press, New York, 411–418.
[17] Lekhnitski, S.G. (1968). *Anisotropic Plates* Gordon and Breach, New York.
[18] Maeda, K. (1992). Effect of crack blunting on dislocation-mobility-controlled brittle-ductile transition. *Scripta Metallurgica et Materialia* 11, 805–809.
[19] Michot, G., Loyola de Oliveria, M.A. and George, A. (1994). Dislocation loops at crack tips: control and analysis of sources in silicon. *Materials Science and Engineering* A176, 99–109.
[20] Mullins, M. and Dokanish, F. (1982). Simulation of the (001) Plane Crack in α-Iron Employing a New Boundary Scheme. *Philosophical Magazine* 46(5), 771–787.
[21] Mullins, M. (1982). Molecular Dynamics Simulation of Propagating Cracks. *Scripta Metallurgica* 16, 663–666.
[22] Mura, T. (1968). The continuum theory of dislocations, Advances in Materials Research (edited by H. Herman H.), Interscience Publ., New York, 1–107.
[23] Narita, N., Higashida, K., Torii, T. and Miyaki, S. (1989). Crack-tip shielding by dislocations and fracture toughness in NaCl crystals. *Materials Transactions, JIM* 30, 895–907.
[24] Noguchi, H. and Furuya, Y. (1997). A method of seamlessly combining a crack tip molecular dynamics enclave with a linear elastic outer domain in simulating elastic-plastic crack advance. *International Journal of Fracture* 87, 309–329.
[25] Rice, J.R. (1974). Limitations to the small scale yielding approximation for crack tip plasticity. *Journal of the Mechanics and Physics of Solids* 22, 17–26.
[26] Rice, J.R. (1992). Dislocation nucleation from a crack tip: an analysis based on the Peierls concept. *Journal of the Mechanics and Physics of Solids* 40, 239–271.
[27] Rice, J.R. and Beltz, G.E. (1994). The activation energy for dislocation nucleation at a crack tip. *Journal of the Mechanics and Physics of Solids* 42, 333–360.

[28] Rice, J.R. and Thomson, R. (1974). Ductile versus brittle behavior of crystals. *Philosophical Magazine* 29, 73–97.
[29] Riedle, J., Gumbsch, P. and Fischmeister, H.F. (1994). Fracture studies of tungsten single crystals. *Materials Letters* 20, 311–317.
[30] Sinclair, J.E., Gehlen, P.C., Hoagland, R.G. and Hirth, J.P. (1978). Flexible boundary conditions and nonlinear geometric effects in atomic dislocation modeling. *Journal of Applied Physics* 49(7), 3890–3897.
[31] Xu, G. and Argon, A.S. (1995). Nucleation of dislocations from crack tips under mixed modes of loading: implications for brittle against ductile behavior of crystals. *Philosophical Magazine* A72, 415–451.
[32] Yang, W., Tan, H. and Guo, T. (1994). Evolution of crack tip process zones. *Modeling and Simulation in Materials Science and Engineering* 2, 767–782.
[33] Zhou, S.J. and Thomson, R. (1991). Dislocation emission at ledges on cracks. *Journal of Materials Research* 6, 639–653.

L. Madej, M. Sitko, K. Perzynski, L. Sieradzki, K. Radwanski, and R. Kuziak

12 Multi scale cellular automata and finite element based model for cold deformation and annealing of a ferritic-pearlitic microstructure

Abstract: Numerical modelling of microstructure evolution during cold rolling and the subsequent annealing of a two phase ferritic-pearlitic sample under an α/γ phase transformation regime is the subject of the present work. The multi scale model based on the digital material representation taking into account exact representation of the microstructure morphology is used in the research to investigate inhomogeneous strain distribution during cold rolling. Obtained results are then incorporated into the discrete cellular automata model of static recrystallization. Data transfer between the finite element and cellular automata models is performed by means of the interpolation method based on the Smoothed Particle Hydrodynamic. Details about the developed cellular automata model of static recrystallization are presented within the paper. The complete multi scale model is finally validated against a series of experimental cold rolling and subsequent annealing operations. Various annealing conditions were used as case studies to prove robustness of the developed numerical approach.

Keywords: static recrystallization, multi scale modelling, cellular automata

12.1 Introduction

The main driving force of innovative experimental and numerical research is the significant need for new metallic materials manifested by the automotive and aerospace industries. Such materials have to meet increasing requirements regarding weight/property ratio, as well as a combination of high strength and high ductility. As a result, dynamic development of the modern steel grades has been observed over the last years. The number of new steel grades which have been developed since the year 2000 has been increased, which is clearly visible in Fig. 12.1. Currently, a wide range of innovative steels (third generation of AHSS (Advanced High Strength Steels), Bainitic, nano-Bainitic, etc.) as well as other metallic materials, e.g. aluminium, magnesium, titanium or copper alloys, is being developed in research laboratories around the world [1–4]. Complex mechanical and thermal cycles are applied to obtain very sophisticated microstructures with a combination of various micro scale features: large grains, small grains, inclusions, precipitates, multi-phase structures, etc. These microstructural features and interactions between them at the micro-scale level during manu-

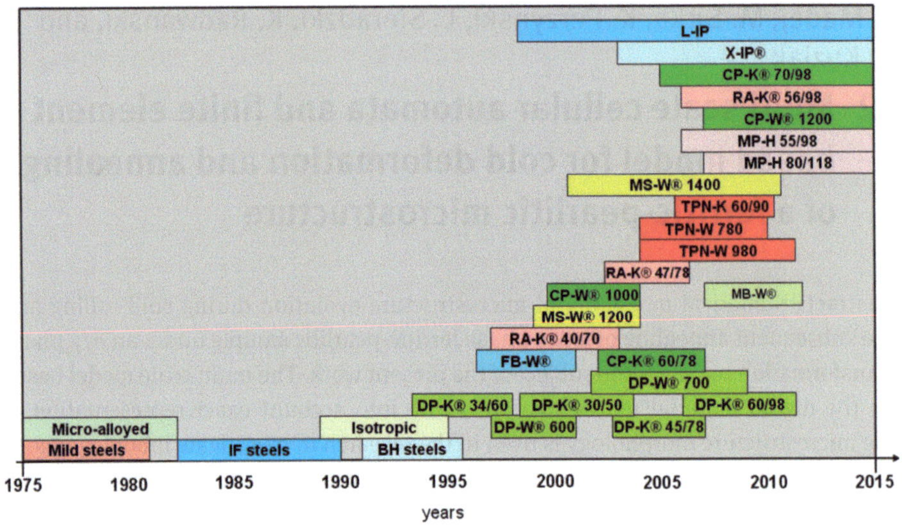

Fig. 12.1: New steel grade development chart in the ThyssenKrupp Steel company (DP – dual phase, BH – baking hardening, IF – interstitial free, CP – complex phases, MS – martensitic, FB – ferrite-bainite, RA – retained-austenite, TPN – three-phase nano, X-IP – iron-manganese TWIP, L-IP – light induced plasticity) [8].

facturing or exploitation stages can eventually lead to superior material properties at the macro-scale level. Finally, some of these innovative steels developed in laboratory conditions find their application in manufacturing auto body components in industrial conditions [5–7].

One of the most important groups of new steel grades is the AHSS group. A major advantage of these steels is that they provide the possibility of reducing automobile weight (increasing fuel efficiency), while maintaining or even increasing their safety under exploitation conditions (crash worthiness). Particular focus in the present research is put on DP (Dual Phase) steel as a representative of the AHSS group. DP thin sheets with tensile strength of 400–1200 MPa have been successfully applied in the production of automobile structural parts because they are characterized by a combination of high strength, good formability, high bake hardenability and crash worthiness. These properties are the result of properly designed microstructure consisting mainly of ferrite matrix (around 70–90 %) and hard martensitic phase islands (around 10–30 %), as seen in Fig. 12.2.

The most direct way of obtaining DP ferritic-martensitic structures is the annealing of steel in the ferrite-austenite ($\alpha + \gamma$) two-phase region, called intercritical annealing, followed by controlled cooling, causing the austenite to transform into martensite [9–11]. A more advanced process of DP microstructure formation, called continuous annealing, involves heating and soaking of cold rolled sheets in the intercritical temperature range, followed by two-stage cooling. The first stage with moderate cooling

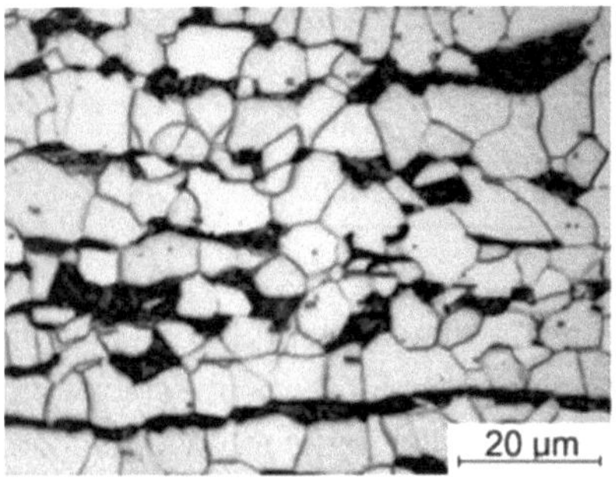

Fig. 12.2: Dual phase steel containing 27 % of martensite: white phase – ferrite, dark phase – martensite; LOM (Light Optical Microscopy).

rate is intended to produce the required volume fraction of ferrite in the microstructure. The aim of the second stage is to transform the remaining austenite into martensite. However, experimental process design is usually time consuming and expensive, which is why much attention has been recently paid to the development of accurate numerical models. Such models can be used to support the design of efficient manufacturing cycles, which allow obtainment of the required DP steel morphology. The investigation of the SRX (Static Recrystallization) phenomenon during the first stage of the process, namely the annealing of a cold deformed ferritic-pearlitic structure to the intercritical temperature range, is the subject of the present work. Firstly, the results of the experimental analysis are presented within the paper. They clearly show the importance of the proper capturing of material inhomogeneities in the numerical model in order to predict real material behavior. Secondly, the developed multi scale model combined with the SRX model is discussed in detail and compared with the obtained experimental results.

12.2 Experimental investigation of static recrystallization

As mentioned, the first stage in manufacturing DP steel is the cold rolling of ferritic-pearlitic microstructure followed by the SRX. This part is analyzed within the present research. A low carbon, cold rolled steel sheet, 1 mm thick was subjected to heating. The progress of the SRX at various temperatures was investigated. The chemical composition of the investigated steel is given in Tab. 12.1. First, the material was casted into a 70 kg ingot using a laboratory vacuum induction furnace. Second, the ingot was forged into bars having a squared cross section of 45 mm × 45 mm. Next, the bars were hot rolled into a 3 mm thick plate with the use of a laboratory reversing mill and cooled in such a way to enable the development of the ferritic-pearlitic microstructure until reaching the ambient temperature. Subsequent cold rolling into a 1 mm thick sheet was also conducted using a laboratory rolling mill.

Tab. 12.1: Chemical composition of experimental steel (wt.%).

C	Mn	Si	P	S	Cr	Ni	Mo	Ti	Al
0.09	1.42	0.1	0.011	0.010	0.35	0.01	0.02	0.001	0.043

The recrystallization process was investigated using the dilatometer DIL 805A/D manufactured by Bchr Termoanalyse GmbH. The kinetics of the SRX was investigated in 1 mm × 7 mm samples cut from the cold rolled sheet parallel to the rolling direction. The initial microstructure of the sheet after cold rolling is presented in Fig. 12.3 (a). The samples were then heated to different temperatures in the range of 600–750 °C at a rate of 3 °C/s. After reaching the defined temperatures, the samples were subjected to fast cooling with nitrogen in order to freeze the microstructure. Selected scanning electron micrographs of the samples after reaching the required temperatures are shown in Fig. 12.3 (b)–(d). Images were taken in the middle of the samples, and this location is further tracked within the experimental as well as numerical investigation.

As can be seen in Fig. 12.3 (a), the microstructure in the cold rolled state is composed of banded, severely deformed, ferrite grains and comparably deformed aggregates of pearlite colonies. The relationship between the band width and length, called aspect ratio, is consistent with the degree of macroscopic deformation in the cold rolling process. During the heating process, the beginning of the SRX is observed at around 600 °C. The recrystallization process is composed of two steps,namely nucleation and growth of nuclei. Looking at the partly recrystallized microstructure of the samples, one may say that the nucleation process is not clearly connected with the favorable sites, such as grain boundaries or deformation bands. Instead, the nucleation starts in the regions of maximum stored energy. Generally, two mechanisms of nucleation can be distinguished in the samples, namely nucleation by sub-boundary migration and preexisting boundary migration. The sub-grain coalescence mechanism was also occasionally observed. Recrystallization is completed at around 750 °C. It is important to note that the cementite particles are still not dissolved at this temperature and have the privileged orientation parallel to the rolling direction (Fig. 12.3 (d). In order to quantitatively examine the state of the samples subjected to annealing, EBSD (Electron Back Scattered Diffraction) analysis was conducted with SEMFEG. The EBSD patterns were acquired using an acceleration voltage of 20 kV. All analysis was conducted on a hexagonal scan grid 180 × 70 μm with a scan step size 0.5 μm. The EBSD data were post-processed by means of the TSL® software. The EBSD analysis can effectively distinguish the recrystallized grains since they do not contain the deformation substructure composed of dislocation arrangements and are surrounded by high-angle grain boundaries. It may be easily distinguished by Kernel Average Misorientation (KAM). This parameter is used in the determination of strain distribution in the material. A KAM value smaller than 0.5 is used to quantitatively determine the volume fractions of recrystallized ferrite and ferrite after phase transformation [12]. The

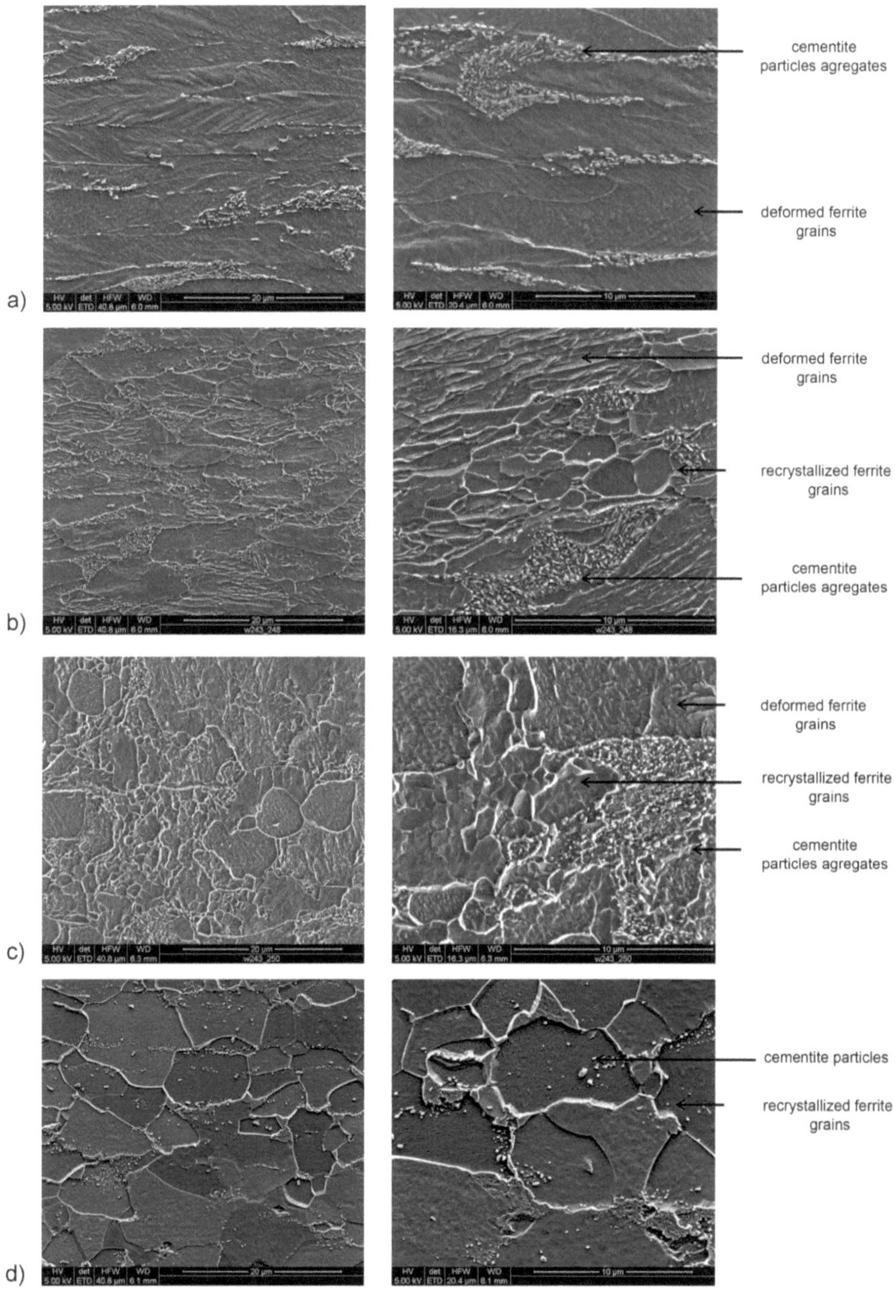

Fig. 12.3: Microstructures of the samples heated to different temperatures: (a) cold rolled, (b) 600 °C, (c) 670 °C, (d) 750 °C; SEMFEG (Field Emission Gun Scanning Electron Microscope). Right column presents results at higher magnification corresponding to the left column.

measurement of the recrystallized grain fraction and the average grain size (average equivalent diameter) were derived from EBSD analysis. The results of the EBSD measurements related to the early stages of recrystallization are given in Fig. 12.4. They are presented in the form of misorientation maps with the grain boundaries marked in dark and gray colors distinguishing high- and low-angle boundaries, respectively.

It is seen in Fig. 12.4 (a) that the elongated grains contain the deformation substructure. At 600 °C, the small, dislocation-free and nearly equiaxed grains appear in the microstructure. However, there is also the recovery of dislocation sub-structure in some deformed grains. As presented in Fig. 12.4 (d) and (e), between 680 and 690 °C, the size of the deformed grains increases, which may indicate that some high-angle boundaries were annihilated. The recrystallized grains are larger compared to the state observed during annealing to temperatures in the range of 600–670 °C (Fig. 12.4 (b) and (c)). This may reflect grain growth without substantial nucleation during annealing to 690 °C. The microstructure obtained after heating to 750 °C was completely recrystallized, which is why it was not presented in Fig. 12.4. The main cause of the observed progress of the microstructure evolution during recrystallization is most likely connected with substantial variability in the strain energy stored in the microstructure and the effect of recovery diminishing the driving force of the recrystallization. As mentioned, the quantification of the microstructures obtained after annealing to different temperatures was conducted using KAM maps that allow for differentiation between recrystallized and non-recrystallized regions of the material (Fig. 12.5). Blue colour in this figure represents KAM values below 0.5, which is characteristic of recrystallized grains.

As presented, significant inhomogeneities in the cold rolled ferritic-pearlitic microstructure is the main feature of material subjected to continuous annealing. These inhomogeneities influence the kinetics of the SRX during annealing cycles. Therefore, in modelling of the SRX, it is essential to adequately describe such inhomogeneities in the microscale. Conventional approaches used for the numerical simulation of macro scale material behavior during cold rolling have some limitations in predicting the influence of the micro scale features such as strain distribution. Thus in order to accurately predict the state of material prior to annealing, a multi scale model based on the DMR (Digital Material Representation) approach was used in the present work. Obtained results accounting explicitly for microstructural features during numerical modelling were then incorporated into the developed CA (Cellular Automata) model for the SRX in ferritic-pearlitic microstructures. The multi scale model based on DMR as well as the SRX model for ferritic-pearlitic steels are described in the following chapters, and the obtained results are then validated against the described experimental research.

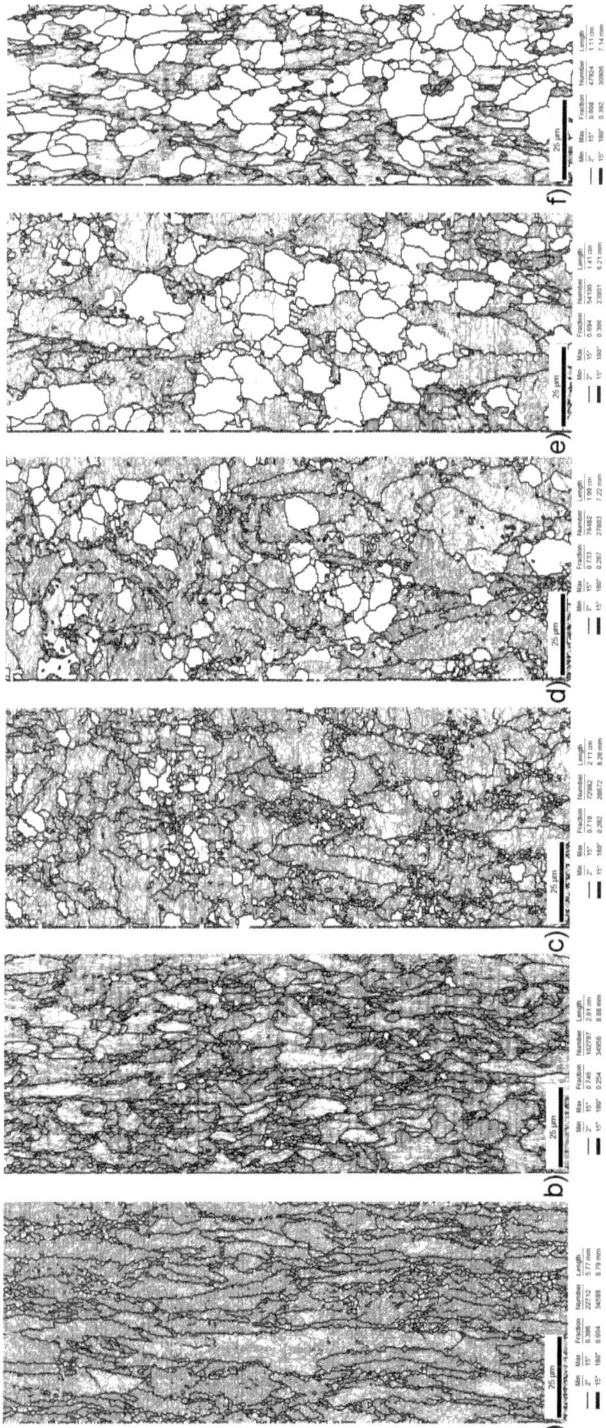

Fig. 12.4: Misorientation maps with 15° random grain boundaries of the microstructure: (a) cold rolled; heated to: (b) 600 °C, (c) 670 °C, (d) 680 °C, (e) 690 °C, (f) 700 °C.

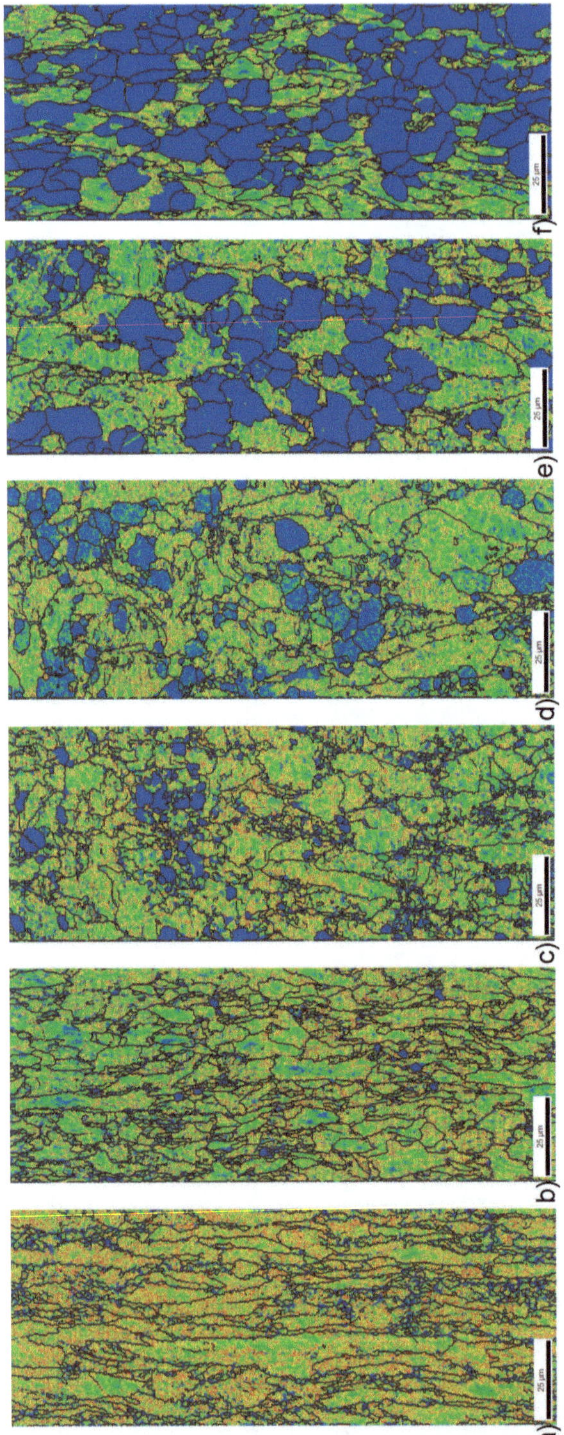

Fig. 12.5: Kernel maps presenting recrystallized fraction of the microstructure: (a) cold rolled; heated to: (b) 600 °C, (c) 670 °C, (d) 680 °C, (e) 690 °C, (f) 700 °C.

12.3 Digital material representation of the ferritic-pearlitic microstructure

The concept of the DMR has been recently proposed and is dynamically evolving [13–17]. The main objective of the DMR is the creation of a digital representation of microstructure with an explicit representation of its features. The DMR basic concept creates the possibility to describe material behavior at various scales. Such an approach offers the gathering and processing of related metallurgical data at different levels of description. The more precisely the DMR is applied, the more realistic are the results of calculations regarding material behavior. Due to that conclusion, the detailed virtual analysis of simulation results can be performed, while errors of calculations are minimized. Various methods can be used to create statistical (e.g., cellular automata, Monte Carlo, sphere growth, inverse analysis) or exact representation of microstructure [17]. To ensure a detailed transfer of the real microstructure into the digital material model, an image processing algorithm described in [18] was used in the present work. As a result, the exact replication of complex shapes of particular phases can be obtained. The procedure of image processing was composed of several steps not only to replicate the shapes of particular phases, but also to distinguish information of subsequent grains. The initial microstructure was subjected to a standard image processing algorithm resulting in binary representation of an image as seen in Fig. 12.6. Then, a colouring algorithm based on cellular automata grain growth method was applied. A single pixel in each grain is selected to represent a grain nuclei. Next, a simple transition rule is applied: when a neighbour of a particular cell in the previous time step is in the state 'already grown', then this particular cell can also change its state. The grains grow with no restrictions until they fill the entire investigated domain. Finally, when each grain has the unique colour identifier, thin grain boundaries are removed. It have to be emphasise that in the present work different colours of subsequent grains do not represent crystallographic orientations, they are just used to distinguished particular grains. Details on the CA grain growth algorithm

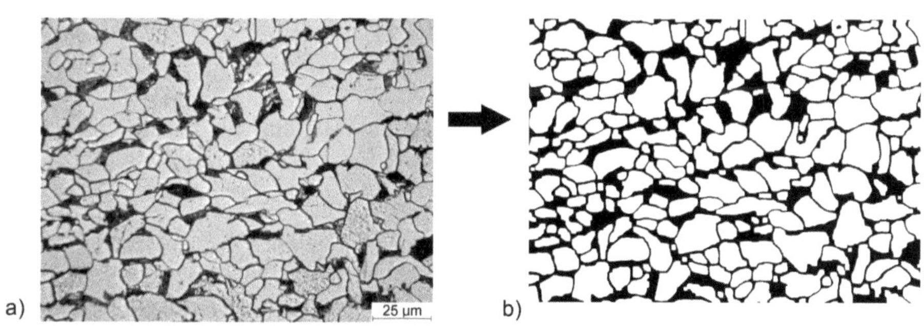

Fig. 12.6: LOM image to binary image conversion of ferritic-pearlitic microstructure: (a) LOM, (b) binary image.

Fig. 12.7: Digital material representation of a two phase ferritic-pearlitic microstructure.

can be found in the authors' earlier work [19]. As a result, a two phase ferritic-pearlitic digital microstructure with separated features is obtained (Fig. 12.7).

To incorporate the obtained digital material representation and to properly capture material behavior along the grain boundaries, specific nonuniform FE (Finite Element) meshes that are refined along the grain boundaries have to be generated. In the commercial FE software, there is a possibility to refine the mesh, but specific areas have to be marked manually. For simple geometry, this approach is usually satisfactory. However, when digital microstructures with sophisticated shapes of grain/phase boundaries are considered, this approach cannot be applied. As seen in Fig. 12.8, the specific triangular mesh was created with a previously developed mesh generation tool and incorporated into the commercial finite element software. Details on this module can be found in [20, 21].

The selected flow curves describing subsequent phases of ferrite and pearlite, respectively, are assigned to particular grains in the microstructure. To capture differences in grain flows due to various crystallographic orientations, a diversification in the flow curves for each grain is introduced using the Gaussian distribution. Thus, each grains is described by slightly different flow stress values [17]. The differences in crystallographic aspects of deformation are not considered in the present model as conventional finite element model is applied. Finally, to provide the capabilities of material behavior not only at the micro scale level, the digital material representation is combined with the macro scale information in the form of the multi scale concurrent model [22, 23].

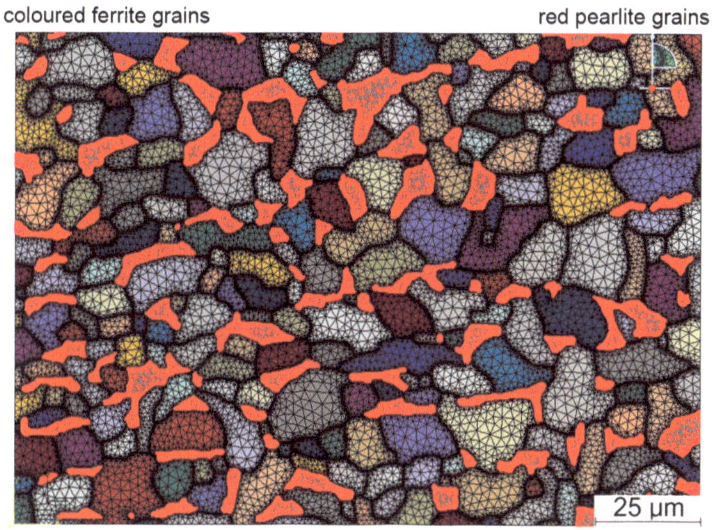

Fig. 12.8: FE mesh imposed on the DMR of the two phase ferritic-pearlitic microstructure.

12.4 Multi scale model of rolling

In the developed multi scale concurrent finite element based model, macro scale material behavior is modelled as a global model with a relatively coarse finite element mesh. At the same time, some regions in this mesh are selected, and finite element meshes created on the basis of the DMR are attached. In this way the model provides information regarding general strain, stress, temperature, etc. at the macro scale level and at the same time very detailed information about microstructure behavior at the micro scale level. The micro model is created using a partition of the macro model.Firstly, the macro scale simulation is carried out, and then the DMR based micro model is resubmitted with displacement boundary conditions taken from the macro scale simulation. Both at macro and micro scales plain strain boundary conditions were applied. It have to be pointed out that assuming columnar grains in the out-of-plane direction may slightly influence equivalent strain distribution at the micro scale level.Numerical results of cold rolling obtained at the macro and micro scale levels using this approach are shown in Figs. 12.9 and 12.10. The obtained inhomogeneous strain distribution and the final geometry of the grains are the input parameters for the CA model of the SRX.

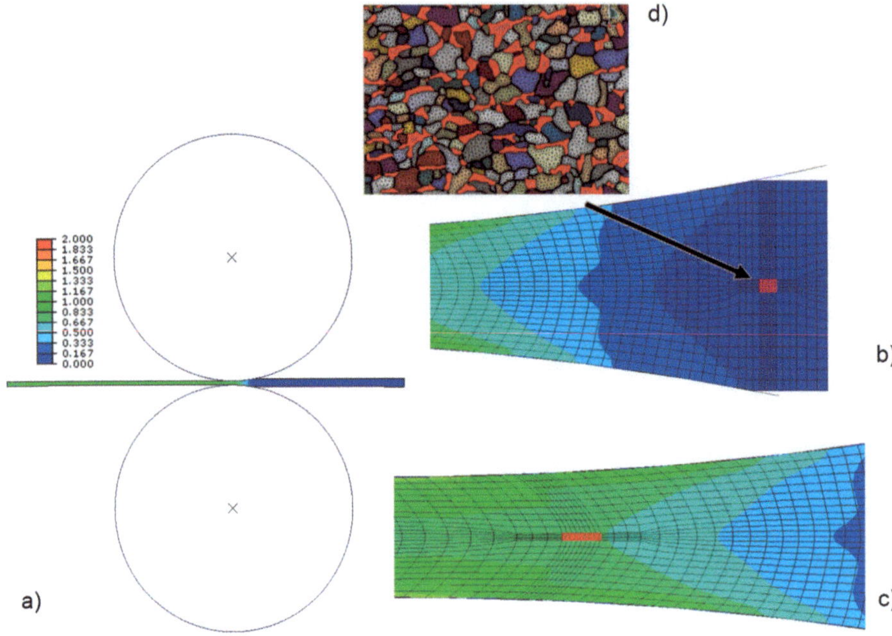

Fig. 12.9: (a) Equivalent plastic strain distribution obtained during macro scale rolling, (b), (c) enlargement of the macro scale model during deformation with clearly visible location of the micro scale model, (d) initial geometry of the micro scale DMR model.

12.5 Cellular automata model of static recrystallization

The data transfer from the FE model into the CA code is based on the SPH (Smoothed Particle Hydrodynamic) interpolation method. The concept of an integral representation of function $f(x)$ at location x in the SPH method is given by an integral of multiplication of the product of the function and an appropriate kernel function W_{ij}. If the value of the $f(x)$ function is known only in a finite set of discrete points, it can be written as:

$$\langle f(x)\rangle \cong \sum_{j=1}^{N} f(x_j) W_{ij}(x - x_j, h_{\text{sm}}) V_j\,, \tag{12.1}$$

where $\langle\cdot\rangle$ is the kernel approximation, W_{ij} the kernel function, h_{sm} the smoothing length of the support domain, and V_j the volume associated with the j-th particle. Eq. (12.1) is a basic equation used in the SPH method. The value of the function at point x is calculated by a summation of the contribution from a set of neighbouring particles (j subscript) from the support domain of the x particle. The Kernel function plays an important role and thus should be required to meet several consistency conditions to be applicable in the SPH method. Several Kernel functions that meet these

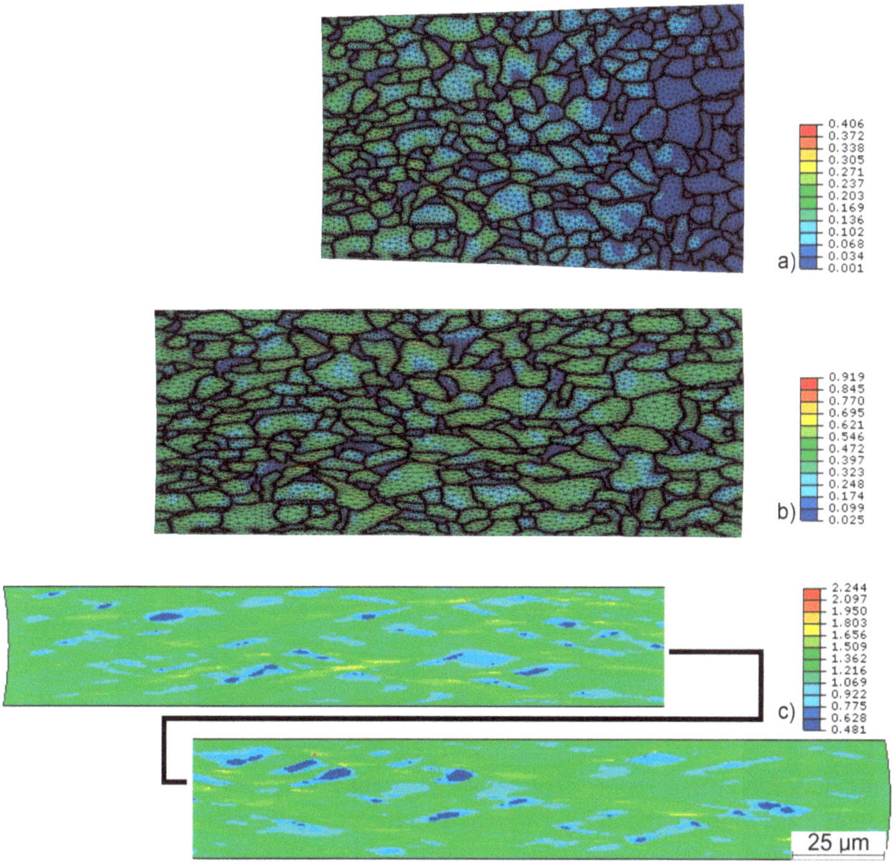

Fig. 12.10: (a), (b) Micro scale model based on the DMR of two-phase steel at various stages of deformation, (c) final equivalent plastic strain distribution without the FE mesh. Dimensions of the analyzed DMR after deformation are 484 × 41 μm.

requirements can be defined. The quintic spline Kernel is used in this work:

$$W(R, h_{\text{sm}}) = \alpha_d \begin{cases} (3-R)^5 - 6(2-R)^5 + 15(1-R)^5 & 0 \le R < 1 \\ (3-R)^5 - 6(2-R)^5 & 1 \le R < 2 \\ (3-R)^5 & 2 \le R < 3 \\ 0 & R > 3 \end{cases} \tag{12.2}$$

with $\alpha_d = 7/478\Pi h_{\text{sm}}^2$. The strain values calculated at every Gaussian integration point are interpolated by the SPH interpolation into the discrete regular CA space. After interpolation and transfer into the CA space, the strain field is converted into the accumulated energy within subsequent grains using the commonly used formula [24, 25]:

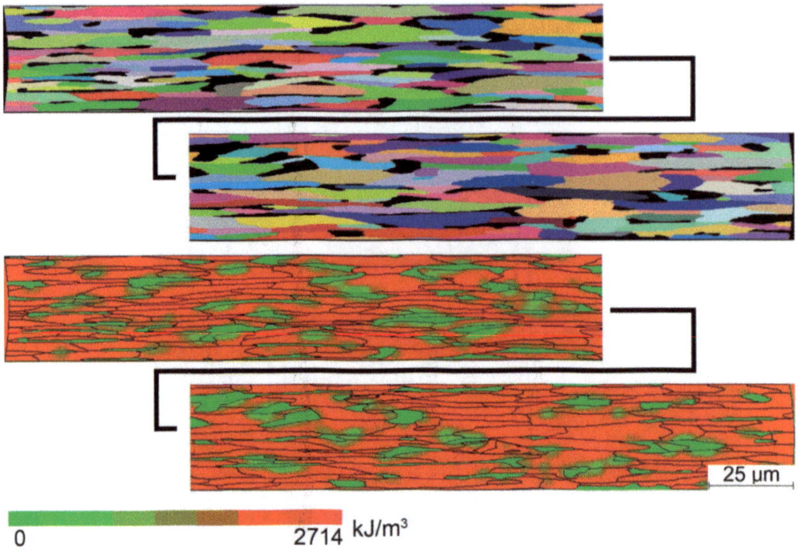

Fig. 12.11: (a) Shape of deformed grains and (b) corresponding accumulated energy distribution in the CA space.

$$H = \frac{\varepsilon}{a\varepsilon + b}\gamma_{\mathrm{LAGB}}\,, \tag{12.3}$$

where ε is the equivalent strain, a and b are coefficients, γ_{LAGB} is the low angle grain boundary energy. The shape of the deformed grains and corresponding accumulated energy distribution in the cellular automata space are presented in Fig. 12.11. Obtained level of accumulated energy is consistent with the literature observations [26, 27].

The obtained data serve as input for the developed CA model dedicated to the SRX simulations. The CA model with all the details on defined CA space, neighborhood and transition rules is described in [19]. Thus, only basic assumptions of the model are summarized below. A two dimensional space with a size of 2114 × 183 CA cells has been created. To correspond with the size of the deformed sample, each CA cell has a physical length equal to 0.229 µm. Only two state variables have been created – a CA cell can either be in the unrecrystallized or recrystallized state. Each CA cell in the unrecrystallized state can become the nucleus with the probability determined by:

$$p = NSt\,, \tag{12.4}$$

where S is the volume in which nucleation can appear, t the time, and N the coefficient computed by:

$$N = M_N \exp\left(-\frac{Q_{\mathrm{a}}}{RT}\right), \tag{12.5}$$

where Q_{a} is the activation energy for nucleation, R the universal gas constant, T the temperature, and M_N the coefficient described as:

$$M_N = C_0(H_i - H^{\mathrm{C}})\,, \tag{12.6}$$

where C_0 is the scaling parameter, H_i the amount of energy in the particular cell, and H^C the critical amount of energy which is necessary to trigger nucleation. After the CA cell becomes a nucleus of a new recrystallized grain, the grain growth process is simulated. It is assumed that the grain growth is controlled by two major factors, namely stored energy and grain boundary curvature. The velocity of a moving grain boundary is calculated as:

$$v = MP\,, \tag{12.7}$$

where M is the grain boundary mobility and P the net pressure on the grain boundary. Grain boundary mobility is defined by:

$$M = \frac{D_0 b_B^2}{kT} \exp\left(-\frac{Q_b}{RT}\right), \tag{12.8}$$

where D_0 is the diffusion constant, b_B is Burger's vector, k the Boltzmann constant, and Q_b the activation energy for grain boundary motion. The value of the net pressure depends on the nature of the driving force causing the grain growth. When the stored energy is under consideration, the net pressure is equal to the value of the stored energy in a particular cell. When grain boundary curvature is considered, the net pressure is expressed as:

$$P = \kappa \gamma_{HAGB} \tag{12.9}$$

where κ is the grain boundary curvature and γ_{HAGB} the high angle grain boundary energy. The methodology of grain boundary curvature computation is described elsewhere [28]. In the present model, the influence of crystallographic orientation on the grain growth characteristic is not taken into account. Finally, the cell coverage by the recrystallization front or another grain is calculated by:

$$\mathrm{RX}_{i,t}^{\mathrm{fraction}} = \mathrm{RX}_{t,t-1}^{\mathrm{fraction}} + \sum_{j=1}^{N} \left(\frac{v_j t_{\mathrm{step}}}{c_s}\right), \tag{12.10}$$

where $\mathrm{RX}_{i,t-1}^{\mathrm{fraction}}$ is the level of coverage of the i-th cell in the previous $(t-1)$ time step, v_j the velocity of the recrystallization front or recrystallized grain boundary, t_{step} the length of time step, and c_s the cellular automata cell size. As mentioned, the developed model was used to simulate the SRX of the cold deformed ferritic microstructure presented in Fig. 12.11. The major model parameters are reported in [19]. To reflect experimental research, the sample was heated up to 600 °C, 670 °C, 680 °C, 690 °C, 700 °C, and 750 °C with a heating rate of 3 °C/s and then cooled down to suppress the grain growth after the SRX. A sample of the grain morphology results are presented in Fig. 12.12. As can be seen, there is grain refinement in the fully recrystallized microstructure. A clear correlation between nucleation sites and regions with high values of the stored energy is observed. That is why it is of importance to precisely predict inhomogeneous energy distribution during numerical calculations. Due to low accumulated energy values, the pearlitic phase did not recrystallize, which has a significant influence on further microstructure evolution during phase transformation

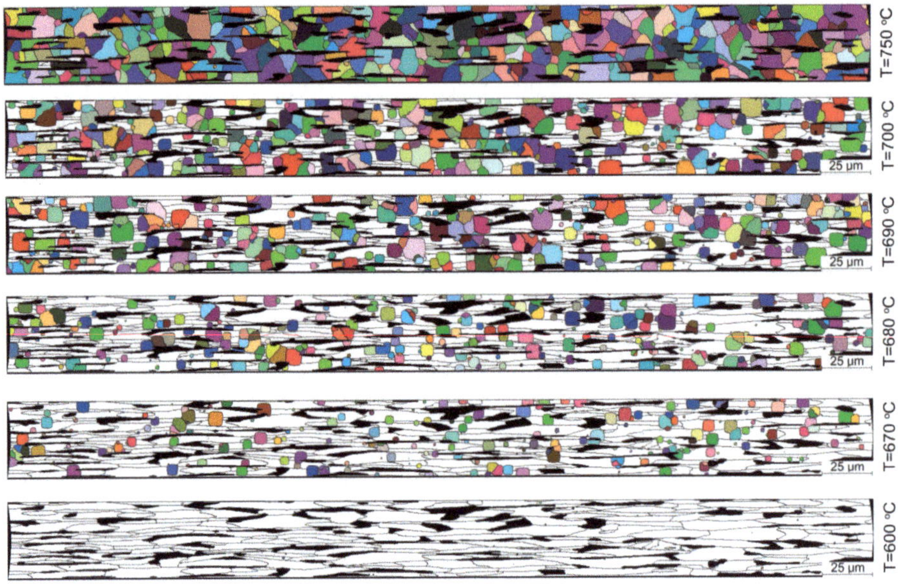

Fig. 12.12: Microstructure morphology after SRX with heating rate 3 °C/s up to (a) 600 °C, (b) 670 °C, (c) 680 °C, (d) 690 °C, (e) 700 °C, (f) 750 °C; black grains – pearlite.

from the ferrite to austenite region. Since the cementite plates are obstacles for the recrystallization process in pearlite, the driving force for recrystallization in this case was arbitrarily set to zero.

The obtained results are in good qualitative agreement with the results presented in Fig. 12.5. Moreover, quantitative agreement is also satisfactory. A comparison between the measured and calculated grain sizes and the recrystallized fractions are presented in Fig. 12.13.

As can be seen in Fig. 12.13, the comparison between the results of measurement and simulation with CA is acceptable. It is noteworthy that the accuracy of the method used for the microstructure quantification has some limitations. The EBSD analysis is a long lasting process, and typically the number of measured microstructural features may not be large enough to accurately characterize the state of the sample. The obtained microstructure morphology as well as information on accumulated energy will become the input data for further numerical modeling of ferrite/pearlite to austenite transformation which is the next step in the manufacturing cycle of dual phase steels. This will be the subject of future works of the authors.

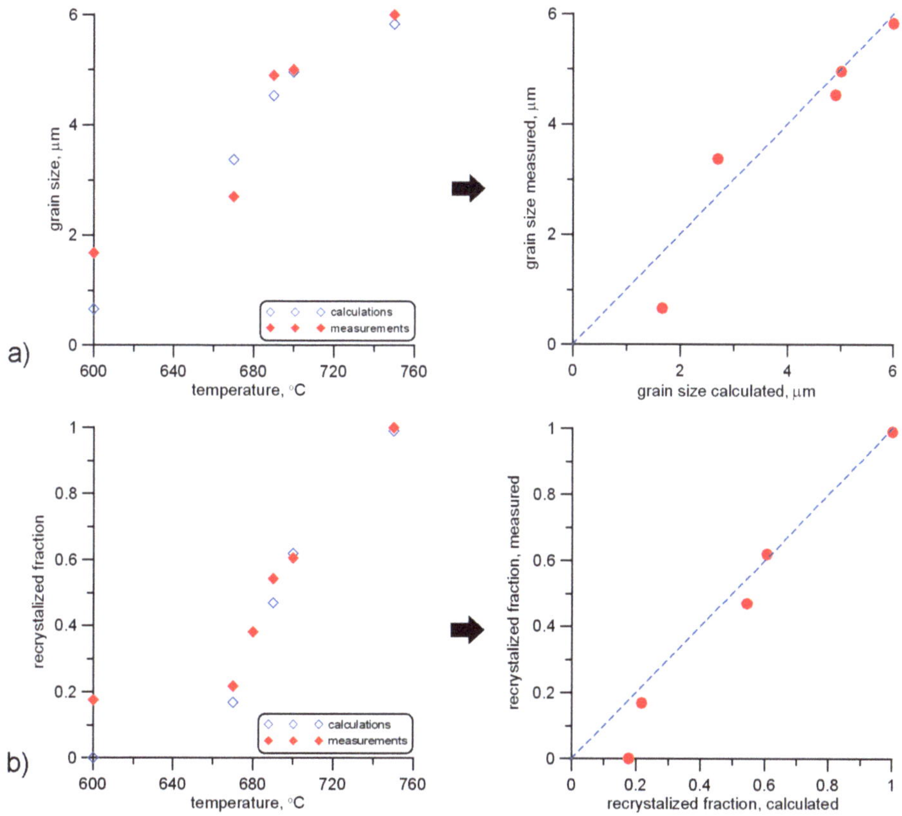

Fig. 12.13: Comparison of the results of measurement and model predictions: (a) average equivalent grain diameter, (b) recrystallized volume fraction.

12.6 Conclusions

The capabilities of the multi scale Cellular Automata Finite Element based model in application to phenomena occurring at the microstructure level during cold rolling and recrystallization of the ferritic-pearlitic phases were investigated within the paper. Based on the presented research the following conclusions can be drawn:

- application ofthe DMR model incorporated into the macro scale FE computations provides a detailed description of microstructure features of dual phase steels. To properly capture the solution gradient occurring along phase boundaries, a refined FE mesh has to be used.
- detailed local strain distribution in ferrite grains can be calculated and applied as the input data for the CA simulation of the SRX of ferrite.
- application of the SPH interpolation method provides the possibility to accurately transfer data between the FE mesh and CA space.

- based on a detailed description of regions with high values of stored energy (preferable sites for nucleation), qualitative and quantitative agreements with experimental results were obtained.
- the nucleation model in the CA approach should be a function of accumulated energy. Other solutions, such as homogeneous site saturated nucleation, may provide inaccurate results.
- based on the model's predictions, grain growth during the SRX is the most intense in the temperature range between 680 ° and 700 °C. In this temperature interval most of the microstructure became recrystallized.
- to model microstructure behavior in higher temperatures – above 750 °C – a phase transformation of ferrite/pearlite to austenite has to be implemented.

The obtained grain morphology of recrystallized ferrite and the remaining pearlite can be the input data for such calculations. Application of the more complex crystal plasticity finite element model [20, 28, 29] to take into account influence of crystallographic orientation on subsequent static recrystallization kinetics will be the subject of future work.

Acknowledgment: The research leading to these results has received funding from the European Community's Research Fund for Coal and Steel (RFCS) under grant agreement no. RFSR-CT-2011-00014. FEM calculations were realized at the ACK CYFRONET AGH MNiSW/IBM_BC_HS21/AGH/075/2010.

References

[1] Beladi H, Adachi Y, Timokhina I, Hodgson P. Crystallographic analysis of nanobainitic steels. Scripta Materialia. 2009;60:455–458.
[2] Robertson L, Hilditch T, Hodgson P. The effect of prestrain and bake hardening on the low-cycle fatigue properties of trip steel. International Journal of Fatigue. 2008;30:587–594.
[3] Sabirov I, Estrin Y, Barnett M, Timokhina I, Hodgson P. Tensile deformation of an ultrafine-grained aluminium alloy: Micro shear banding and grain boundary sliding. Acta Materialia. 2008;56:2223–2230.
[4] Timokhina I, Hodgson P, Ringer S, Zheng R, Pereloma E. Precipitate characterisation of an advanced high-strength low-alloy (hsla) steel using atom probe tomography. Scripta Materialia. 2007;56:601–604.
[5] Liedl U, Traint S, Werner E. An unexpected feature of the stress-strain diagram of dual-phase steel. Computational Materials Science. 2002;25:122–128.
[6] Saleh M, Priestner R. Retained austenite in dual-phase silicon steels and its effect on mechanical properties. Journal of Materials Processing Technology. 2001;113:587–593.
[7] Bayram A, Uguz A, Ula M. Effects of microstructure and notches on the mechanical properties of dual-phase steels. Materials Characterization. 1999;43:259–269.
[8] Jaroni U, Imlaau K, Osburg B. Neue losungsansatze fur innovative produkte und umformtechnologien im automobilbau. In: Conf. Proc. ASK25. pp. 3–14.
[9] Speich G, Miller R. Fundamentals of dual-phase steels. Metallurgical Society/TMS-AIME. 1981; 503:279–304.

[10] Speich G, Miller R. Structure and properties of dual-phase steels. Metallurgical Society/TMS-AIME. 1979;145:62–90.
[11] Qu J, Dabboussi W, Hassani F, Nemes J, Yue S. Effect of microstructure on static and dynamic mechanical property of a dual phase steel studied by shear punch testing. The Iron and Steel Institute of Japan (ISIJ International)E. 2005;45:1741–1746.
[12] Zaefferer S, Romano P, Friedel F. Ebsd as a tool to identify and quantify bainite and ferrite in low alloyed al-trip steels. Journal of Microscopy. 2008;230:499–508.
[13] Bernacki M, Resk H, Coupez T, Loge R. Finite element model of primary recrystallization in polycrystalline aggregates using a level set framework. Modelling and Simulation in Materials Science and Engineering. 2009, 17.
[14] Dawson P, Miller M. The digital material - an environment for collaborative material design; 18 December 2009. Available from: http://anisotropy.mae.cornell.edu.
[15] Li W, Zabaras N. A virtual environment or the interrogation of 3d polycrystalline microstructures including grain size effect. Computational Material Science. 2009;44:1163–1177.
[16] Rollett A, Saylor D, Frid J, et al. Modelling polycrystalline microstructures in 3d. In: Proc. Conf. Numiform, eds S. Ghosh. pp. 71–77.
[17] Madej L, Rauch L, Perzynski K, Cybulka P. Digital material representation as an efficient tool for strain inhomogeneities analysis at the micro scale level. Archives of Civil and Mechanical Engineering. 2011;11:661–679.
[18] Rauch L, Madej L. Application of the automatic image processing in modelling of the deformation mechanisms based on the digital representation of microstructure. International Journal for Multiscale Computational Engineering. 2010;8:343–356.
[19] Sieradzki L, Madej L. A perceptive comparison of the cellular automata and monte carlo techniques in application to static recrystallization modeling in polycrystalline materials. Computational Materials Science. 2013;67:156–173.
[20] Madej L, Perzynski K, Wajda W, Paul H, Cybulka P, Rauch L. Numerical analysis of the texture evolution on the basis of the cpfe and digital material representation. Steel Research International, Special Edition. 2011;67:854–859.
[21] Madej L, Kruzel F, Cybulka P, Perzynski K, Banas K. Generation of dedicated finite element meshes for multiscale applications with delaunay triangulation and adaptive finite element - cellular automata algorithms. Computer Methods in Material Science. 2012;12:85–96.
[22] Madej L, Mrozek A, Kus W, Burczynski T, Pietrzyk M. Concurrent and upscaling methods in multi scale modelling - case studies. Computer Methods in Material Science. 2008;8:1–15.
[23] Madej L, Hodgson P, Pietrzyk M. Development of the multi-scale analysis model to simulate strain localization occurring during material processing. Archives of Computational Methods in Engineering. 2009;16:287–318.
[24] Song X, Rettenmayr M. Modeling recrystallization in a material containing fine and coarse particles. Computational Materials Science. 2007;40:234–245.
[25] Song X, Rettenmayr M, Muller C, Exner H. Modeling of recrystallization after inhomogeneous deformation. Metallurgical and Material Transactions A. 2001;32:2199–2206.
[26] Choi S, Jin Y. Evaluation of stored energy in cold-rolled steels from ebsd data. Metallurgical and Material Transactions A. 2004;371:149–159.
[27] Raabe D, Hantcherli L. 2d cellular automaton simulation of the recrystallization texture of an if sheet steel under consideration of zener pinning. Metallurgical and Material Transactions A. 2005;34:299–313.
[28] Delannay L, Jacques P, Kalidini S. Finte element modeling of crystal plasticity with grains shaped as trunced octahedrons. Journal of Plasticity. 2006;22:1879–1890.
[29] Zhang P, Balint D, Lin J. An integrated scheme for crystal plasticity analysis: Virtual grain structure generation. Computational Materials Science. 2011;50:2854–2864.

J. Wiedmaier, W. Verestek, U. Weber, and S. Schmauder

13 Multiscale simulation of the mechanical behavior of nanoparticle-modified polyamide composites

Abstract: In this paper nano silicate modified polyamide 6 (PA 6) is investigated. Micromechanical modeling is proposed to predict the mechanical behavior/response of this material. A RVE (representative volume element) model is chosen to simulate the quasi static mechanical tensile behavior of the PA 6-composite. The model is modified with the consideration of debonding between the nano silicate platelets and the PA 6 matrix. The interaction between silicate platelets and PA 6 matrix, which is not accessible experimentally, is determined numerically via molecular dynamics (MD) simulations. The knowledge gained from the MD simulation as quantitative description of interface debonding (traction separation law) is used as input data for the micromechanical RVE models. The numerical predictions of mechanical properties are in very good agreement with the experimental results.

Keywords: polyamide 6-composite, layered silica, mechanical properties, micromechanical modeling, molecular dynamic simulation

13.1 Introduction

Ever since the 40's, polyamides are one of the major engineering thermoplastics in many applications, e.g. in packaging, automotive and electrical components [1]. With respect to plastics, they offer a unique combination of high mechanical strength and stiffness, high melting point and low wear due to a good chemical resistance to non-polar solvents [2]. Nowadays, more and more polyamide materials are used instead of metallic materials in the automotive industry. In order to increase the strength of polyamide further, the incorporation of a second dispersed phase of nanoscale layered silicate is required. The modified polyamide 6 can, therefore, meet higher demands, especially if high strength is required. To manufacture these new materials inexpensively and better understand their mechanical behavior, numerical simulations are essential. Based on experimental investigations on the morphology and the mechanical properties the real microstructures are modeled. The findings from the simulations can then be used for economical production, which represents an effective way to achieve the desired mechanical properties (high strength with sufficient ductility).

13.2 Used Materials

The matrix material polyamide 6 (PA 6, Ultramid B40) was provided by BASF. Nanoscale layered silica has been manufactured by ion exchange of sodium montmorillonite (MMT) and an organic modifier (OM). The morphology und mechanical properties of these materials, which are used as input for the simulations described here, have been discussed extensively in [3].

The plastic behavior of the composite has been characterized by the Johnson-Cook-Model, which is able to model the deformation with different strain rates for many materials. The Johnson-Cook-Law is based on the von-Mises-Plasticity-Model, but takes the "hardening" law as well as the strain rate dependency into account in an analytical form, see Eq. (13.1):

$$\bar{\sigma} = [A + B(\bar{\varepsilon}^{\mathrm{pl}})^n]\left[1 + C\ln\left(\frac{\dot{\bar{\varepsilon}}^{\mathrm{pl}}}{\dot{\varepsilon}_0}\right)\right](1 - \hat{\theta}^m)\,, \tag{13.1}$$

where $\bar{\sigma}$ is the yield stress; A, B, C, n, m, $\dot{\varepsilon}_0$ are material parameter at or below glass transition temperature, $\bar{\varepsilon}^{\mathrm{pl}}$ is the plastic strain, and $\dot{\bar{\varepsilon}}^{\mathrm{pl}}$ the plastic strain rate.

$$\hat{\theta} \equiv \begin{cases} 0 & \text{for } \theta < \theta_{\mathrm{tr}} \\ \frac{\theta - \theta_{\mathrm{tr}}}{\theta_{\mathrm{melt}} - \theta_{\mathrm{tr}}} & \text{for } \theta_{\mathrm{tr}} \le \theta \le \theta_{\mathrm{melt}} \\ 1 & \text{for } \theta > \theta_{\mathrm{melt}}\,, \end{cases} \tag{13.2}$$

where θ is the current temperature, θ_{tr} the glass transition temperature, and θ_{melt} the melting temperature.

These parameters were derived by iterative, inverse simulation with the software PolyUmod and based on the Johnson-Cook-Model. The reference curve for the inverse simulation is the true stress strain curve of this composite, which has been obtained by ARAMIS measurements of a tensile test [4]. All parameters were varied in a predefined interval until the simulated and the reference curve agree to a certain tolerance (Fig. 13.1). The parameters that have been determined are listed in Tab. 13.1.

Tab. 13.1: Derived Johnson-Cook parameters.

Parameter	*A*	*B*	*n*	*m*	*C*
Composite	0.044	0.034	0.141	1.409	0.057

13.3 RVE model – tensile test

The elastic-plastic material behavior of the toughened PA 6 composite has been modeled by a two dimensional representative volume element (2D RVE) model. Using optical, local strain measurements (ARAMIS) during the tensile test, the true stress strain

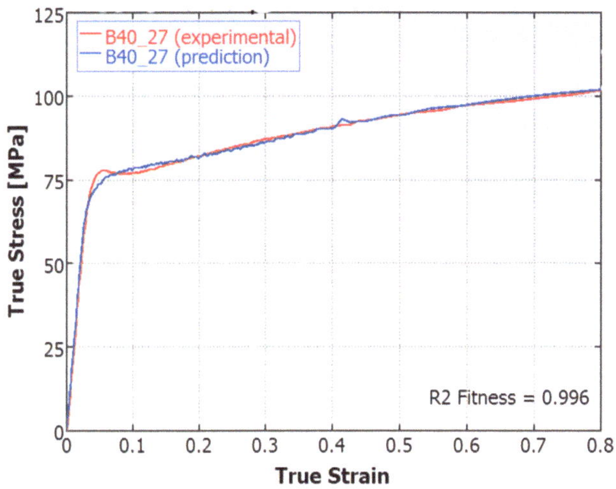

Fig. 13.1: Inverse simulation of the stress strain curve of PA 6 (red: experimental reference curve; blue: simulated curve).

behavior of the material is determined. The characteristic values for the PA 6 matrix as well as for MMT are used as input for the modeling.

To define the correct RVE size the minimum number of layered silicate platelets was investigated on the basis of the work of Gitman et al. [5] with the Chi-square criterion.

Due to the manufacturing process the tensile specimen can be divided into three areas (Fig. 13.2) [6]. Close to the surface friction between mold and cast during the injection molding process leads to an orientation of the silicate platelets parallel to the surface. In contrast there is no predominant direction of orientation at the core structure of the tensile specimen.

The experimental data obtained with ARAMIS depicts the deformation of the surface. Therefore, the models for comparison with the experiment will represent the oriented area close to the surface.

The geometry of the RVEs is modeled with the software Digimat/FE. The matrix as well as the inclusion is generated separately at the first stage. The morphological parameters of the silicate platelets (orientation and position, aspect ratio, exfoliation grade und volume fraction) can be adjusted very accurately.

The interface interaction (interface delamination) between silicate platelet and PA 6 matrix is modeled by cohesive elements (“cohesive surface”). The properties of these cohesive surfaces are described by a traction separation law. The characteristics of the interface are hardly accessible in experiment due to the nano scale size of the silicate platelets. Hence, the traction separation law is derived from atomistic simulations. This multi-scale approach is depicted in Fig. 13.3.

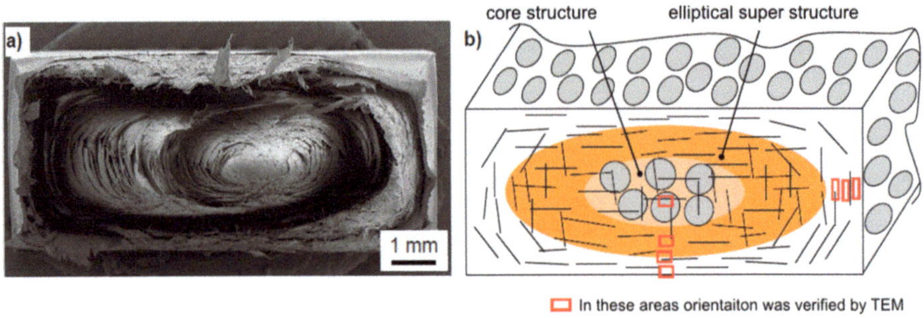

Fig. 13.2: (a) REM picture of the fracture surface of the tensile bar (b) Model of the silicate platelet orientation [6].

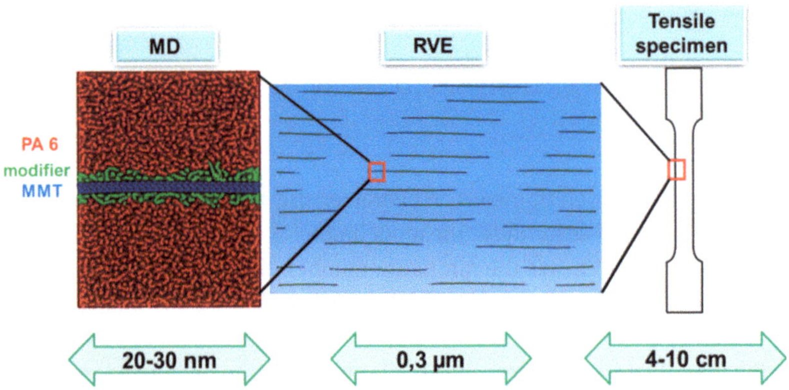

Fig. 13.3: Representation of the multi scale approach (a) molecular dynamics (MD) model, (b) 2D micro mechanical finite element model (RVE), and (c) macroscopic model of a tensile specimen.

13.4 Molecular dynamics simulations: Derivation of the traction separation law

For the MD simulations a PA 6 structure was generated by a self-avoiding random walk algorithm. In this self-avoiding random walk algorithm for each molecule a path on a cubic face centered (fcc) lattice is created. In the creation process the orientation of each step is independent of all previous ones and no site is allowed to be occupied twice. For this purpose a preexisting code from [7] for "united atom" Polyethylene structures (Hydrogen is not explicitly modeled) was modified so a certain number of PA 6 molecules with a predefined polymerization grade as an "all atom" structure (including explicitly modeled Hydrogen) is created. Organically modified (OM) MMT is inserted into the atomistic PA 6 structure. For the structure developed here water resp. humidity is not considered. The presented structure is periodic in all three dimensions; visualization of the atomistic structures was done with Ovito [8].

Due to the self-avoiding random walk the initial structure is far from equilibrium. This has to be covered by initial simulations to relax and equilibrate the system. Therefore a sequence of NPT (constant number of particles (N), constant pressure (P), constant temperature (T)) and NVT (constant number of particles (N), constant volume (V), constant temperature (T)) simulations was performed. During the NVT simulations the structure was heated up and cooled down ("simulated annealing") to accelerate the equilibration process and overcome energetic barriers more easily, while pressure relaxation happened in the NPT simulations. As MD-Code LAMMPS [9] was used. The INTERFACE Force Field [10] was used to describe the interaction between the atoms. And a time step of 0.25 fs was chosen.

Using the equilibrated structure as initial structure a nano tensile test has been simulated using a true strain rate of $5 \cdot 10^{-7}$/fs for a temperature of 293.15 K, see Fig. 13.4. PA 6 is depicted in red, the organic modifier in green and the MMT platelet in blue. As can be observed in Fig. 13.4 delamination happens between PA 6 and organic modifier. In Fig. 13.5 the resulting traction separation curve of the MD simulation is shown from which the adhesive damage behavior of the cohesive surface elements is derived. The curve depicts the resulting stress between the PA 6 matrix and the organically modified MMT. Tab. 13.2 lists the resulting values which are used as input for the traction separation law in the RVE model.

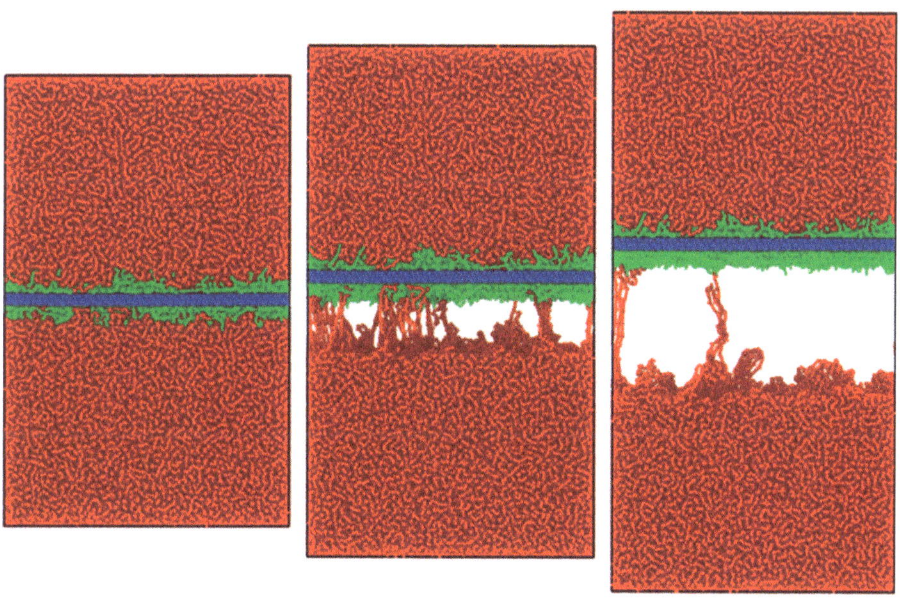

Fig. 13.4: Simulation of a nano tensile test (red = PA 6, green = organic modifier, blue = MMT).

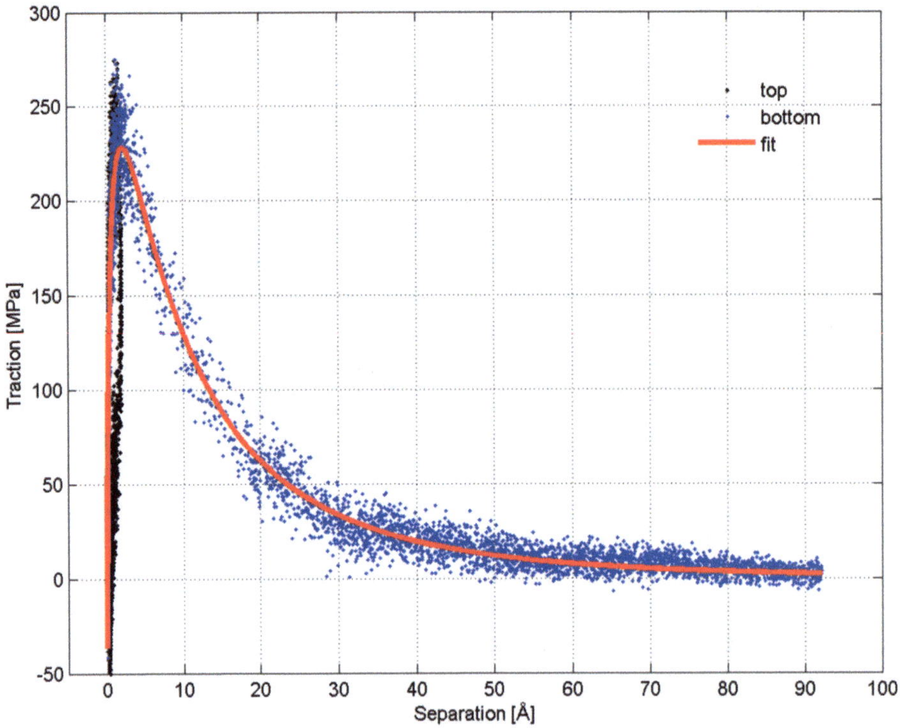

Fig. 13.5: Traction separation curve at a true strain rate of $5 \cdot 10^{-7}$/fs and 293.15 K (top = Interaction PA 6 – MMT/OM above MMT, bottom = Interaction PA 6 – MMT/OM below MMT, fit = approximation of the traction separation curve).

Tab. 13.2: Input parameter of the adhesive damage behavior (traction separation law).

Max. Traction [GPa]	Adhesive energy [J/m^2]
0.22	0.387

13.5 Results and discussion

Fig. 13.6 shows the true stress strain curves of the composite from experiment and simulation of the 2D RVE model. Experiment and simulation are in good agreement in the elastic region. In the plastic region the experimental and simulated stress strain curves show a similar trend. However, the simulated curves lie slightly above the real measured ones. This discrepancy has two reasons. In the present model the orientation of the silicate platelets was assumed to be perfectly parallel. In reality the silicate platelets in the surface area have a tendency to orient parallel to the surface, but there is a statistical influence which has not been covered in the current model. On the other

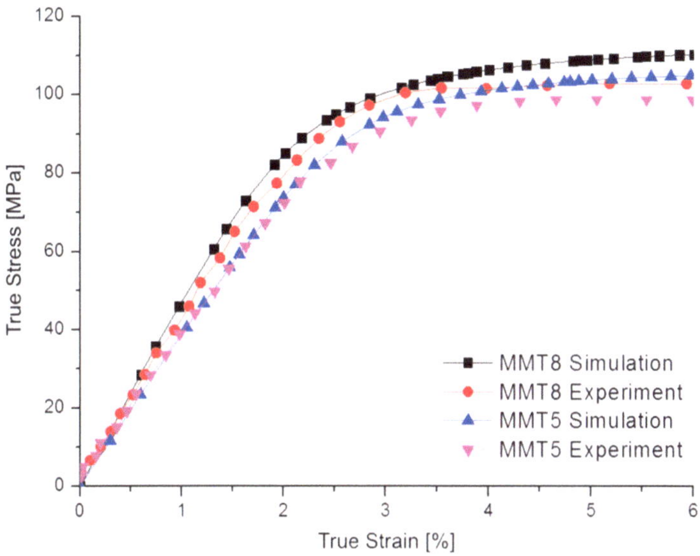

Fig. 13.6: Comparison of numerical and experimental stress strain curves of the tensile test (ARAMIS), (MMT5: 1.24 vol. % MMT; MMT8: 1.99 vol. % MMT).

hand the simulation reproduces solely the surface area whereas in the real tensile specimen three different orientations are observable, see Fig. 13.2. Furthermore the simulated strain rates of the MD simulations are too fast compared to reality. For this reason the derived values for the traction separation law are too high too.

13.6 Conclusion and outlook

Using numerical simulations, quasi static (tensile test) mechanical properties of nano particle modified Polyamide 6 was numerically investigated. Employing a multiscale approach the 2D RVE model is able to describe the behavior of the composite very well and reliably predicts the mechanical properties of the composite. The findings from the simulation can be used to optimize the composite which results in decreasing costs and time during the manufacturing process.

To predict better the influence of the strain rate on the adhesive behavior of the interface between MMT and PA 6 matrix more simulations with different strain rates are planned. Furthermore all three orientations of the silicate platelets should be considered in the RVE model. The results of MD and RVE simulations will be used as input for simulating the tensile test on a macroscopic scale.

The application of the described simulation models on binary material systems reveals extensive relations between micro structure and mechanical properties. The expected insights are supposed to help understanding the transition from ductile to stiff material properties and thereby optimize the toughness/stiffness balance of the material system.

Acknowledgment: The research was supported by Deutschen Forschungsgemeinschaft (DFG) in the projects SCHM 746/74-2 and GO 579/11-2. The MD simulations were performed on the Hermit computational cluster of HLRS at University of Stuttgart.

References

[1] Page IB. Polyamides as engineering thermoplastics materials. Tech. Rep. 11(1), RAPRA Technology Ltd., Shawbury; 2000.
[2] Michler GH, Balta-Calleja FJ. Mechanical Properties of Polymers based on Nanostructure and Morphology. Taylor & Francis Group; 2005.
[3] Huang J, Geier S, Weber U, Schmauder S. Modeling stiffness of nanolayered silicate-modified polyamide 6 via fem micromechanical modeling and analytical composite models. Journal of Applied Polymer Science. 2012;125(6):4416–4420.
[4] Huang J, Schmauder S, Weber U, Geier S. Micromechanical modeling of the elastoplastic behavior of nanodispersed elastomer particle-modified pa 6. Computational Material Science. 2012;52:107–111.
[5] Gitman IM, Askes H, Sluys LJ. Representative volume: Existence and size determination. Engineering Fracture Mechanics. 2007;74(16):2518–2534.
[6] Geier S. Optimierung von Steifigkeit/Zähigkeits-Eigenschaften nanoskaliger Polyamid 6-Verbundwerkstoffe durch Analyse von Struktur/Eigenschafts-Korrelationen. Phd thesis, Universität Stuttgart; 2011.
[7] Hossain D, Tschopp MA, Ward DK, Bouvard JL, Wang P, Horstemeyer MF. Molecular dynamics simulations of deformation mechanisms of amorphous polyethylene. Polymer. 2010;51:6071–6083.
[8] Stukowski A. Visualization and analysis of atomistic simulation data with ovito – the open visualization tool. Modelling and Simulation in Materials Science and Engineering. 2009;18: 015012.
[9] Plimpton S. Fast parallel algorithms for short-range molecular dynamics. Journal of Computational Physics. 1995;117(1):1–19.
[10] Heinz H, Lin TJ, Kishore Mishra R, Emami FS. Thermodynamically consistent force fields for the assembly of inorganic, organic, and biological nanostructures: The INTERFACE force field. Langmuir. 2013;29(6):1754–1765.

Part III: Multiscale simulations of biological and bio-inspired materials, bio-sensors and composites

C.-S. Chen*, Y.-C. Shih, C.-C. Chou, S.-W. Chang, and C. Liou

14 Multiscale Modeling of Nano-Biosensors

Abstract: Chemo-bio-mechanical phenomena in biosensing offer an enticing opportunity for researchers in the field of computational mechanics to investigate these phenomena with methodologies that have not previously been exploited, and to provide new insights into the design of next-generation ultrasensitive nano-biosensors. In this chapter, we review the top-down and bottom-up schemes inspired by information passage from one scale to another to predict the surface stress and strain distribution of microcantilevers through the microscopic fidelity of molecular interactions. The top-down scheme couples a continuum description with first principles density functional theory (DFT) calculations or classical molecular dynamics/statics (MD) simulations and links atomic contributions with kinematic constraints imposed by continuum mechanics. The bottom-up scheme relies on detailed DFT or MD calculations in a representative volume at the microscopic level and uses this information to solve the desirable macroscopic response with finite element (FE) analysis.

Keywords: nano-biosensor, microcantilever, multiscale modeling, density functional theory, molecular dynamics, finite element method

Ultrasensitive nano-biosensor related research is a vital and rapidly developing field owing to its potential social and economic impact. For example, the technology shows promise in detecting disease in an early stage; monitoring contaminants and bioweapons in air, water and soil; reducing food borne infection; and benefiting healthcare, biomedical technology, and clinical analysis. Chemo-bio-mechanical phenomena in biosensing offer an enticing opportunity for researchers in the field of computational mechanics to investigate these phenomena with methodologies that have not previously been exploited, and to provide new insights into the design of next-generation ultrasensitive nano-biosensors.

The microcantilever is one of the most promising ultrasensitive nano-biosensors [1]. It is highly portable and supports label-free molecular recognition measurement with an ultra-high sensitivity. With the advance of the micro/nano electromechanical systems (MEMS/NEMS) technology, variously shaped microcantilevers can be now routinely fabricated in batches (Fig. 14.1). The device has been used to detect biomolecules (such as DNA and proteins associated with cancer or other diseases), chemicals (such as explosives and glucose), and ionic species (such as calcium ions). It is thus rather surprising to note that the theoretical description and predictive modeling of these devices are not well developed, and lag behind advances in fabrication and applications [2–4].

(a)

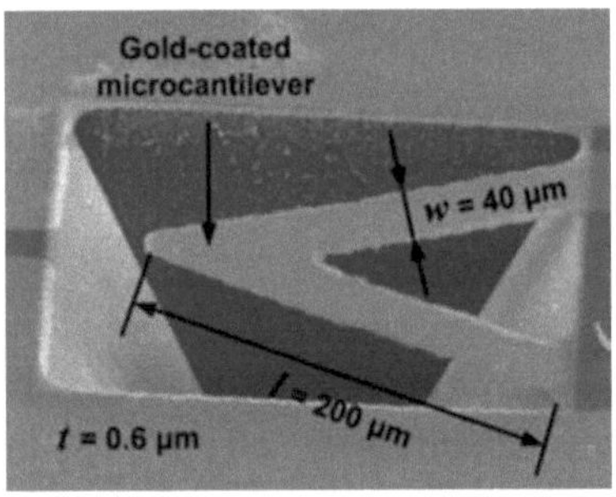

(b)

Fig. 14.1: Scanning electron micrograph of various shapes of a gold-coated microcantilever fabricated using MEMS/NEMS technology.

At its simplest, the key components of biosensors consist of a sensing layer, a transducer and a recorder (Fig. 14.2). The sensing layer is normally a receptor which can bind or interact with a target analyte molecule. The transducer transforms the signals and the recorder is a read-out system. For example, microcantilever biosensors can translate bio-recognition events into mechanical motion ranging from a few to hundreds of nanometers. The theoretical modeling challenge thus lies in how to couple molecular interactions with a continuum description: a daunting but rewarding intellectual challenge with great potential. In this article, we will highlight some recent progress made in predicting the macroscopic response of microcantilevers through the microscopic fidelity of molecular interactions.

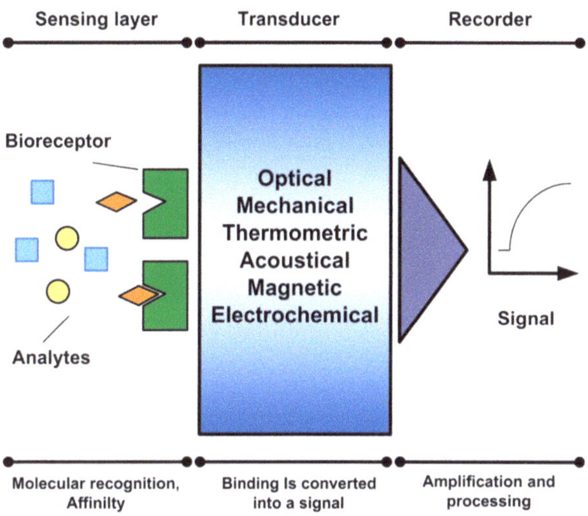

Fig. 14.2: The three key components of biosensors: a sensing layer for molecular recognition, a transducer for signal translation, and a recorder for processing.

14.1 Top-down Information Passage

One simple way to couple a continuum description with first principles density functional theory (DFT) calculations or classical molecular dynamics/statics (MD) simulations is to link atomic contributions with kinematic constraints imposed by continuum mechanics [2, 3]. For example, if we consider a differential element in a microcantilever subjected to a uniform curvature, we can take the derivative of the total potential energy with respect to curvature and directly relate the Stoney formula for surface stress to atomic positions and atomic forces originating from the local rearrangement of surface atoms as well as charge redistribution induced by ligand adsorption.

Fig. 14.3 shows a calculated charge distribution of an alkanethiol (SC_4H_9) adsorbed on a gold film. Detailed DFT calculations allow us to capture the binding process between the sulfur and gold atoms. Binding of the sulfur atom alters the atomic positions as well as charge distribution of the gold atoms in a few surface layers. Consequently, it induces surface stresses and bends the microcantilever.

Similarly, we can use MD simulations to capture bio-molecular recognition with length and time scales beyond the reach of DFT calculations. For example, consider a cyclic shrimp anti-lipopolysaccharide factor (cSALF), an antimicrobial peptide that can affect the infection of grouper nervous necrosis virus [5]. Fig. 14.4 shows snapshots of cSALF adsorbed onto thiol group linkers from molecular dynamics simulations. Again the induced surface stress can be predicted from the scheme imposed with a uniform curvature constraint [2].

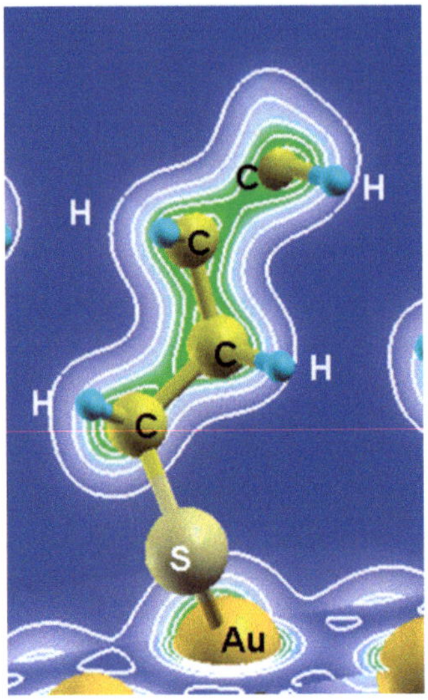

Fig. 14.3: Charge distribution from DFT calculations of an alkanethiol (SC_4H_9) adsorbed on a gold film.

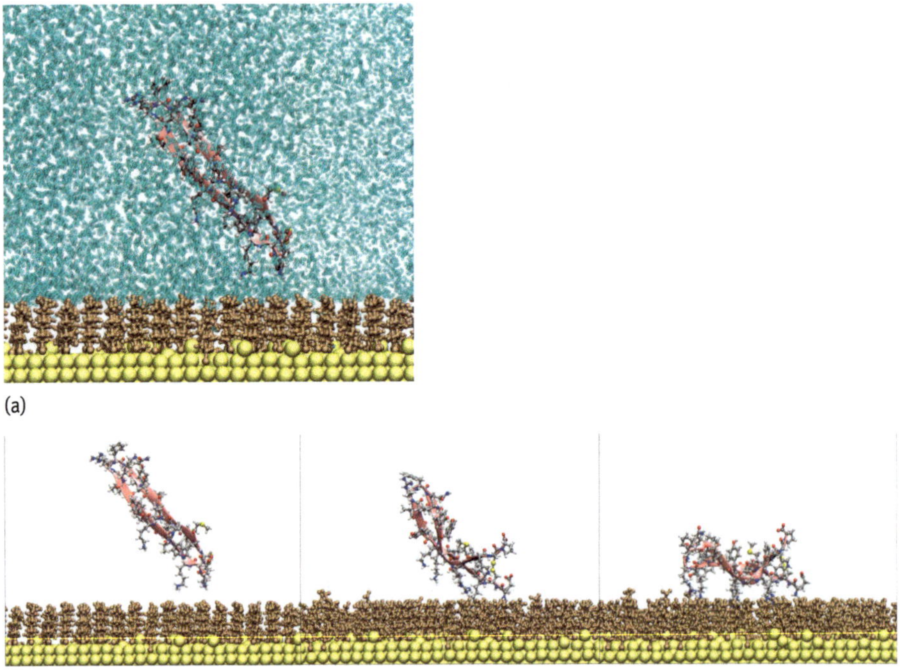

(a)

(b)

Fig. 14.4: Adsorption of cSALF onto a microcantilever surface. (a) cSALF (pink ribbon) above linkers (ochre) and gold surface (yellow) surrounded by solution (cyan) and (b) snapshots of the adsorption process (the solution is not shown for clarity).

14.2 Bottom-up Information Passage

The simple kinematically constrained method provides a pathway to study rich physics for adsoption-induced surface stress. However, we sometimes do not know or do not want to assume the deformation field a priori. For example, to maximize the portability of the microcantilever biosensors, an embedding piece of piezoresistive material or a metal-oxide semiconductor field effect transistor (MOSFET) is often used to detect the deflection-induced strain due to biomolecular binding. As they are directly embedded in the microcantilever, there is no need for an extra instrument for deflection detection. Strain distributions in a microcantilever are not homogeneous and cannot be resolved by simply imposing a uniform curvature deformation. We thus need to perform detailed DFT or MD calculations in a representative volume at the microscopic level and use this information to solve the desirable macroscopic response with finite element (FE) analysis.

Let us take DFT and FE coupling to illustrate this bottom-up information passage scheme. By using DFT calculations, we can obtain the spatial distribution of electrons in a representative volume before and after molecular adsorption. Fig. 14.5 shows a charge difference slicing through the surface gold atoms upon methanethiolate (SCH_3) adsorption. Charge removal occurs in the blue region where the sulfur atom binds to the surface, indicating an induced compressive surface stress. Using the Hellmann-Feynman theorem, all the forces and stresses in the representative volume can be obtained from the DFT calculations. These prescribed forces and stresses can then be applied to a finite element model of a microcantilever. Fig. 14.6 shows the deformation and the strain distribution of a 3D microcantilever with these prescribed stresses and forces. It is not surprising to observe that the maximum strain occurs at the clamped-end of the microcantilever, indicating a good location to embed a piezoresistor or MOSFET. Using the bottom-up information passage scheme, we open a new gateway for simulation-based modeling and design to optimize the signal to noise (S/N) ratio of nano-biosensors.

14.3 Conclusion

We have shown two simple schemes inspired by information passage from one scale to another to predict the surface stress and strain distribution of microcantilevers through the microscopic fidelity of molecular interactions. We finally remark that these scales should be seamlessly coupled and our preliminary work using a 3D quasicontinuum method (Fig. 14.7) has shown a promising future for concurrent coupling of nano-biosensor modeling.

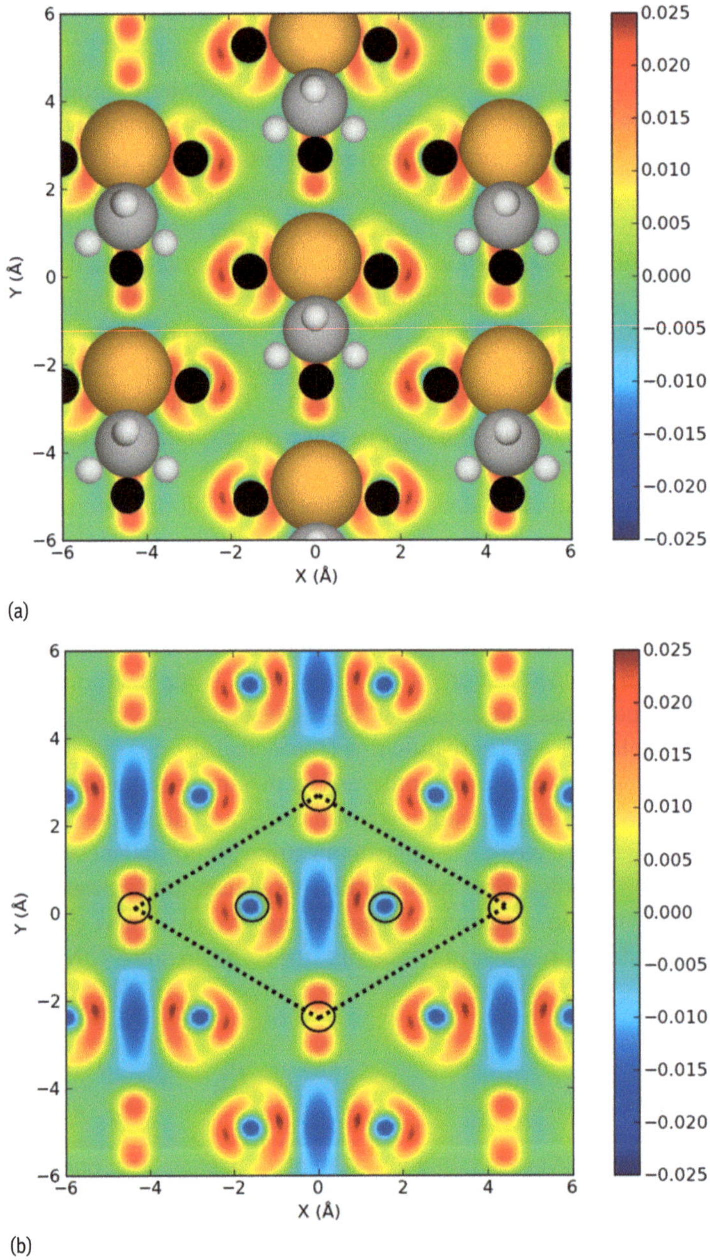

Fig. 14.5: Charge difference slicing through the surface gold atoms upon methanethiolate (SCH_3) adsorption. In (b) methanethiolate is not shown for clarity.

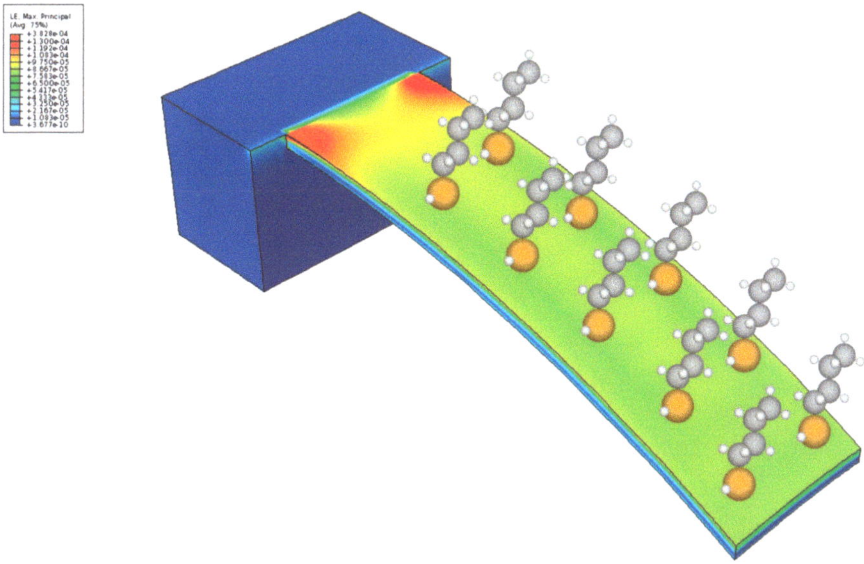

Fig. 14.6: Deformation and strain distribution of a microcantilever with prescribed stresses and forces, obtained from DFT calculations.

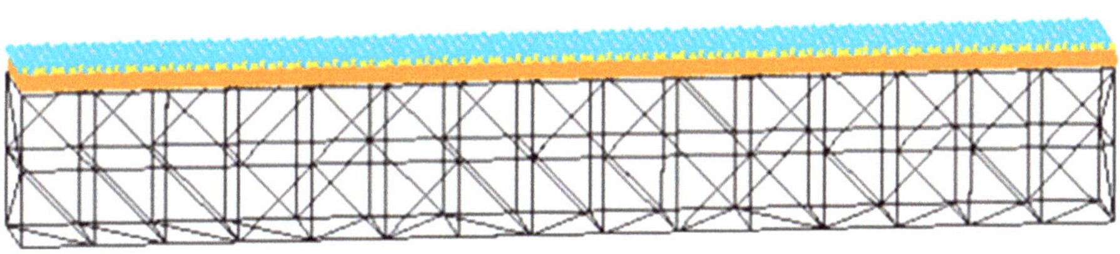

Fig. 14.7: Simple 3D quasicontinuum model for concurrent coupling of nano-biosensor modeling.

References

[1] Heath JR, Davis ME, Hood L. Nanomedicine targets cancer. Sci Am. 2009;300:44–51.

[2] Chen CS, Chou CC, Chang SW. Multiscale analysis of adsorption-induced surface stress of alkanethiol on microcantilever. J Phys D-Appl Phys. 2013;46:035301Cover image.

[3] Chen CS, Kuan S, Chang TH, Chou CC, Chang SW, Huang LS. Microcantilever biosensor: sensing platform, surface characterization, and multiscale modeling. Smart Struct Syst. 2011;8(1):17–37.

[4] Shih YC, Chen CS, Wu KC. Chain length effect on surface stress of alkanethiolates adsorbed onto $\sqrt{3} \times \sqrt{3}$ R30° Au(111) surface: A van der Waals density functional study. J Mech. 2013;30(3): 241–246.

[5] Kuan S, Chi SC, Cheng YJ, Chia TJ, Huang LS. Binding kinetics of grouper nervous necrosis viruses with functionalized antimicrobial peptides by nanomechanical detection. Biosens Bioelectron. 2012;31(1):116–123.

D. Weidt and Ł. Figiel*

15 Finite strain compressive behaviour of CNT/epoxy nanocomposites

2D versus 3D RVE-based modelling

Abstract: The macroscopic finite strain compressive behaviour of CNT/epoxy nanocomposites at quasi-static and high strain-rates was predicted and compared using 2D and 3D RVE approaches, a nonlinear and rate-dependent model for the epoxy matrix and nonlinear numerical homogenisation. Effects of CNT orientation, CNT volume fraction, CNT aspect ratio on the nanocomposite response were predicted and analysed, to compare 2D and 3D modelling approaches and obtain an insight into the nonlinear finite compressive behaviour of CNT/epoxy systems. It was found that the nanocomposite nonlinear compressive stress–strain response cannot be accurately captured by 2D RVEs (assuming the plane strain condition), when compared to 3D RVEs, primarily because of the stress transfer effect and the interparticle interaction accurately captured only in 3D. In general, 3D models predicted that the increasing CNT aspect ratio, CNT volume fraction and CNT alignment enhance the nonlinear finite strain compressive response by increasing the yield peak true stress and changing the post-yield deformation behaviour from softening to hardening. Also, the CNT alignment was identified as the major factor for enhancing the nonlinear stress–strain response at both quasi-static and impact rates of strain.

Keywords: CNT/epoxy nanocomposites, compression, finite strains, strain-rate dependence, finite element modelling, representative volume element

15.1 Introduction

Epoxy resin-based carbon fibre-reinforced polymer (CFRP) laminates are widely used in light-weight structures in the aerospace and wind-turbine industries. They became very attractive due to their good specific stiffness/strength, improved fatigue life and corrosion resistance, when compared to metals. All these unique material characteristics promise to meet current demands for the reduction of fuel consumption, efficient energy generation and the long-term structural and environmental durability. However, one of the remaining issues, which still compromise the use of epoxy-based laminates in high-performance applications, is their relatively brittle behaviour when exposed to in-service impact events such as those caused by foreign objects. Therefore, improvement of the energy absorption characteristics of those materials is an important research goal for material scientists and engineers working with composite materials.

Over the last two decades, attention in composites research has been shifted to the use of nanofillers, such as carbon nanotubes (CNTs), to improve the toughness of epoxy matrices and their composites. The exceptional mechanical properties of CNTs (e.g. stiffness, resilience) [1, 2] and their specific surface area which is several orders of magnitude larger than conventional microfillers [3] offer means to enhance the mechanical performance of epoxy at low CNT concentrations. This also includes the enhancement of energy absorption characteristics, which are brought by two main deformation mechanisms: (1) enhanced nonlinear finite deformation of epoxy in compression due to the presence of CNTs [4] and (2) CNT crack bridging mechanisms [5, 6] accompanied by CNT pull-out and interface debonding in tension. This work addresses the first of those two deformation mechanisms.

In contrast to the brittle behaviour of epoxy in tension, a prominent feature of epoxy undergoing compressive loading is its ductility [7]. Three distinct regimes can be recognised within the stress–strain curve. The linear elastic regime and yield is followed by a strain softening regime and a subsequent strain hardening regime. The nonlinear deformation of CNT/epoxy in compression depends on the manufacturing method-induced nanocomposite morphology (e.g. CNT distribution and CNT orientation), material composition (e.g. CNT volume fraction) and the deformation rate applied. Theoretically, CNTs aligned and optimally distributed within the epoxy matrix promise to provide the maximum enhancement in mechanical properties in the direction of CNT alignment. However, obtaining homogenous CNT dispersion and CNT alignment within polymeric matrices remains a significant research challenge. Despite some successful attempts of controlling the CNT alignment [8–11], only few studies investigated the effect of aligned CNTs in epoxy on the nonlinear compressive response (see for example [12]). Likewise, the nonlinear compressive response of CNT/polymer composites at high strain rates has only been sparsely researched [13–15].

Advanced material modelling can complement experimental work, and assist experimentalists in exploring systematically the relationship between the process-induced nanocomposite morphology and rate-dependent nonlinear behaviour of CNT-based nanocomposites. Despite its practical importance, only a few numerical studies have considered the nonlinear compressive response of CNT/epoxy or CNT/polymer composites. Nanocomposite behaviour (e.g. CNT instability and damping behaviour) under compressive loading was predicted using molecular dynamics (MD) and multiscale finite element (FE) modelling [16–18]. The proposed models did not however consider sufficiently large deformations (important as an energy absorbing mechanism), strain-rate dependence (occurring during impact), and nanocomposite morphological effects (e.g. CNT orientation) for the prediction of the nanocomposite mechanical response.

Therefore, this work seeks to determine the influence of material composition, nanocomposite morphology, and strain rate on the nonlinear compressive response of CNT/epoxy nanocomposites using the representative volume element (RVE) concept and nonlinear FE-based modelling. The RVE concept was followed in this work to account explicitly for the nanocomposite material composition and morphology, and study the influence of CNT volume fraction, CNT aspect ratio and CNT orientation on the nanocomposite response. A nonlinear finite strain constitutive model, developed and validated previously for the epoxy resin [19], was integrated with RVEs and implemented into the FE framework (ABAQUS) to capture the rate-dependent behaviour of the nanocomposite matrix ranging from quasi-static to impact strain rates. Two- and three-dimensional (2D and 3D) RVE-based computational models were developed to study their accuracy and computational-efficiency in capturing the nonlinear, finite strain response of the nanocomposites.

This paper is structured as follows. Modelling methodology and its numerical implementation are introduced first. Then, results of a mesh convergence analysis and RVE size study are shown. Subsequently, 2D and 3D RVE-based predictions of the finite strain behaviour of CNT/epoxy nanocomposites are presented and discussed with respect to nanocomposite morphological characteristics, strain rate, computational effort and experimental results.

15.2 Framework of modelling

The applied modelling approach (see Fig. 15.1) involves four length scales of the nanocomposite: (1) the nanoscale, where CNT properties were derived using molecular mechanics; (2) the microscale, where the discrete CNT and discrete epoxy were represented by a homogenised solid cylinder and continuous medium, respectively; (3) the mesoscale, where the morphology of the nanocomposite was reconstructed using RVEs under assumptions of global periodicity; and (4) the macroscale at which the nanocomposite response (stress–strain curves) was predicted. These stress–strain curves might serve as further input in a bottom-up multiscale approach for the prediction of the impact performance of CNT/epoxy-based composite systems. The parameters captured by each scale are presented in Tab. 15.1.

Continuum theory was assumed to hold down to the scale at which the nanocomposite morphology can be captured with an RVE. This offers a compromise between accuracy and computational time. The RVE, as a building block of the macroscopic nanocomposite, allows for the prediction of the macroscopic behaviour using a numerical nonlinear homogenisation technique.

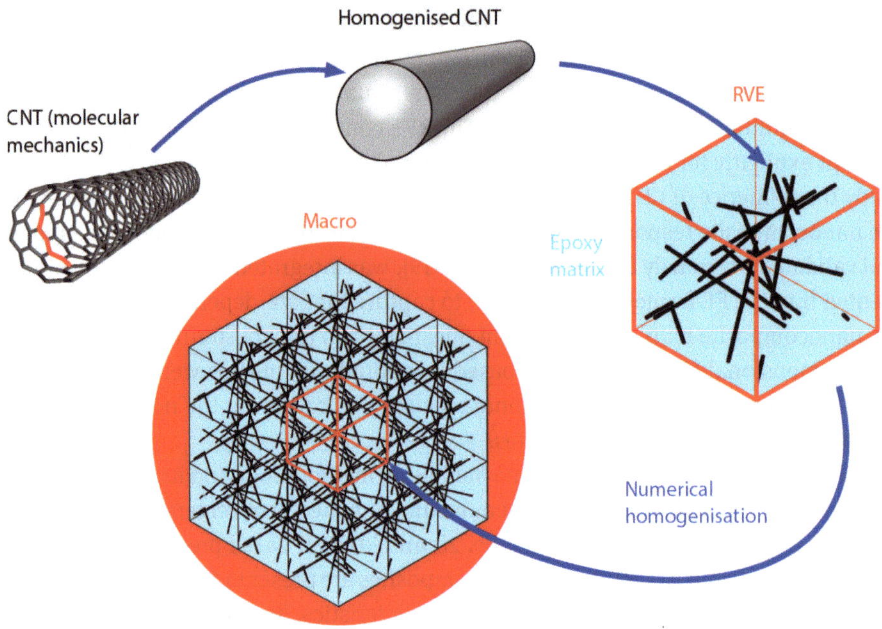

Fig. 15.1: Modelling methodology.

Tab. 15.1: Effective parameters captured by each scale (following Shokrieh and Rafiee [20]).

Effective scale	Effective length	Parameters
Nano	nm	CNT chirality, atomistic configuration, bond deformations
Micro	nm/μm	Mechanical properties of CNT (linear elastic, isotropic), CNT aspect ratio, elasto-visco-plastic behaviour of epoxy, relaxation (epoxy), adiabatic heating (epoxy), rejuvenation (epoxy)
Meso	μm	Nanocomposite morphology, CNT volume fraction, CNT orientation, CNT distribution
Macro	mm	Mechanical properties of nanocomposite

15.2.1 Representative volume elements (RVEs)

An RVE is defined here as the minimum size of a heterogeneous material model that ensures a sufficiently accurate representation of the macroscopic material behaviour. Hence, it has to contain a minimum number of CNTs and deformation mechanisms to represent statistically the response of the macroscopic nanocomposite.

The morphology of CNT/epoxy nanocomposites is quite complex and its reconstruction using RVEs poses significant challenges even when CNTs are simplified as straight and non-bundled short fibres of different distributions and orientations, as is the case in this work. 3D RVEs are a natural choice to capture such morphologies, but they require a large number of finite elements and computational resources. On the other hand, 2D RVEs are computationally more efficient than 3D RVEs, but they can compromise the accuracy of their predictions.

Therefore, two types of RVEs were considered and compared here: (a) 2D RVEs assuming the plane strain condition (see Fig. 15.2 (a) and (b)) and (b) 3D RVEs (see Fig. 15.2 (c) and (d)). 3D RVEs were chosen because of their ability to accurately capture the CNT geometry with two 'nanometre' dimensions in the plane of the circular cross-section and one dimension in 'micrometre' scale in CNT axis direction. As a consequence, 3D RVEs exhibit the correct ratio of number of CNTs to RVE volume for a given CNT volume fraction. On the other hand, 2D RVEs were chosen for their computational efficiency on the account of the following assumptions: (a) the third dimension of 2D RVEs under plane strain is neglected (set to unity here), and only one 'nanoscale' dimension is captured (i.e. thickness of a CNT) – hence, the CNT shape is captured accurately in the x–y plane only, and the ratio of the number of CNTs to 2D RVE volume for a given CNT volume fraction depends here on the scaling of the CNT dimensions, and (b) the well-known constraint imposed on the out-of-plane strains (in the z-direction here), which is expected to provide a stiffer macroscopic response.

Both, 2D and 3D RVEs assumed two different CNT orientations: (1) aligned (in the x-direction) (see Fig. 15.2 (a) and (c)), and (2) misaligned (random) (see Fig. 15.2 (b) and (d)). The sides of each of the RVEs used possess equal lengths, except for 3D RVEs of randomly distributed and aligned CNTs (see Fig. 15.2 (c)), where a cuboidal cell was chosen over a cube in order to save computational costs related to the required number of CNTs. For example, a cubic RVE with a minimum edge length of two times the CNT length, a CNT volume fraction of 0.01 and an aspect ratio of 50, would require 255 CNTs. In contrast, cuboidal RVEs would need to be populated with a significantly smaller number of CNTs depending on the ratio of the longer to shorter RVE edge. The ratio of CNT dimensions to RVE dimension is much smaller in the transverse direction. Hence, the long side of the cuboidal RVE was assumed to be crucial for its size. The length of the short edges of the cuboidal RVE were set equal to the short edges in an RVE of 20 CNTs that exhibits a long edge length of $2.5 \times$ CNT length.

In-house developed scripts were used to build 2D and 3D RVEs of the nanocomposite. One of the key-functions included in these codes was the random positioning and the random alignment of CNTs in a squared, cubic or cuboidal geometric cell representing the RVE. A typical acceptance-rejection algorithm was employed to populate RVEs with cylinder-like objects representing homogenised CNTs (see Sec. 15.2.1.1). For this, checks for intersections with previously generated objects were carried out either resulting in the deletion or inclusion of the newly generated object. This procedure continued until the requested number of CNTs was incorporated. The assumption of morphological periodicity imposed additional conditions on the objects intersecting the boundaries of the RVE – those objects were divided into inner and outer parts,

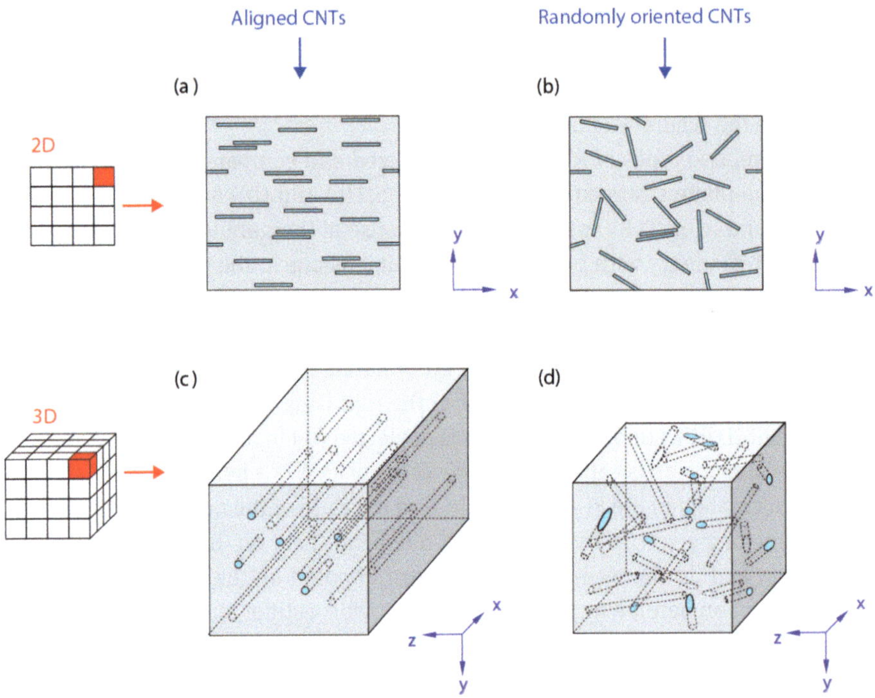

Fig. 15.2: Schematic sketches of RVEs used in this work: (a) squared 2D RVE of randomly distributed and aligned CNTs, (b) squared 2D RVE of randomly distributed and randomly oriented CNTs, (c) cuboidal 3D RVE of randomly distributed and aligned CNTs, (d) cubic 3D RVE of randomly distributed and randomly oriented CNTs.

and the latter ones were translated to the inside of the opposite RVE edges. A review on numerical generation of random fibre composites may be found in [21].

A non-contacting distribution of CNTs was enforced by assuming a minimum distance between CNT surfaces, which was set equal to the CNT diameter. Accounting for overlapping and connection of CNTs may affect the stiffness of the nanocomposite as shown by Makvandi and Öchsner [22]. Additionally, the effect of CNT agglomerations is not considered, although the level of CNT agglomeration was found to have a significant effect on the nonlinear stress–strain response of CNT nanocomposites [23]. However, CNT agglomerations may be minimised by using an appropriate dispersion method [5], surface functionalisation [24], low CNT volume fractions [25] and low CNT aspect ratios [25, 26].

The meshing of the RVEs was performed using the ABAQUS built-in free meshing algorithm. Periodic boundary conditions require that relevant nodal coordinates of finite elements lying on opposite faces (or edges) of an RVE are the same before and during deformation. In the case of 3D RVEs, the matrix material was discretised by 10-node quadratic tetrahedron elements (C3D10) and the CNT by 20-node quadratic brick elements (C3D20). 6-node quadratic plane strain triangle elements (CPE6) were used in 2D RVEs for both, CNT and matrix discretisation.

15.2.1.1 Carbon nanotube representation in RVEs

The concept of the effective (homogenised) representations of single-walled CNTs (SWCNTs) [27] was followed by modelling SWCNTs as homogenised solid fibres of the given aspect ratio. This SWCNT representation is computationally more efficient when compared with a discrete, tubular representation of a SWCNT. In general, the effective CNT is transversely isotropic with five independent elastic constants. In the present work, it was assumed for simplicity that the material behaviour of each homogenised CNT is linear elastic and isotropic, i.e. it can be described by only two elastic constants: Young's modulus E^{eff} and Poisson's ratio ν, which were determined from the following equations for zig–zag SWCNTs based on harmonic potentials [27]:

$$E^{\mathrm{eff}} = \frac{4\sqrt{3}C_\rho}{(R/2)[9 + 3(C_\rho r_0^2/C_\theta)/(2\eta)]}, \tag{15.1}$$

$$\nu = \frac{-1 + (C_\rho r_0^2/C_\theta)/(2\eta)}{3 + (C_\rho r_0^2/C_\theta)/(2\eta)}, \tag{15.2}$$

$$\eta = \frac{14 + 12\cos(\pi/n) - 2\cos^2(\pi/n)}{10 + 4\cos(\pi/n) - 6\cos^2(\pi/n)}, \tag{15.3}$$

where $C_\rho = 742\,\mathrm{nN/nm}$ [27] is the bond stretching force constant, $C_\theta = 1.42\,\mathrm{nN\,nm}$ [27] is the bond angle variation constant, $r_0 = 0.1421\,\mathrm{nm}$ is the C–C bond length, $R \approx 5\,\mathrm{nm}$ is the CNT radius and $n = 128$ is the corresponding chirality of a zig–zag SWCNT. Effective Young's modulus and Poisson's ratio were calculated to be equal to 143.62 GPa and 0.16, respectively.

15.2.1.2 Constitutive model for the epoxy matrix in the RVE

The physically-based constitutive 'glass–rubber' model proposed and experimentally-validated for thermosetting resins [19] was chosen to represent the nonlinear strain rate-dependent behaviour of the epoxy matrix. The model assumes two sources of energy arising from perturbation of interatomic potentials (relaxed by thermally activated flow events) and from entropy-elastic perturbation of molecular conformations (arising from preferred molecular alignment) [19]. As a result, there are two deviatoric stress components: (1) bond-stretching stress and (2) conformational stress (described in terms of isotropic hyperelasticity). The model is able to capture strain-rate dependence of the yield peak, strain softening, and to account for the adiabatic heating (active upon impact rates of deformation). The constitutive model incorporates those features by applying the Eyring and fictive temperature concepts, and a representation of the plastic-strain-induced rejuvenation [19].

The model was described within the framework of nonlinear continuum, and hence its kinematics is driven by the deformation gradient tensor, while its rate-dependency is driven by the viscous part of the deformation rate tensor. The model was implemented into the nonlinear FE framework using a user subroutine UMAT (ABAQUS/Standard). Model parameters obtained for the standard bisphenol A resin were used in these simulations [19].

15.2.1.3 CNT/epoxy interface/interphase in the RVE

The CNT and epoxy were assumed to be perfectly bonded in this work. However, this is only an approximation to the real picture of the interface behaviour in CNT/epoxy nanocomposites, where the strength of interactions between CNT and epoxy varies depending on the CNT surface functionalisation [28]. Also, it was assumed that the epoxy properties in the vicinity of CNTs are the same as in the bulk. This is again an approximation as the presence of CNTs can change the epoxy dynamics and/or disrupt its network near the CNT [29]. Incorporation of an interface/interphase can significantly change the stress transfer efficiency [30–32] and affect the nonlinear behaviour of the nanocomposite [23, 33].

15.2.2 Computational homogenisation: RVE-to-macro transition

Computational homogenisation was applied to calculate macroscopic stresses from each RVE. The nanocomposite was assumed to be globally periodic, i.e. consisting of the same repeatable RVE. The macroscopic deformation was considered as strain-controlled and it was assumed that the applied macroscopic strain $\boldsymbol{\varepsilon}_{\mathrm{M}}$ is uniform such that each RVE experienced the same strain. The strain within each RVE boundary consisted of the linear contribution resulting from the macroscopic strain, and the periodic component, which arises from the nanocomposite morphology. Therefore, the resulting displacements in the RVE are given as:

$$\boldsymbol{u}_{\mathrm{RVE}} = \boldsymbol{u}_{\mathrm{M}} + \boldsymbol{u}_{\mathrm{per}} , \tag{15.4}$$

where the displacement due to the macroscopic strain $\boldsymbol{\varepsilon}_{\mathrm{M}}$ is given through the well-known linear relation:

$$\boldsymbol{u}_{\mathrm{M}} = \boldsymbol{\varepsilon}_{\mathrm{M}} \boldsymbol{x}_{\mathrm{RVE}} , \tag{15.5}$$

where $\boldsymbol{x}_{\mathrm{RVE}}$ denotes the relevant position within the RVE, and the periodic boundary conditions were applied to link the opposite boundaries (faces and/or edges) of the RVE to ensure they undergo the same displacements according to:

$$\boldsymbol{u}^{+}_{\mathrm{RVE}} - \boldsymbol{u}^{-}_{\mathrm{RVE}} = \boldsymbol{u}^{+}_{\mathrm{M}} - \boldsymbol{u}^{-}_{\mathrm{M}} = \boldsymbol{\varepsilon}_{\mathrm{M}}(\boldsymbol{x}^{+}_{\mathrm{RVE}} - \boldsymbol{x}^{-}_{\mathrm{RVE}}) , \tag{15.6}$$

where symbols '+' and '–' represent corresponding points on two opposite faces and/or edges of the RVE boundary.

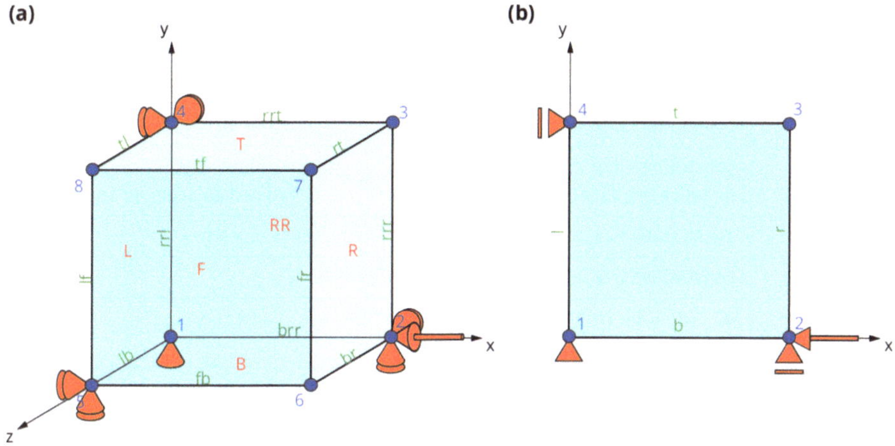

Fig. 15.3: Schematics of boundary conditions in 3D RVE (a) and 2D RVE (b) with notations of faces: top (T), bottom (B), left (L), right (R), front (F), rear (RR); edges: top (t), bottom (b), left (l), right (r), topleft (tl), topfront (tf), bottomright (br), bottomrear (brr), leftbottom (lb), leftfront (lf), righttop (rt), rightrear (rrr), frontbottom (fb), frontright (fr), reartop (rrt), rearleft (rrl); and corner nodes $(1, 2, \ldots, 8)$.

In this work, the RVE domain was separated into corner nodes ($I = 1$–8), faces (Top, Bottom, Left, Right, Front, Rear) and edges. The corner node 1 was constrained in all directions to prevent rigid body movement and the Eqs. (15.5) and (15.6) were applied to the relevant RVE regions to simulate the macroscopic uniaxial compression of the nanocomposite in the x-direction. Fig. 15.3 and Tab. 15.2 show the displacements and constraints applied to 2D and 3D RVEs.

The macroscopic Cauchy stresses $\boldsymbol{\sigma}_{\mathrm{M}}$ were calculated from the forces resulting from the surface tractions caused by the displacements imposed on the RVE boundaries (see Eq. (15.4)):

$$\boldsymbol{\sigma}_{\mathrm{M}} = \frac{1}{V_{\mathrm{RVE}}} \int_{V_{\mathrm{RVE}}} \boldsymbol{\sigma}_{\mathrm{RVE}} \, \mathrm{d}V = \frac{1}{V_{\mathrm{RVE}}} \int_{\Gamma_{\mathrm{RVE}}} \boldsymbol{t}_{\mathrm{RVE}} \otimes \boldsymbol{x}_{\mathrm{RVE}} \, \mathrm{d}\Gamma \,. \tag{15.7}$$

For the boundary conditions applied in this work (see Fig. 15.3 and Tab. 15.2), Eq. (15.7) results in the macroscopic stress component $\sigma_{xx} = V_{\mathrm{RVE}}^{-1} f_{x(2)} x_{x(2)}$ where $f_{x(2)}$ and $x_{x(2)}$ denote force and position of the node 2 (see Fig. 15.3).

15.3 Results and discussion

The developed 2D and 3D computational models were used to predict the large strain compressive response of CNT/epoxy for different material compositions and nanocomposite morphologies. The results are presented and discussed in this section. First a mesh convergence study is addressed followed by the determination

Tab. 15.2: Boundary conditions for 3D and 2D RVEs applied to relevant nodes.

Nodes	3D	2D
Node 1:	constrained in x, y, z	constrained in x, y
Node 2:	constrained in y, z; negative displacement applied in x	constrained in y; negative displacement applied in x
Node 4:	constrained in x, z	constrained in x
Node 5:	constrained in x, y	
Face nodes:	$\mathbf{u}_F^+ - \mathbf{u}_F^- = \mathbf{u}_I$, with $I = 2, 4, 5$ for case R–L, T–B, F–RR	
Edge nodes:	$\mathbf{u}_E^+ - \mathbf{u}_E^- = \mathbf{u}_I$, with $I = 2$ for case 'rrr–rrl', 'br–lb', and 'rt–tl'; $I = 4$ for case 'rrt–brr', 'tf–brr', and 'tl–lb'; $I = 5$ for case'fr–rrl','lf–rrl', and 'fb–brr'	$\mathbf{u}_E^+ - \mathbf{u}_E^- = \mathbf{u}_I$, with $I = 2, 4$ for case 'r–l', 't–b'
Corner nodes:	$\mathbf{u}_J^+ - \mathbf{u}_J^- = \mathbf{u}_I$, with $I = 2$ for 3–4, 6–5, and 7–8; $I = 4$ for case 8–5	$\mathbf{u}_J^+ - \mathbf{u}_J^- = \mathbf{u}_I$, with $I = 2$ for 3–4

of the RVE size and ensemble size. Subsequently, the results section is focused on the investigation of the 2D and 3D modelling approach with respect to the effects of CNT aspect ratio, CNT volume fraction and strain-rate effects on the large strain nanocomposite response. For this purpose, stress–strain parameters introduced in Sec. 15.3.2 were analysed. Furthermore the computational effort of 2D and 3D RVEs is presented and finally the predictions are compared against experimental data from literature.

15.3.1 Mesh convergence

The mesh density can influence FE results significantly, especially when nonlinear FE simulations are performed. Thus, the sensitivity of the stress–strain behaviour to the density of the FE mesh was studied for 2D and 3D models (CNT aspect ratio of 50, CNT volume fraction of 1 %). In order to minimise the computational time, relatively small models incorporating only five CNTs were used in the mesh study. 2D models allowed for constant and fine mesh densities due to their computational efficiency (relatively small number of degrees of freedom and element integration points). Therefore, the mesh was refined globally by reducing the average element size. For the sake of computational cost reduction, 3D models were only refined in regions of interest, i.e. the mesh density of the matrix domain close to the CNTs was enhanced by using more elements along the CNT circumference and also along the CNT axis.

These mesh refinement strategies are denoted here by: (A) one number in 2D, e.g. 11, which stands for the element number along the 2D CNT, and (B) a set of two numbers, e.g. 17; 4 – the first number refers to the element number along the 3D CNT axis and the second number specifies the element number along the circumference of the 3D CNT.

Additionally, and for completeness, two special cases of mesh refinement were also considered for the 3D model. First, the effect of the size growth parameter for ABAQUS built-in free meshing algorithm on the result was investigated. Apart from the element seeding at the geometric edges (e.g. RVE edges), the user can adjust a size growth parameter for interior elements (`sizeGrowth`), which affects the element edge lengths inside the geometric volume. The parameter was decreased from `sizeGrowth=MAXIMUM` to `sizeGrowth=MODERATE` (case 17; 4*). Secondly, the element edge length at the model cube edges was treated as an additional factor and it was changed from 0.108 × CNT length to 0.077 × CNT length (case 17; 4**).

The mesh density effect on 2D stress–strain curves are not shown here as the true stress-true strain curves of the two mesh extremes, i.e. mesh densities with 50 and 11 elements along the CNT axis, were coincident. Hence, Fig. 15.4 shows only the results of mesh study for 3D models. Moreover, as it was found that the linear elastic regime and yield peak stress are relatively unaffected by the tested mesh densities in the 3D model, the analysis is focused on the nonlinear portion of the 3D stress–strain behaviour.

In particular, the post-yield behaviour showed a mesh dependence. The number of elements along the CNT axis (see Fig. 15.4 (a)) and along the CNT circumference (see Fig. 15.4 (b)) influences the post-yield behaviour significantly. The effect of size growth parameter and the element edge length at the model cube edges is negligible. The stress–strain curves converge for the case 50; 4 as further mesh refinements with more elements along the CNT axis (case 100; 4) or along the CNT circumference (case 50; 8) brought negligible effects. As a summary, Fig. 15.5 shows the convergence of the overall nanocomposite stress value (for 2D and 3D models at 15 % of the overall true strain) when plotted as a function of element number.

It is noteworthy to mention that premature termination of FE analyses becomes a serious problem when relatively fine meshes are used, as for example obtained for the mesh refinement cases 50; 4, 100; 4, and 50; 8. This is caused by the excessive magnitude of local stresses in the matrix (in the vicinity of CNTs) resulting from the enhanced nonlinear hyperelastic response of the constitutive model, which captures the strain hardening behaviour of the epoxy matrix. This resulted in local stresses rising rapidly with a small increase of strains, leading to convergence problems in the numerical solution. The numerical convergence may be improved by the development of a more robust constitutive tangent. However, as those local stresses exceeded the strength of pure epoxy matrix under macroscopic compression [7], continuation of the analysis would lead to unphysical results. Instead, introduction of a damage variable into the constitutive model for the epoxy matrix will be explored in the future. There-

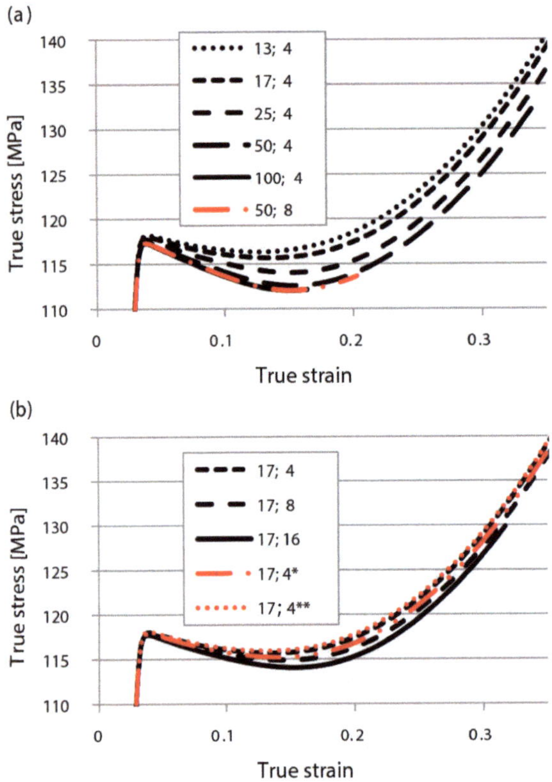

Fig. 15.4: Nonlinear portion of the stress–strain behaviour for various mesh refinements for 3D models: mesh refinement by (a) changing the number of elements along the CNT axis (first 5 black curves), and mesh refinment by (b) changing the number of elements along the CNT circumference (first 3 black curves); * mesh refinement via element size growth parameter; ** mesh refinement through reduction of element edge length at the model cube edges.

fore, case 17; 4 for 3D models was finally chosen as a compromise between accuracy and achievable finite strains, which are nearly twice as much as for the case 50; 4.

15.3.2 RVE size and ensemble size

RVE size and the ensemble size (number of RVE realisations per case) are important for accurate predictions of the nanocomposite response. Deterministic size effects specify the minimum volume of the RVE, provided that they converge with increasing RVE size, while statistical size effects specify the number of RVE realisations according to the required accuracy. Thus, studies with 2D and 3D RVEs composed of different numbers of CNTs were carried out to investigate deterministic and statistical RVE size effects.

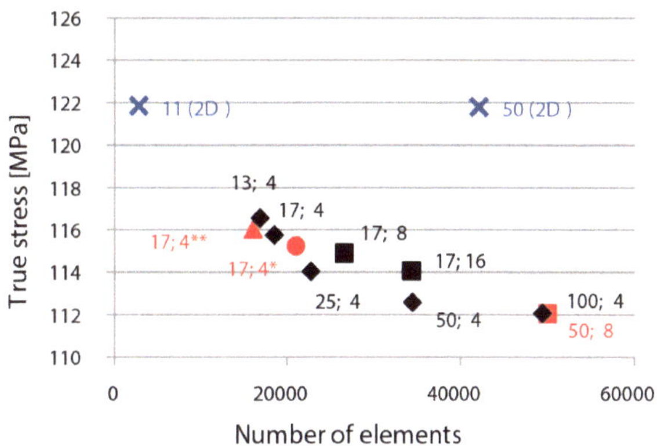

Fig. 15.5: True stress of the nanocomposite at 15 % of true strain as a function of number of elements for 2D and 3D models; (×) in blue are datapoints obtained from 2D simulations.

Specific parameters representing subsequent stages of the stress–strain behaviour of CNT/epoxy nanocomposites were selected for the analysis of the RVE size and ensemble size: (a) the Young's modulus E was defined here as initial slope of the stress–strain response of both, 2D and 3D models; (b) the yield peak stress σ_Y (start of material softening) or, in absence of a pronounced yield peak, the yield strength σ_{p1} with an offset of 1 % of plastic strain; (c) the minimum tangent modulus T_{min} of the whole curve; (d) the softening strain limit ε_S (the true strain value representing the end of the softening regime and the start of the hardening regime, corresponding to a stress minimum between softening and hardening); (e) the initial hardening modulus H (only for 2D RVEs) defined here as secant modulus between the point (in the hardening regime), where the tangent modulus is equal to $1.5 \cdot 10^8$ Pa, and the point on the graph determined by adding 0.05 to the true strain value of the first point.

All RVE-size studies were carried out for the CNT volume fraction of 1 %, which represents the highest volume fraction considered in this work and thus accounts for the maximum CNT–CNT interactions. The influence of the CNT aspect ratio on the RVE size was studied for 2D (aspect ratios of 25, 50, and 100) and for 3D (aspect ratios of 25 and 50) models. Two nanocomposite morphologies were considered: (1) 2D and 3D, with randomly oriented and randomly distributed CNTs, and (2) 3D with aligned and randomly distributed CNTs. RVE-size studies were performed for the quasi-static strain rate of $1 \cdot 10^{-3}\ s^{-1}$, as it was assumed that the conclusions will also hold for higher strain rates.

The volume element (VE) size was varied indirectly by changing the number of incorporated CNTs. Thus, the objective was to determine a minimum number of CNTs, and to choose a number of RVE realisations (RVE ensemble size) that ensures an acceptable width of the confidence interval. All stress–strain parameters mentioned above were investigated and the number of realisations varied between 10 (for small VEs) and 3 (for large VEs).

As an example, Fig. 15.6 (a) and (b) shows the evolution of E and T_{min} as a function of CNT number, i.e. VE size. The standard deviations of the investigated parameters of VE ensembles are represented by the error-bars. Deterministic size effects refer to variation of the means across VE sizes, while statistical size effects refer to variation of the standard deviations across VE sizes. Thus, potential deterministic size effects are indicated by effects of CNT number on the investigated parameters.

Due to the limited number of VE realisations, parametric Welch's tests (with and without rank transformation) [34] were performed to conclude upon the statistical significance of potential deterministic size effects. Unequal variances were assumed due to statistical size effects that cause the standard deviations to decrease with increasing VE size. The null hypotheses (H_0) on the equality of the means of two VE sizes were tested at two significance levels of $\alpha = 0.05$ (significant) and $\alpha = 0.01$ (highly significant). The results of the Welch's tests are shown in Tab. 15.3. Fig. 15.6 (c) shows how a deterministic size effect shows up in a larger slope in the post-yield regime for 3D VEs with aligned CNTs (CNT aspect ratio of 50), and previously indicated as the effect on T_{min} (see Tab. 15.3).

Tab. 15.3: Results of Welch's test for 2D and 3D models of various CNT aspect ratios (AR)*.

Model	AR	Randomly oriented CNTs	Aligned CNTs
2D	25	–	Not tested
	50	H: 30–60 (0.05,nrt,rt)	Not tested
	100	E: 10–30 (0.05,nrt), 10–60 (0.01,nrt; 0.05,rt); σ_Y: 10–20 (0.05,nrt,rt); ε_S: 10–30 (0.05,nrt; 0,01,rt), 10–40 (0.05,nrt,rt), 10–60 (0.05,rt)	Not tested
3D	25	–	Not tested
	50	–	E: 9–12 (0.05,rt), 9–20 (0.05,nrt; 0.01,rt); T_{min}: 9–20 (0.05,nrt; 0.01,rt)

* Notation: e.g. 'T_{min}: 9–20 (0.05,wrt; 0.01,rt)', where T_{min} is the stress–strain parameter affected by deterministic size effect, 9 and 20 are the two numbers of CNTs corresponding to the two VE sizes that are compared, 0.05,nrt stands for the significance level (0.05) tested without rank transformation (nrt) of data, and 0.01,rt stands for the significance level tested using rank transformed data (rt); '–' means that no deterministic size effect was observed.

Investigation of the strain-hardening behaviour of the nanocomposites was only possible for 2D VEs of CNT aspect ratio of 25 and 50 because of the premature termination of simulations at early stages of the hardening response (see comment in Sec. 15.3.1).

The results suggest that the minimum number of randomly oriented CNTs of aspect ratio of 50 in 2D RVEs is 60 CNTs or higher due to a deterministic size effect affecting the hardening modulus H. A convergence study of the deterministic size

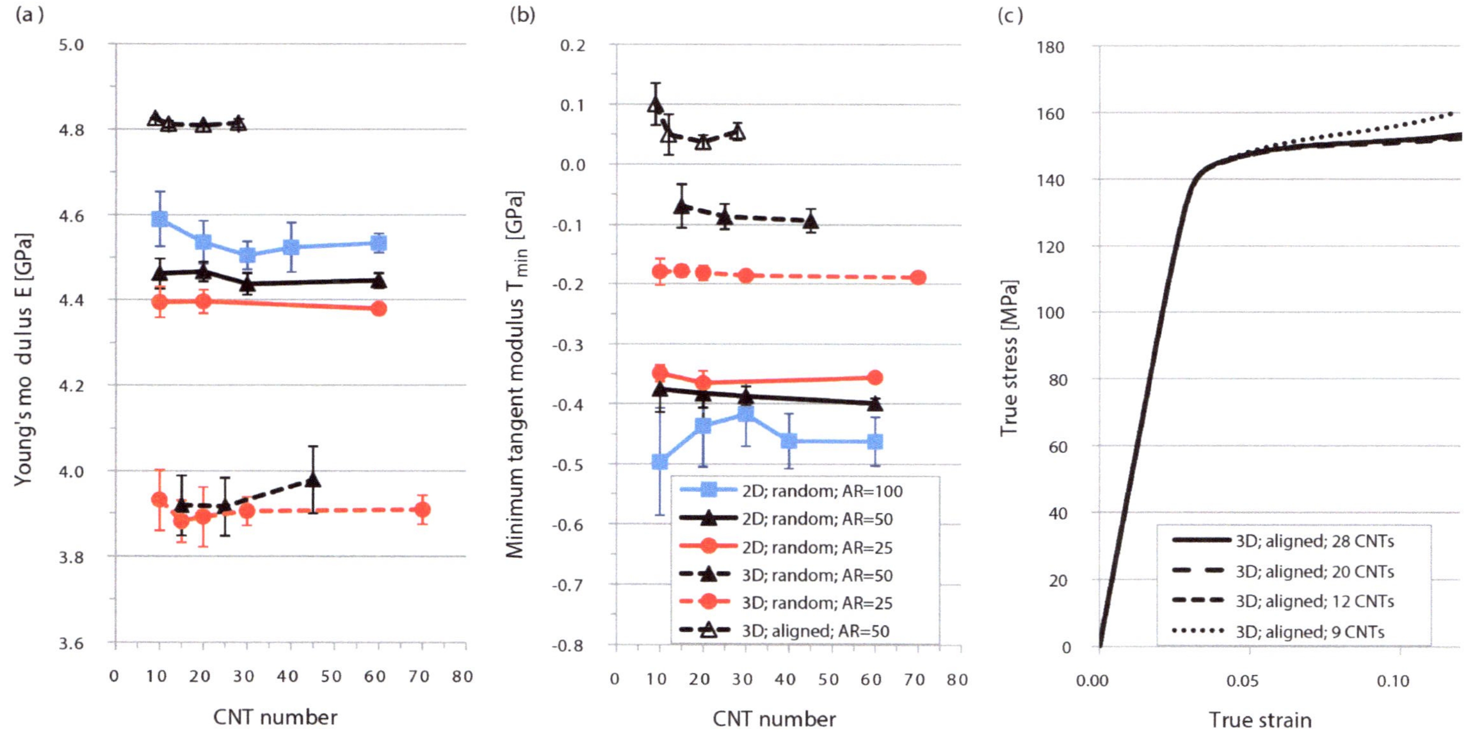

Fig. 15.6: Evolution of (a) Young's modulus and (b) minimum tangent modulus with increasing CNT number for 2D and 3D models of various CNT aspect ratios (AR); CNT volume fraction of 1 %; strain rate of $1 \cdot 10^{-3}\ s^{-1}$; standard deviations are represented by error-bars; (c) initial portion of stress–strain curves for 3D RVEs incorporating aligned CNTs with an aspect ratio of 50.

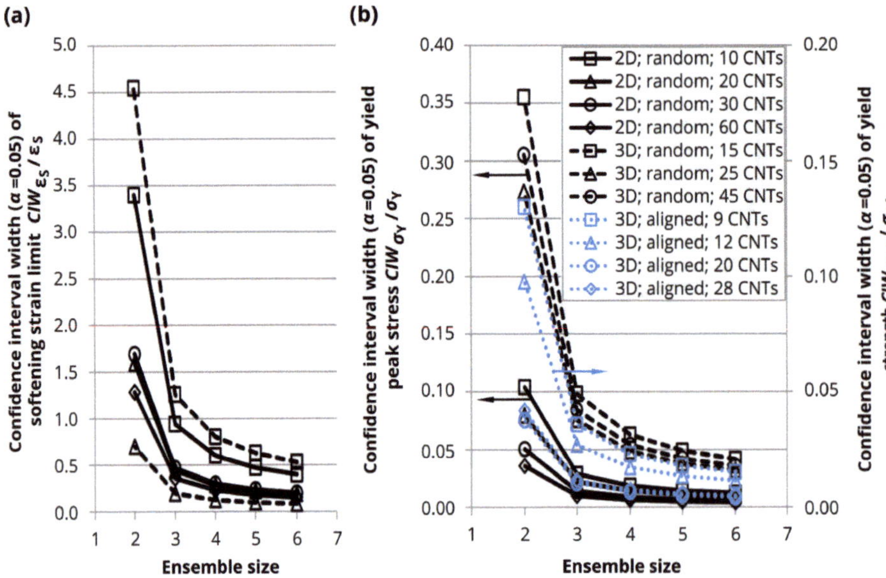

Fig. 15.7: Normalised width of the confidence interval of (a) softening strain limit, (b) yield peak stress or yield strength respectively, as indicated by arrows, as a function of ensemble size; CNT aspect ratio of 50, CNT volume fraction of 1%; strain rate $1 \cdot 10^{-3}$ s^{-1}.

effect on *H* was not carried out due to computational constraints. However, in the context of comparison of 2D and 3D RVE approaches, an investigation of the hardening modulus was omitted due to premature termination of simulations of 3D RVEs. Hence, the selection of a smaller RVE size was deemed to be reasonable. The minimum number of randomly oriented CNTs with an aspect ratio of 100 in 2D RVEs is 20 CNTs. For 3D cuboidal RVEs (with aligned CNTs) a minimum number of 12 CNTs is suggested.

The confidence interval widths (*CIW*) were calculated for the stress–strain curve parameters for various ensemble sizes (i.e. the number of VE realisations). Ensemble standard deviations were assumed to be equal for various ensemble sizes. Student's *t*-distribution was used due to small ensemble sizes. The results for two stress–strain curve parameters are shown in Fig. 15.7.

For 3D RVEs composed of randomly oriented CNTs a significant reduction of the *CIW* of the parameter ε_S was obtained by increasing the CNT number from 15 to 25 CNTs. An RVE populated with 25 CNTs possesses a normalised *CIW* of the parameter σ_Y of 5% for an ensemble size of 5. Minor improvements were obtained for this parameter by increasing the ensemble size. Hence, an RVE containing 25 CNTs of an acceptable ensemble size of 5 was chosen in the 3D case. In a similar manner, RVEs populated with 20 CNTs and subject to 5 realisations were chosen for the 2D random orientation case, while RVEs containing 20 CNTs with 3 realisations were selected for the 3D aligned case.

15.3.3 2D versus 3D RVE-based analyses of finite strain compressive behaviour of the nanocomposite

Finite strain compressive behaviour was predicted using the computational model (see Sec. 15.2) to elucidate effects of CNT aspect ratio, CNT volume fraction, and strain rate on stress–strain parameters (as introduced in Sec. 15.3.2). Parameter results for the single constituents are presented in Tab. 15.4. It is important to point out that the Young's modulus was defined here as initial tangent modulus, which defines the slope of the linear portion of the initial stress–strain behaviour in 2D (plane strain) and 3D. Different assumptions invoked by the 2D (plane strain) and 3D models resulted in the deviation between the initial tangent modulus E in 3D and its counterpart E_{pe} in 2D (see Tab. 15.4), which can be described through the relationship $E = (1 - \nu^2)E_{pe}$ [35], where ν denotes Poisson's ratio. Throughout this paper we use the term Young's modulus to define the initial tangent modulus of the 2D (plane strain) and 3D stress–strain curves.

Tab. 15.4: Mechanical properties of CNT phase and pure epoxy (2D and 3D) (Young's moduli for pure epoxy were obtained from the initial slope of stress–strain curves).

Parameter		CNT	2D epoxy	3D epoxy
Young's modulus E	[GPa]	143.62	4.20	3.63
Yield peak stress σ_Y	[MPa]		138.5	110.2
Yield strength σ_{p1}	[MPa]			108.6
Softening strain limit ε_S	[%]		20.6	28.3

Previously selected RVE sizes (in terms of CNT numbers) and RVE ensemble sizes for randomly oriented CNTs (see Section 3.2) were used for RVEs populated with aligned CNTs. For 2D RVEs, the RVE size for randomly oriented and aligned CNTs was assumed to be the same.

15.3.3.1 Effect of CNT aspect ratio

The effect of CNT aspect ratio is shown in Figs. 15.8 and 15.9. Both figures show results obtained for RVEs containing randomly distributed and randomly oriented CNTs volume fraction of 1 % and subjected to quasi-static strain rates ($1 \cdot 10^{-3}$ s^{-1}). Typically, aspect ratios for CNTs are claimed to be larger than 100. However, as CNTs frequently occur in a wavy form, the effective aspect ratios (i.e. those, which allow for an efficient stress transfer) are actually much smaller – thus, three aspect ratios were investigated: 25 (2D and 3D), 50 (2D and 3D), and 100 (2D case only).

As expected, incorporation of CNTs (and their increasing aspect ratio) leads to an increase of Young's modulus E of the nanocomposite of around 4 % (5.5 %, 8 %) for

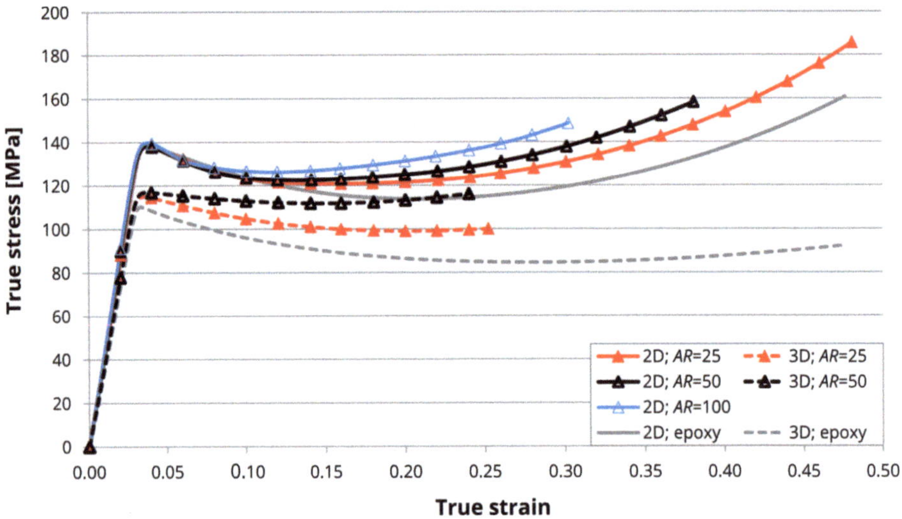

Fig. 15.8: Examples of stress–strain curves from 2D and 3D RVEs incorporating CNTs of various aspect ratios (*AR*); reference stress–strain curves of pure epoxy in 2D and 3D; randomly oriented CNTs, CNT volume fraction of 1 %; strain rate $1 \cdot 10^{-3}\ s^{-1}$.

2D RVEs of aspect ratio of 25 (50, 100) and 7.5 % (9.5 %) for 3D RVEs of aspect ratio of 25 (50) (see Fig. 15.9 (a)). 2D models predicted a stiffer response (in the linear elastic regime) both for pure epoxy and its nanocomposite, when compared to 3D models (see Fig. 15.8). Therefore, it is concluded that the cause of the stiffer 2D response is the well-known effect of the constraint on the out-of-plane strain caused by the plane strain assumption. The normalised Young's modulus in the 3D case is slightly larger than in the 2D case, probably caused by the better stress transfer due to larger CNT surface area in 3D and the smaller average CNT distance due to the 3D effect.

Values of the parameter σ_Y (yield peak stress) show a difference between 2D and 3D models, which is qualitatively similar to those obtained for pure epoxy under plane strain (2D) and 3D conditions (see Fig. 15.8). However, the CNT presence and the increase in CNT aspect ratio had a negligible effect for 2D RVEs, whereas for 3D RVEs a significant increase from around 4.5 % to 7 % in σ_Y was predicted for the increase in CNT aspect ratio from 25 to 50 (see Fig. 15.9 (b)). The local matrix behaviour of the 2D and 3D models was compared in order to determine the reason for this behaviour. In the averaged true stress-applied strain curves of the matrix one can observe a slight decrease of the yield peak stress in 2D models and the increase of the yield peak stress in 3D models. This can be explained by comparing the CNT distributions in 2D and 3D models. For the same volume fraction a denser CNT distribution is generated in 3D with a much smaller average CNT to CNT distance – a 3D scaffold of densely packed CNTs is compared against well separated CNTs in the 2D plane (see Fig. 15.10). The dense CNT networks in the 3D RVEs limit the expansion in the lateral directions when

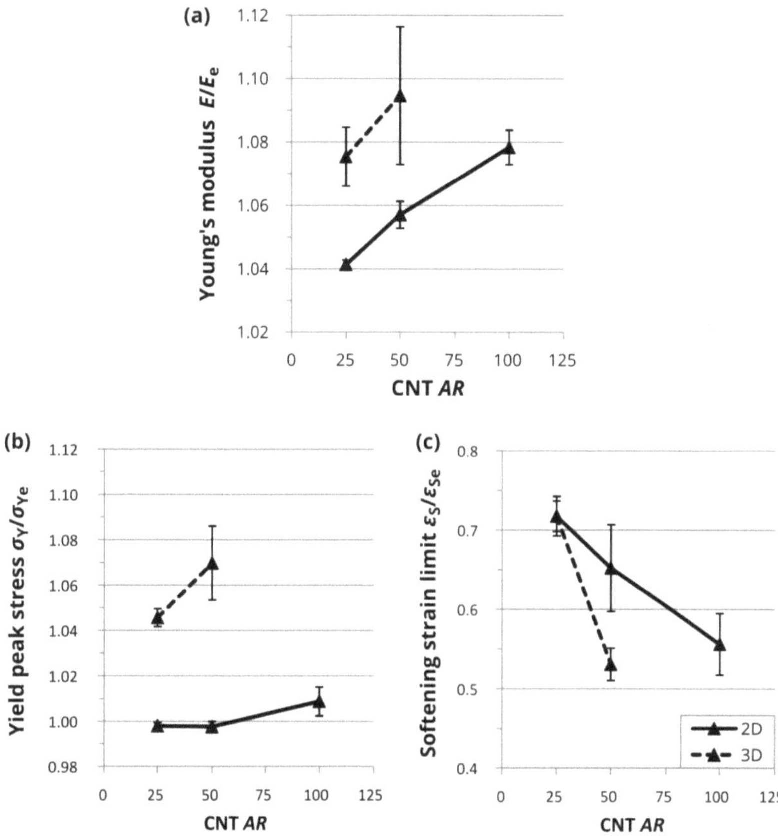

Fig. 15.9: Effect of CNT aspect ratio (*AR*) and RVE approaches (2D, 3D) on the normalised stress–strain parameters (a) Young's modulus, (b) yield peak stress, and (c) softening strain limit; randomly oriented CNTs, CNT volume fraction of 1 %; strain rate $1 \cdot 10^{-3}\ s^{-1}$; standard deviations are represented by error-bars.

the 3D RVEs are subjected to uniaxial loading. Within the matrix, this promotes a bigger volume change as well as an increase in mean stress when compared to pure epoxy. This led to bigger averaged stresses in the matrix and the nanocomposite – a similar effect is imposed by the plane-strain assumption. In contrast, for 2D models with a bigger average CNT to CNT distance, the average mean stress within the matrix around the yield peak was slightly smaller than for the pure epoxy.

Subsequently, a decrease of the magnitude of softening modulus (defined by a reduced softening slope) with increasing CNT aspect ratio was found for 3D RVEs (see Fig. 15.8), while a small increase of the magnitude of softening modulus was predicted for 2D RVEs. Two main effects were identified to contribute to the reduced softening response in 3D. First, the CNT to CNT distance is reduced (as explained earlier), which causes the increase of the averaged mean stress in the matrix. Secondly, the averaged

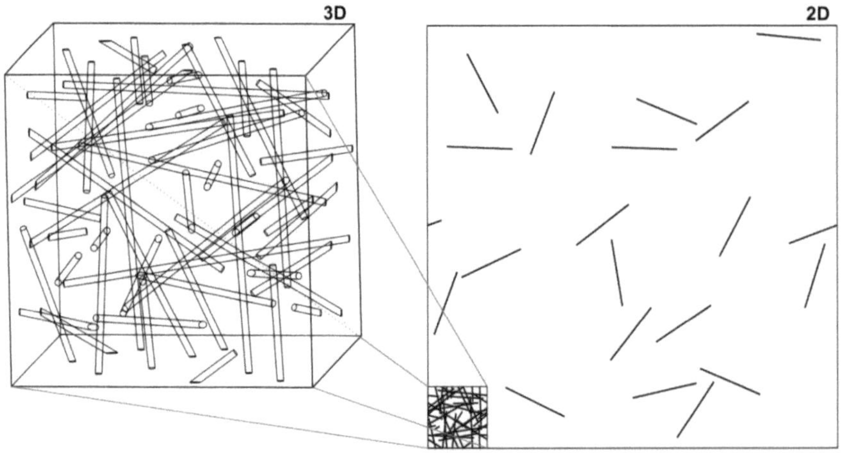

Fig. 15.10: Proportion of 2D to 3D RVE size and underlying nanocomposite morphology; randomly oriented CNTs, CNT aspect ratio of 50, CNT volume fraction of 1 %.

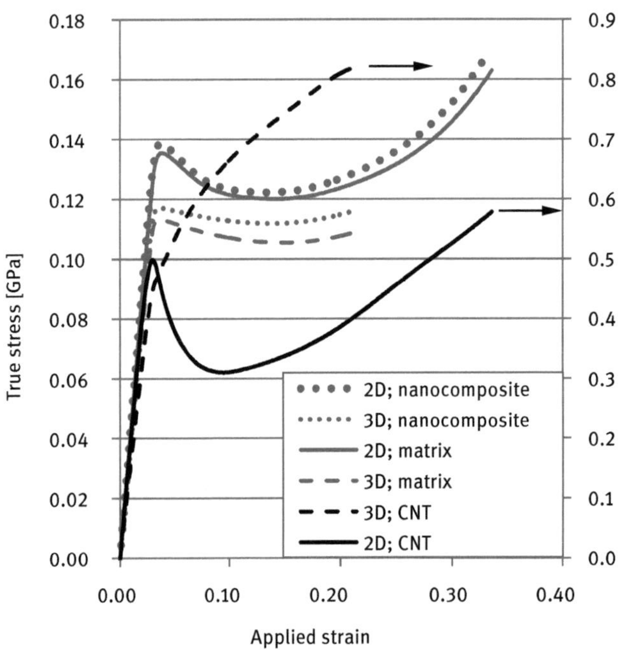

Fig. 15.11: Averaged stress–strain curves for the nanocomposite, CNT phase (true stress on secondary vertical axis, as indicated by arrows) and matrix phase in 2D and 3D RVEs; randomly oriented CNTs, CNT aspect ratio of 50, CNT volume fraction of 1 %.

stress in the CNTs increases well beyond the yield peak within the matrix for the 3D RVE, as shown in Fig. 15.11. In contrast, the averaged stress within the CNTs for 2D RVEs resembles closely the average response of epoxy with regimes such as yield, softening and hardening. The large CNT to CNT distance in 2D RVEs limits the inter-particle interaction, and thus the averaged stress response of CNTs is mainly governed by the behaviour of epoxy. It is believed that the increased inter-particle interaction in 3D RVEs (because of the smaller CNT to CNT distances) changes an epoxy-like behaviour for the CNTs and therefore for the nanocomposite.

It was predicted that increasing CNT aspect ratio accelerates the onset of strain hardening, as indicated by the decreased strain softening limit ε_S (see Fig. 15.9 (c)). Changing the aspect ratio from 25 to 50 reduced ε_S both for 2D and 3D models around 20 % for 3D, and 7 % for 2D models. This difference between 2D and 3D models is a further consequence of the difference in nanocomposite morphology, as explained above.

15.3.3.2 Effect of CNT volume fraction for aligned and random CNT orientations

The effect of CNT volume fraction on the compressive true stress-true strain curves for aligned CNTs and randomly oriented CNTs is shown in Figs. 15.12 and 15.13 using 2D and 3D RVEs. In each case, the predictions are for RVEs with CNT volume fractions of 0.5 % and 1 % and aspect ratio of 50. They were obtained for the quasi-static strain rate of $1 \cdot 10^{-3}\,s^{-1}$.

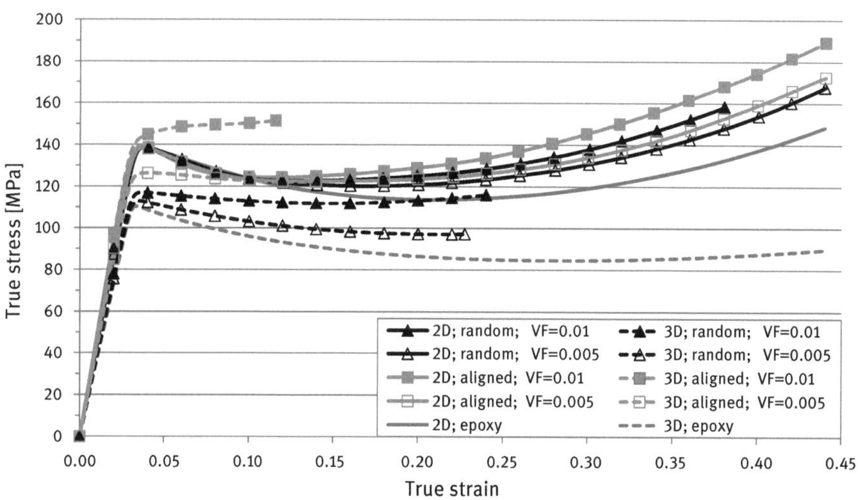

Fig. 15.12: Examples of stress–strain curves from 2D and 3D RVEs incorporating CNTs of various volume fractions (*VF*); reference stress–strain curves of pure epoxy in 2D and 3D; randomly oriented and aligned CNTs, CNT aspect ratio of 50; strain rate $1 \cdot 10^{-3}\,s^{-1}$.

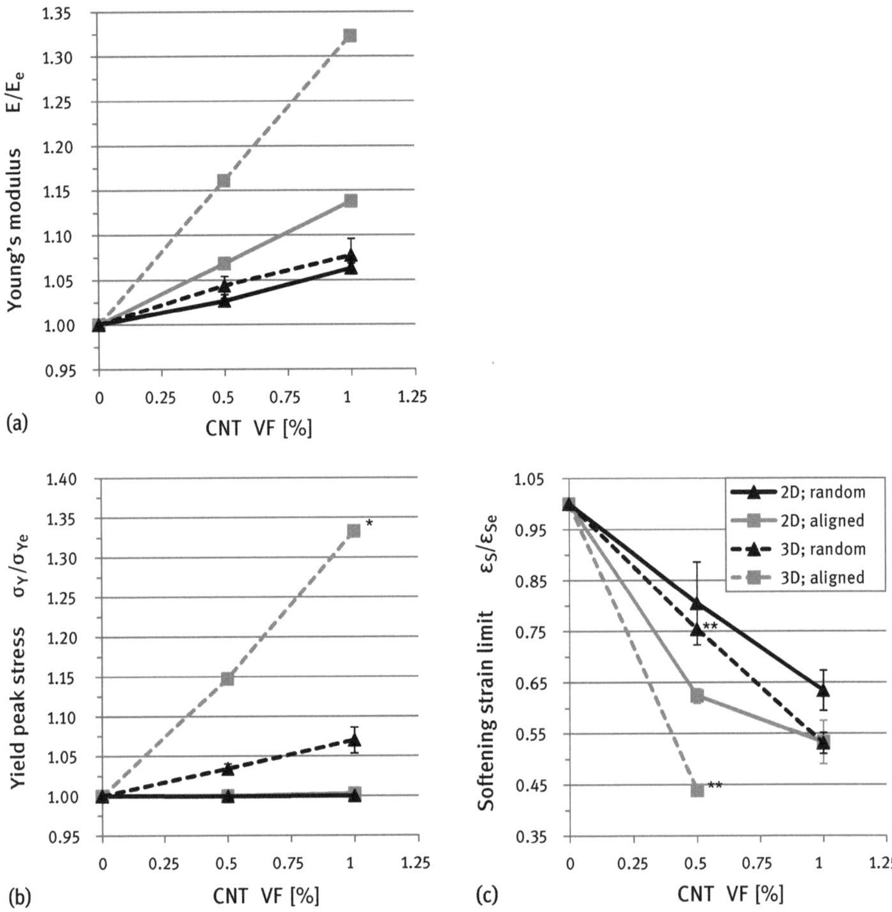

Fig. 15.13: Effect of CNT volume fraction (*VF*) and RVE approaches (2D, 3D) on the normalised stress–strain parameters (a) Young's modulus, (b) yield peak stress, and (c) softening strain limit; RVEs with randomly oriented and aligned CNTs; CNT aspect ratio of 50; strain rate of $1 \cdot 10^{-3}\ s^{-1}$; standard deviations are represented by error-bars; * value of the normalised yield stress σ_{p1}/σ_{p1e} at the strain offset 0.01; ** only one realisation considered due to premature termination of simulations.

In general, the CNT volume fraction has a qualitatively similar effect on the stress–strain parameters, as the CNT aspect ratio. As expected, the Young's modulus increased linearly with increasing CNT volume fraction (see Fig. 15.13 (a)). The largest enhancement is associated with 3D nanocomposite morphologies of aligned CNTs with an improvement of 32 % for CNT volume fraction of 1 %, while less than half this value was obtained for its 2D counterpart (i.e. 14 % increase of Young's modulus for CNT volume fraction of 1 %). This is again caused by the improved stress transfer due to the larger surface area in 3D and the smaller average CNT distance due to the 3D effect. Significantly smaller differences for the Young's modulus between 2D and 3D RVEs were found for random CNT orientations.

An interesting result was observed regarding the effect of volume fraction on the nanocomposite yield peak. In particular, the volume fraction does not affect the yield peak parameter σ_Y in 2D models (see Fig. 15.13 (b)) (similarly to the effect of aspect ratio in 2D). However, for 3D models there is a linear increase in σ_Y with increased CNT volume fraction. Larger effects were found for 3D nanocomposite morphologies of aligned CNTs (rather than random) with an increase from 15 % to 33 %. This matches the magnitude of the effect of CNT volume fraction on the Young's modulus of the nanocomposite. It is noteworthy to mention that in contrast to 3D nanocomposite morphologies of randomly oriented CNTs, this effect does not arise from the increase in the nanocomposite mean stress. Rather the mean stress within the matrix was actually smaller than for the pure epoxy at applied strains corresponding to the yield peak stress. Instead, it is believed that the major contribution here is brought by the larger stress, which was carried by the 3D CNTs, as shown in Fig. 15.14, in terms of averaged true stress-applied strain curves for the reinforcement. It is expected that a larger surface area of CNTs within the 3D models contributes to that by promoting a much better stress transfer.

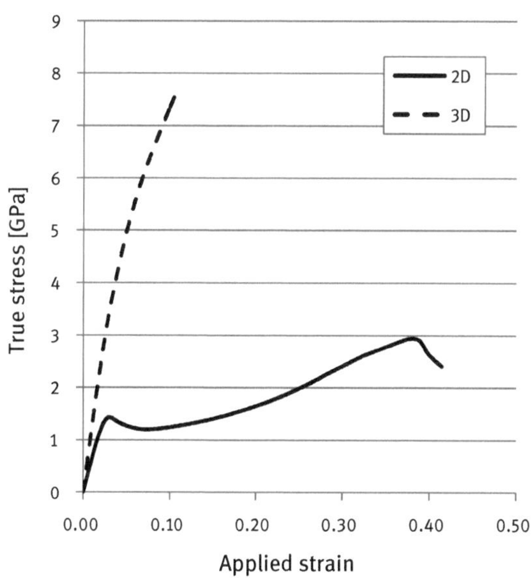

Fig. 15.14: Average stress–strain curves for the CNT phase in 2D and 3D RVEs; aligned CNTs, CNT aspect ratio of 50, CNT volume fraction of 1 %.

Additionally, it must be mentioned that there is a second averaged CNT stress peak followed by the stress drop located in the hardening regime (at around 35–40 % of applied nominal strain) as predicted by 2D RVEs (see Fig. 15.14). It originates from intensive bending and rotation of 2D CNTs, which reduces the stress transfer.

Then, a significant decrease in the magnitude of softening modulus (case CNT volume fraction of 0.5 %) and the disappearance of the softening behaviour (case CNT volume fraction of 1 %) were predicted by 3D RVEs with aligned CNTs. It is believed that this results from the significant increase of the averaged CNT stress in 3D RVEs, as discussed above and shown in Fig. 15.14.

The softening strain limit ε_S was predicted to decrease with increasing CNT volume fraction (see Fig. 15.13 (c)). The corresponding true strain values were found to be larger for 3D RVEs than for 2D RVEs, and also larger for aligned CNTs than for randomly oriented CNTs: (a) for CNT volume fraction of 0.5 %, a reduction of 20–25 % in ε_S was obtained for 2D and 3D approaches of randomly oriented CNTs, (b) decrease of 40 % (2D) and 55 % (3D) was predicted for aligned CNTs.

15.3.3.3 Effect of strain rate for aligned and random CNT orientations

The effect of strain rate for aligned CNTs and randomly oriented CNTs is shown in Figs. 15.15 and 15.16 for strain rates of $1 \cdot 10^{-3}$ (quasi-static strain rate) and $1 \cdot 10^{3}\ s^{-1}$ (impact rate of strain). Both figures show predictions of RVEs containing CNTs of aspect ratio set to 50 and volume fraction equal to 1 %. The stress–strain parameters in Fig. 15.16 were normalised against corresponding values of pure epoxy subjected to quasi-static strain rate.

For pure epoxy, the increased strain rate shifted the yield peak stress and the corresponding strain to larger values (see Figs. 15.15 and 15.16 (a)). A conventional explanation is that the time window of material deformation becomes smaller than the relaxation time of the material. Similar trends were predicted for the nanocomposite using 2D and 3D RVEs with aligned and randomly oriented CNTs. As recognised in the previous section, nanocomposite morphologies of aligned CNTs in the 3D case exhibit a larger increase of σ_Y or σ_{p1} than morphologies of randomly oriented CNTs. However, the magnitude of this enhancement was larger for impact rates of strain (~ 55 MPa) than for the quasi-static strain rates (~ 35 MPa) (see Fig. 15.15).

Considering the softening strain limit parameter ε_S, its normalised value did not show a significant dependency on the strain rate as predicted by 2D RVEs – for pure epoxy it shifted to slightly smaller values of strain with increasing strain rate.

For 3D approaches, the effect of strain rate on ε_S is shown only for nanocomposites of randomly oriented CNTs because the softening regime is no longer present after CNT alignment. In that case, the normalised parameter ε_S was significantly reduced from 0.53 to 0.43 when increasing the strain rate from $1 \cdot 10^{-3}$ to $1 \cdot 10^{3}\ s^{-1}$ (see Fig. 15.16 (b)). For the 3D case with aligned CNTs a significantly larger slope of the curve in the post yield regime with increased strain rate was predicted (see Fig. 15.15). It is believed that the enhanced inter-particle interaction in 3D RVEs and CNT alignment cause an improved matrix to CNT stress transfer in the post-yield regime with the increased strain rate. This is confirmed by the increase of average stresses within CNTs, extracted from RVE models, as shown in Fig. 15.17.

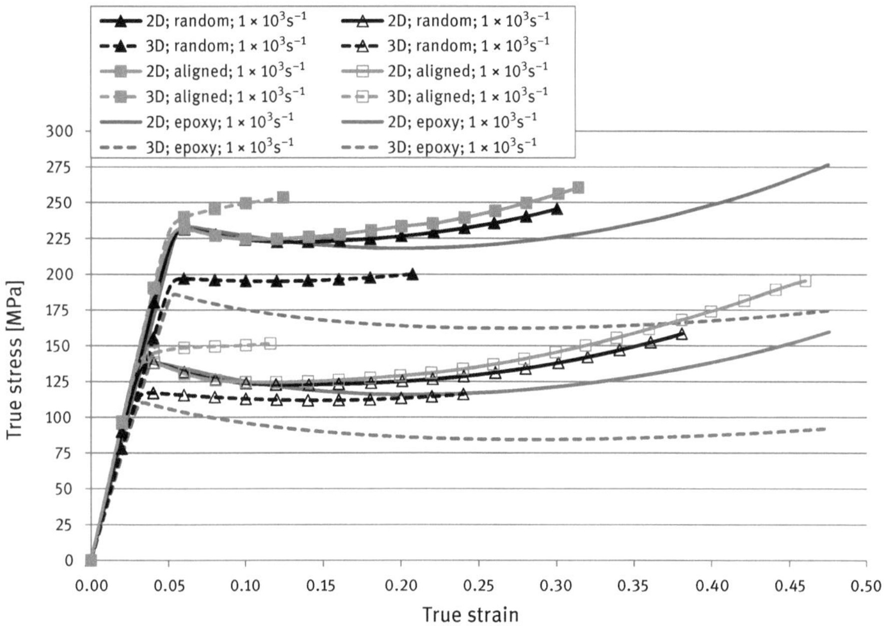

Fig. 15.15: Examples of stress–strain curves from 2D and 3D RVEs; RVEs subjected to quasi-static ($1 \cdot 10^{-3}\ s^{-1}$) and impact rates of strain ($1 \cdot 10^{3}\ s^{-1}$); reference stress–strain curves of pure epoxy in 2D and 3D; randomly oriented and aligned CNTs, CNT aspect ratio of 50, CNT volume fraction of 1 %.

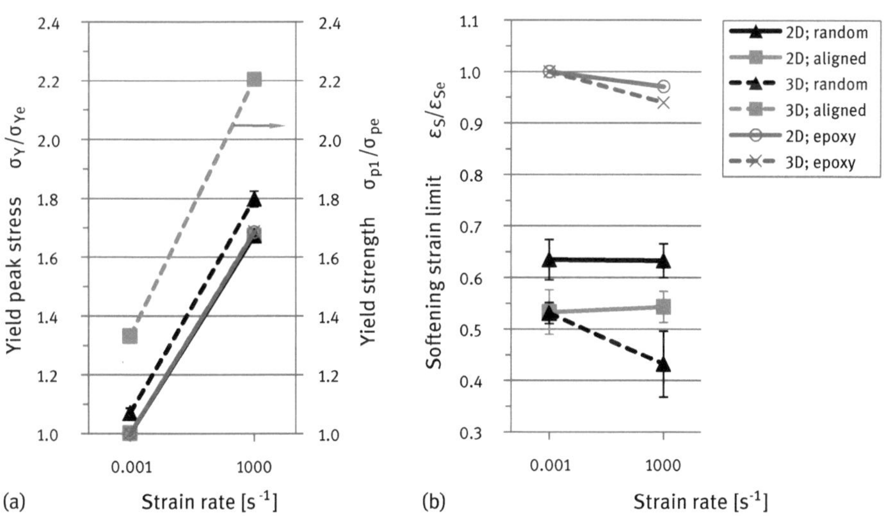

Fig. 15.16: Effect of strain rate and RVE approaches (2D, 3D) on the normalised stress–strain parameters (a) yield peak stress or yield strength respectively, as indicated by the arrows, and (b) softening strain limit; randomly oriented and aligned CNTs, CNT aspect ratio of 50, CNT volume fraction of 1 %.

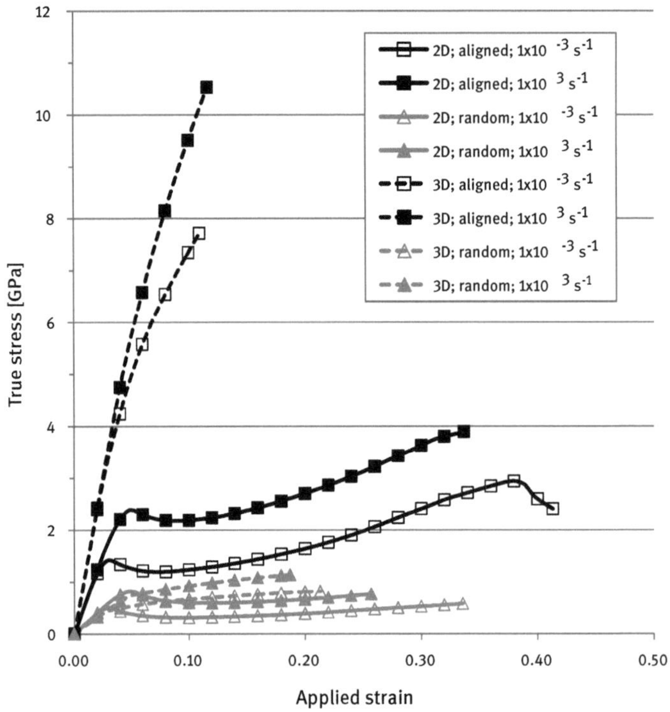

Fig. 15.17: Average stress–strain curves for the CNT phase; aligned and randomly oriented CNTs, CNT aspect ratio of 50, CNT volume fraction of 1 %; quasi-static ($1 \cdot 10^{-3}\ s^{-1}$) and impact rates of strain ($1 \cdot 10^{3}\ s^{-1}$).

15.3.4 Computational time

All simulations were performed using a multi-processor supercomputer. 3D RVEs containing 25 randomly oriented CNTs consisted of around 100,000 elements and 800,000 degrees of freedom. Typical 3D simulations took around 43 h of wall-clock time with 60 CPUs and a total CPU time reported by ABAQUS of 364 h (abort at around 21 % of true strain). In contrast, 2D RVEs incorporating 20 randomly oriented CNTs consisted of around 14,000 elements and 56,000 degrees of freedom. In this case the wall-clock time was 1.5 h using 24 CPUs and a total CPU time reported by ABAQUS was 7 h (abort at around 36 % of true strain).

15.3.5 Comparison with experiments

Some of the predicted stress–strain trends were qualitatively compared with the available experimental data from the literature. Experimental results of Dassios et al. [12] were selected, where the authors investigated the large-strain response of aligned CNT mats impregnated with epoxy under quasi-static compressive loading. Similarly to the predictions using 3D RVE based models with aligned CNTs, the authors observed an increase in Young's modulus and yield peak, and a shift of yielding to larger values of strain for the CNT/epoxy systems, compared to pure epoxy. Due to the relatively large CNT volume fraction of 0.27 % and the special arrangement of CNTs in the mat (thickness of the mat is equal to the CNT length), their compression test result represents an upper bound for the current 3D models of aligned CNTs.

15.4 Conclusion

RVE-based modelling of aligned and misaligned (with random orientations) morphologies of randomly distributed CNTs in an epoxy matrix was carried out to investigate the compressive nonlinear, rate-dependent, finite strain response of CNT/epoxy nanocomposites. A particular goal of this work was to compare the 2D RVE-based approach against the 3D RVE-approach to study its accuracy in capturing the behaviour of the nanocomposite and their computational efficiency.

Relatively large differences between predictions of the nonlinear compressive response were obtained from the 2D and the 3D modelling approach. In general, the 2D RVE approach suffers from a lack of inter-particle interaction due to a larger average CNT to CNT distance and the smaller surface area available for the stress transfer from the matrix to CNTs, when compared to their 3D counterparts. As a consequence (1) the 2D RVE expanded easier in the lateral direction which resulted in an inaccurate prediction of the average mean stress in the matrix, and (2) the average CNT stress in 2D RVEs was found to be significantly smaller than for 3D RVEs. The larger surface area in 3D RVEs promotes more stress to be transferred into CNTs, which better exploits their mechanical properties. Additionally, in 2D RVEs the averaged stress-applied strain response of CNTs resembled that of the pure matrix. In contrast, a smaller CNT to CNT distance in 3D RVEs increased inter-particle interaction, which resulted in an increase of the averaged stress in the CNT component during the strain softening phase of the matrix. Thus, the smaller CNT to CNT distance in 3D models shields the CNTs from the behaviour of the matrix.

As a result, the 2D RVEs underestimated the effect of the morphological factors (CNT aspect ratio, CNT volume fraction and CNT orientation) on the salient parameters of stress–strain curves. In particular, the yield peak stress was found to be nearly unaffected by the presence of CNTs and various changes of the morphological factors in the 2D case. In contrast, 3D models predicted significant effects of morphological factors on the yield peak stress. Therefore, despite obvious demands on substantial computational resources (approximately a 70-fold increase in computational time for 3D models), it is nevertheless advisable to use 3D RVEs for accurate prediction of the nanocomposite response.

The 3D modelling approach predicted that incorporation of CNTs into the epoxy matrix leads to an enhancement of the nonlinear deformation that results in: (1) the increase of yield stress with the CNT aspect ratio, CNT volume fraction and CNT alignment, (2) the change from softening to hardening with the CNT alignment, and (3) the shift of the onset of the hardening regime to smaller stages of strain with the CNT aspect ratio and the CNT volume fraction. The CNT alignment was found to have the most significant enhancement effect, especially at high strain rates, which highlights the reinforcement potential of CNTs for impact-resistant applications.

Some of the predicted stress–strain trends obtained in this work were qualitatively compared with the available experimental data from the literature. Similarly to the predictions using 3D RVE-based models with aligned CNTs, those experimental results show an increase in Young's modulus and yield peak, and a shift of yielding to larger values of strain for the CNT/epoxy systems compared to pure epoxy [12].

However, it has to be mentioned that in reality, morphologies of CNT based nanocomposites show CNT waviness, CNT agglomerations and frequently also an imperfect bonding between CNT and polymer matrix. Future modelling has to account for those aspects in order to overcome the often observed deviation between experiments and modelling.

Acknowledgment: The PhD scholarship from the Irish Research Council (IRC) is gratefully acknowledged. Computational facilities and support were provided by the Irish Centre for High-End Computing (ICHEC). The second author would like to acknowledge the financial support from the National Centre for Science (Poland) through the OPUS programme (Grant No. DEC-2011/01/B/ST8/06492).

References

[1] Treacy MMJ, Ebbesen TW, Gibson JM. Exceptionally high Young's modulus observed for individual carbon nanotubes. Nature. 1996;381:678–680.

[2] Cooper CA, Young RJ, Halsall M. Investigation into the deformation of carbon nanotubes and their composites through the use of raman spectroscopy. Compos Part A: Appl S. 2001; 32(3–4):401–411.

[3] Peigney A, Laurent C, Flahaut E, Bacsa RR, Rousset A. Specific surface area of carbon nanotubes and bundles of carbon nanotubes. Carbon. 2001;39(4):507–514.

[4] Weidt D, Figiel L, Buggy M. Prediction of energy absorption characteristics of aligned carbon nanotube/epoxy nanocomposites. IOP Conf Ser Mater Sci Eng. 2012;40(1):12028.

[5] Gojny FH, Wichmann MHG, Köpke U, Fiedler B, Schulte K. Carbon nanotube-reinforced epoxy-composites: enhanced stiffness and fracture toughness at low nanotube content. Compos Sci Technol. 2004;64(15):2363–2371.

[6] Zhang W, Picu RC, Koratkar N. Suppression of fatigue crack growth in carbon nanotube composites. Appl Phys Lett. 2007;91(19):193109.

[7] Buckley CP, Harding J, Hou JP, Ruiz C, Trojanowski A. Deformation of thermosetting resins at impact rates of strain. part i: Experimental study. J Mech Phys Solids. 2001;49(7):1517–1538.

[8] Ci L, Suhr J, Pushparaj V, Zhang X, Ajayan PM. Continuous carbon nanotube reinforced composites. Nano Lett. 2008;8:2762–2766.

[9] Thostenson ET, Chou TW. Aligned multi-walled carbon nanotube-reinforced composites: processing and mechanical characterization. J Phys D Appl Phys. 2002;35(16):L77.

[10] Martin CA, Sandler JKW, Windle AH, Schwarz MK, Bauhofer W, Schulte K, et al. Electric field-induced aligned multi-wall carbon nanotube networks in epoxy composites. Polymer. 2005; 46(3):877–886.

[11] Kimura T, Ago H, Tobita M, Ohshima S, Kyotani M, Yumura M. Polymer composites of carbon nanotubes aligned by a magnetic field. Adv Mater. 2002;14(19):1380–1383.

[12] Dassios KG, Musso S, Galiotis C. Compressive behavior of mwcnt/epoxy composite mats. Compos Sci Technol. 2012;72(9):1027–1033.

[13] Jindal P, Pande S, Sharma P, Mangla V, Chaudhury A, Patel D, et al. High strain rate behavior of multi-walled carbon nanotubes-polycarbonate composites. Compos Part B - Eng. 2013;45: 417–422.

[14] Al-Lafi W, Jin J, Xu S, Song M. Performance of mwcnt/hdpe nanocomposites at high strain rates. Macromol Mater Eng. 2010;295(6):519–522.

[15] Mantena PR, Al-Ostaz A, Cheng AHD. Dynamic response and simulations of nanoparticle-enhanced composites. Compos Sci Technol. 2009;69(6):772–779.

[16] Namilae S, Chandra N. Role of atomic scale interfaces in the compressive behavior of carbon nanotubes in composites. Compos Sci Technol. 2006;66(13):2030–2038.

[17] Rafiee R, Moghadam RM. Simulation of impact and post-impact behavior of carbon nanotube reinforced polymer using multi-scale finite element modeling. Comp Mater Sci. 2012;63(0): 261–268.

[18] Li C, Chou TW. Multiscale modeling of compressive behavior of carbon nanotube/polymer composites. Compos Sci Technol. 2006;66(14):2409–2414.

[19] Buckley CP, Dooling PJ, Harding J, Ruiz C. Deformation of thermosetting resins at impact rates of strain. part 2: constitutive model with rejuvenation. J Mech Phys Solids. 2004;52(10):2355–2377.

[20] Shokrieh MM, Rafiee R. Stochastic multi-scale modeling of cnt/polymer composites. Comp Mater Sci. 2010;50(2):437–446.

[21] Pan Y, Iorga L, Pelegri AA. Numerical generation of a random chopped fiber composite rve and its elastic properties. Compos Sci Technol. 2008;68(13):2792–2798.

[22] Makvandi R, Öchsner A. A refined analysis of the influence of the carbon nanotube distribution on the macroscopic stiffness of composites. Comp Mater Sci. 2013;77(0):189–193.

[23] Barai P, Weng GJ. A theory of plasticity for carbon nanotube reinforced composites. Int J Plasticity. 2011;27(4):539–559.

[24] Ma PC, Kim JK, Tang BZ. Effects of silane functionalization on the properties of carbon nanotube/epoxy nanocomposites. Compos Sci Technol. 2007;67(14):2965–2972.

[25] Bai JB, Allaoui A. Effect of the length and the aggregate size of mwnts on the improvement efficiency of the mechanical and electrical properties of nanocomposites—experimental investigation. Compos Part A: Appl S. 2003;34(8):689–694.

[26] Hernández-Pérez A, Avilés F, May-Pat A, Valadez-González A, Herrera-Franco PJ, Bartolo-Pérez P. Effective properties of multiwalled carbon nanotube/epoxy composites using two different tubes. Compos Sci Technol. 2008;68(6):1422–1431.

[27] Shen L, Li J. Transversely isotropic elastic properties of single-walled carbon nanotubes. Phys Rev B. 2004;69:045414.

[28] Gojny FH, Wichmann MHG, Fiedler B, Schulte K. Influence of different carbon nanotubes on the mechanical properties of epoxy matrix composites – a comparative study. Compos Sci Technol. 2005;65(15–16):2300–2313.

[29] Putz KW, Palmeri MJ, Cohn RB, Andrews R, Brinson LC. Effect of cross-link density on interphase creation in polymer nanocomposites. Macromolecules. 2008;41:6752–6756.
[30] Wan H, Delale F, Shen L. Effect of cnt length and cnt-matrix interphase in carbon nanotube (cnt) reinforced composites. Mech Res Commun. 2005;32(5):481–489.
[31] Hernández-Pérez A, Avilés F. Modeling the influence of interphase on the elastic properties of carbon nanotube composites. Comp Mater Sci. 2010;47(4):926–933.
[32] Li C, Chou TW. Multiscale modeling of carbon nanotube reinforced polymer composites. J Nanosci Nanotechnol. 2003;3(5):423–430.
[33] Yang S, Yu S, Ryu J, Cho JM, Kyoung W, Han DS, et al. Nonlinear multiscale modeling approach to characterize elastoplastic behavior of cnt/polymer nanocomposites considering the interphase and interfacial imperfection. Int J Plasticity. 2013;41:124–146.
[34] Ruxton GD. The unequal variance t-test is an underused alternative to student's t-test and the mann–whitney u test. Behav Ecol. 2006;17(4):688–690.
[35] Hutchinson JW, Suo Z. Mixed mode cracking in layered materials. Adv Appl Mech. 1991;29: 63–191.

I. Schäfer*, G. Lasko, T. A. Do, J. Pleiss, U. Weber, and
S. Schmauder

16 Peptide–zinc oxide interaction

Finite element simulation using cohesive zone models based on molecular dynamics simulation

Abstract: In this study, a multiscale simulation approach of coupling molecular dynamics (MD) and finite element method (FEM) simulations was established to investigate the mechanical properties of a ZnO-peptide material. MD simulations of a single 6-mer peptide adsorbed on the polar ZnO(0001)–O surface were performed to calculate the adsorbed peptide conformations and their adsorption force parameters, which were used to estimate mechanical properties of the ZnO-peptide composite material in three point bending tests using FEM simulations. The results from the multiscale simulations revealed that the influence of the Elastic modulus of the peptide on the material properties of the composite differs depending on the elastic properties of the cohesive zone. For developing a nanocomposite based on ZnO and a peptide, this dependency should be carefully considered and used to create stronger nanocomposites. Based on these simulation results, a set of binding affinities of the peptide and mechanical properties like the crack opening displacement of ZnO-peptide material could be predicted.

Keywords: biomimetic, bio-inspired, zinc oxide (ZnO), peptide, nano composite, MD simulation, FEM simulation, multiscale, COD.

16.1 Introduction

In material science, especially for metals, multiscale simulations have been recently applied to use the results of molecular dynamics (MD) simulations as input data for other simulation tools like finite element method (FEM) simulation or phase field simulations. Molnar et al. reported on multiscale simulations on the coarsening of Cu-rich precipitates in α-Fe using kinetic Monte Carlo, molecular dynamics and phase-field simulations [1]. Similar multiscale simulations were also performed on metal/ceramic materials where fracture on the metal/ceramic interface was analyzed [2]. In the present work, a method is developed to combine MD simulations of peptide binding to a ceramic surface with FEM simulations. Coupling molecular dynamics simulations to FE simulations is important for bridging the gap between nanoscale and micro- or macroscale. MD simulations have been successfully applied to study atomistic effects in nanoscale of biological molecules [3, 4], although they are usually limited to microsecond timescale simulations for typical systems (10^6–10^5 atoms) [5], while FEM simulations in the framework of conventional continuum mechanics are used for macroscale material behavior [6]. A combination of the two simulation methods is able to predict mechanical properties of organic-inorganic nanocomposites as shown in [7].

A virtual ceramic composite material which consists of zinc oxide (ZnO) and peptides with a high binding affinity for ZnO was examined for its mechanical properties. The molecular binding mechanism of peptides to ZnO was studied by MD simulations and the binding affinities were evaluated.

The material properties of layered structures of peptides and zinc oxide are of high interest for the industry. The use of ZnO is based on its favorable properties making it well suited for many applications [8], which leads to the need to understand the mechanical properties of peptides adsorbed on the ZnO surface material. The use of ZnO includes construction of solar cells, luminescent materials and acoustic devices [8–11]. Within the last decade, many attempts have been made to study the adsorption of peptides to the ZnO surface to prepare nanostructured hybrid materials [12–17]. It was also demonstrated that ZnO-binding peptides could be used as bioactive layers for surface functionalization and modification [18] and that nano-sized structures of ZnO have unique material properties and a remarkable performance in electronics, optics, and photonics [19]. One major disadvantage of ZnO is that it is a relatively soft material [20]. A reinforcement of the mechanical properties of these nanomaterials could lead to longer life spans and an improved usability of these systems in environments where resistance against mechanical stress is essential [21]. An exciting hybrid material with superior mechanical properties can be found in nature: nacre, the inner layer of mollusk shells. Nacre contains a relatively brittle mineral phase that is combined with an organic microstructure resulting in an increase of toughness of the composite compared to the mineral phase alone which is reinforced through microstructures

Nacre is hierarchically structured, it consist of a layered brick (95 % aragonite – platelets) and mortar structure (5 % organic components like proteins) [22]. Viewed in more detail, nano-mineral-bridges between the platelets get visible. Artificial nacre with material properties that are optimized for the respective application would be highly appreciated in engineering [23, 24]. Based on a mechanical model derived from multiscale simulations, optimal peptides can be designed as a basis of future ZnO-peptide hybrid materials with optimized mechanical stabilities.

ZnO-binding peptides (6- to 12-mer) are usually selected through biopanning techniques of phage display [13, 16] or bacterial surface display [25–27] peptide libraries. In a previous study [16], the adsorbed conformations and the adsorption free energy of peptide 31 (amino acid sequence HHGHSPTSPQVR) was investigated using MD simulations. Peptide 31 was previously identified by phage display [17]. It was found that the peptide preferred the negatively charged ZnO(0001)–O surface. To investigate peptide binding to this surface, peptide 31-6 with amino acid sequence HHGHSR was designed which was derived from peptide 31. In a first step, the adsorption properties of peptide 31-6 were studied by MD simulations to obtain mechanical properties on the atomic level. In a second step, the thus obtained nanoscopic mechanical properties were used for a three point bending test of a ZnO-peptide composite to characterize the macroscopic mechanical properties, thus bridging the gap between MD simulations and FEM simulations as within a multiscale approach.

The necessary parameters for FEM simulations were generated from the results of the MD simulations. Because the Elastic modulus (EM) of the peptide is not exactly known, a sensitivity analysis with different elastic moduli for the peptides layer was made for the peptide conformation.

In our study, crack propagation at the interface between ZnO and a peptide layer was studied, assuming that the crack will not primarily propagate within the peptide layer. Crack propagation in the peptide layer could be prevented by crosslinking the peptides or by using longer polypeptides rather than short peptides as in the model presented here.

16.2 Material and Methods

16.2.1 Using MD simulations to estimate the adsorption affinity of the peptide

The adsorption force of peptide 31-6 on the ZnO surface were determined from the MD simulations of a single peptide adsorbed and desorbed on the negative polar ZnO(0001)–O surface. The MD simulation procedure involves three stages: first, the peptide was pulled slowly towards the surface using steered MD simulations. Second, the pulling force was removed after the peptide contacted the surface, and the peptide was equilibrated on the ZnO surface. Third, the peptide was pulled away from the surface in the desorption process using steered MD simulations, as described previously [16]. All MD simulations were performed in explicit water at constant temperature (300 K) and pressure (1 bar) using Berendsen's thermostat implemented in the simulation package GROMACS 4.5.3 [28]. Further details of the simulation can be found in the appendix.

To obtain different binding conformations of the peptide, three independent steered MD simulations were performed where peptide 31-6 was pulled towards the ZnO(0001)–O slab model and equilibrated at the surface. The last snapshot of each equilibrium simulation was chosen as the adsorbed conformation of the peptide.

To calculate the adsorption force, peptide 31-6 was pulled away from the ZnO surface in the desorption process using steered MD simulation. The external force applied in the steered MD simulation was monitored to measure the adsorption force. Three different conformations of the adsorbed peptide were used as starting conformations for the steered MD simulations. In order to evaluate the dependence of the potential of mean force (PMF) on different pulling velocities, the peptide was pulled with 4 different velocities (0.25, 0.5, 1.0, and 1.5 nm/ns). Each simulation was repeated 10 times with different initial velocities.

16.2.1.1 MD simulation details

All MD simulations were performed using the standard AMBER force field [29, 30] for the peptide and TIP3P model for water [29]. Histidines and arginines were simulated in their protonated state, while the N-terminus and the C-terminus were simulated in their protonated and deprotonated state, respectively. The ZnO slab was modeled based on a wurtzite crystal structure with lattice parameters of $a = 0.3249$ nm and $c = 0.5207$ nm [16, 31]. The force field parameters for ZnO were used based on initial computations by Raymand [32] using atomic partial charge values of 1.026 e and –1.026 e for Zn and O atoms, respectively. To reduce the dipole moment of the polar ZnO(0001) surface, 25 % of the negatively charged oxygen atoms on the O-terminated side and 25 % of the positively charged Zinc atoms on the Zn-terminated side were removed, as described by Kornherr [33]. The negative surface charge of the O-terminated side is in agreement with the experimentally determined zeta potential at neutral pH [17]. In aqueous environment, water adsorbed to the negatively charged ZnO(0001)–O surface forming ordered water layers.

After equilibration of peptide 31-6 in explicit water for 5 ns at a distance of 6 nm from the ZnO surface, the adsorption was simulated by pulling the center of mass of the peptide along the z-axis towards the ZnO surface in a steered MD simulation of 20 ns, in which a spring constant of 3000 kJ mol^{-1} nm^{-1} and a decrease in pulling velocity from 0.5 to 0.1 nm ns^{-1} were used. When the peptide contacted the ZnO surface within a center of mass distance of 1.1 nm, the peptide was simulated for 5 ns without external forces to reach an equilibrium state for the system. The peptide conformation at the end of this equilibration time was used as an adsorbed conformation for the next MD simulations in the adsorption process. The adsorption force (f) during a steered MD simulation is monitored and can be represented at time t by the following equation:

$$f(t) = k[(Z_{\text{peptide}}(0) + vt) - Z_{\text{peptide}}(t)]\,, \tag{16.1}$$

where k is the spring constant, v is the pulling velocity, and $Z_{\text{peptide}}(t)$ and $Z_{\text{peptide}}(0)$ are the z coordinates of the center of mass of peptide 31-6 at time t and initial time 0, respectively. The rupture event during the desorption process is recognized when an anchor residue is detached from the ZnO surface meanwhile the force reaches a peak in the adsorption force profile.

16.2.2 FEM simulations

FEM simulations of a three point bending test of a ZnO/peptide composite were conducted. The peptide was simulated as one averaged peptide, which was calculated from the results obtained from three adsorbed conformations of the MD simulations. In a real bulk material, which would consist out of the three or even more peptide confirmations, the material properties of all peptide conformations would influence the results of the composite material and so an averaged peptide conformation will be used in the FEM simulations (compare Tab. 16.3). FEM simulation was made with ABAQUS 6.11. To model the adhesive like bonding between ZnO and peptide, so called "cohesive elements" were used. The nature of the mechanical constitutive response or damage is based on the traction-separation law of the interface [34].

16.2.2.1 Variations of material properties

The material properties of the two components of the composite have to be defined. The properties of ZnO are described in Tab. 16.1. However, the Young's modulus and the Poisson ratio of the peptide are not exactly defined. Therefore, a parameter study was performed to analyze the influence of these parameters. The isothermal compressibility (β_T) of the peptide, which is reciprocal to the factor K (bulk modulus), has the value of 0.04–0.116 GPa based on the MD simulation. Because only a specific loading geometry is considered here (normal to the ZnO surface) not the word Young's modulus is used but Elastic modulus.

Tab. 16.1: Young's modulus and Poisson ratio for ZnO used for FEM simulations.

Material	Poisson ratio	Young's modulus [GPa]	Source
ZnO	0.356	35	[35, 36]

The Elastic modulus is defined through the following formula [37]:

$$E = 3K \cdot (1 - 2V)\,, \tag{16.2}$$

where V is the Poisson ratio and K the bulk modulus.

The Poisson ratio normally has a value in between 0 and 0.5. A typical value for the Poisson ratio of amorphous materials is 0.3 [38] and does not vary widely for different amorphous polymers [39]. The value of an amorphous polymer like polystyrene, which has value of 0.3 [40] can therefore be used as a value for the Possion ratio of the protein. A sensitivity study including K and E was performed (Tab. 16.2) to study the influence of Poisson ration, isothermal compressibility and Elastic modulus, which might influence the results of the simulation as described previously [41].

16.2.2.2 Determination of cohesive element values for the peptide conformations

Based on the results of the MD simulations the input values for the FEM simulation have been determined. The adsorption force curves were visualized as stress/displace-

Tab. 16.2: Calculated values for Elastic modulus of the peptides based on isothermal compressibility.

Poisson ratio	Isothermal compress-ibility (1/*K*) [GPa]	Elastic modulus [GPa]	
0.01	0.04	73.5	used in simulation
0.01	0.116	25.35	
0.2	0.04	45	
0.2	0.116	15.52	
0.25	0.04	37.5	used in simulation
0.25	0.078	19.23	used in simulation
0.25	0.116	12.93	used in simulation
0.3	0.04	30	used in simulation
0.3	0.078	15.38	used in simulation
0.3	0.116	10.35	used in simulation
0.39	0.04	16.5	
0.39	0.078	8.46	used in simulation
0.39	0.116	5.69	used in simulation
0.44	0.12	3.10	used in simulation
0.49	0.04	1.5	used in simulation
0.49	0.116	0.52	used in simulation

ment curves, which represent the binding capabilities of the peptide. One example for these curves is shown in Fig. 16.1. The curve exhibits several maxima, which are characterized by the loss of a connection between single amino acids and the ZnO-surface. From these curves, the required values for the FEM-analyses were extracted (see next chapter) as well as the correct geometric properties of the analyzed specimen (see Tab. 16.4, which shows the contact area used in the MD simulations).

The resulting stress/displacement curve (Figure 16.1) from the MD simulations was recalculated to start at zero (A). The first maximum describes the first detachment of one of three anchor amino acids (B). The last maximum describes the moment when the peptide is fully detached (C). These values were used to determine the required input values for cohesive traction separation simulation in ABAQUS. The results are presented in Tab. 16.3.

16.2.2.3 Three point bending probe simulations

The results of the MD simulations are used to simulate a FEM model of layered material later on. The model consists of a soft part containing many peptides and bulk ZnO. The specimen was investigated in a three point bending test and the results of the three different peptide conformations with different Elastic moduli (EM) were compared. It was shown in literature [42], that the traction separation law which is used in ABAQUS to describe the detachment and crack propagation, is independent of the geometry of the specimen. Based on this work, the calculated values for traction separation in

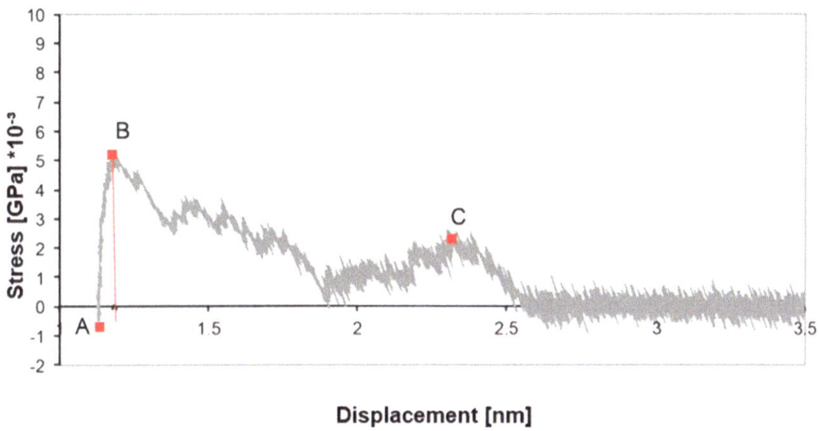

Fig. 16.1: The three points on the load/displacement curve obtained from the adsorption stress profile of the conformation 1 at the pulling velocity of 0.25 nm/ns. Important events for the cohesive element calculation are marked with square symbols: A – offset value, B – first maximum (initiation point), and C – last maximum.

Tab. 16.3: Input parameters for the cohesive element function of ABAQUS for the three different conformations. The sketch show the responding points on a traction-separation curve.

		Conformation 1	**Conformation 2**	**Conformation 3**	**Averaged**
δ_{init} (*x*-value of point B in Fig. 16.1)	[nm]	0.040868	0.189105	0.319635	0.183203
N_{max} (*y*-value of point B in Fig. 16.1, red line)	[GPa]	0.0051486	0.0056713	0.0054042	0.005408
EM (N_{max}/δ_{init})	[GPa/nm]	0.125982	0.02999	0.016907	0.02952
δ_C (Maximal displacement at failure) (*x*-value of point C in Fig. 16.1)	[nm]	1.170144	0.994595	1.172075	1.112271

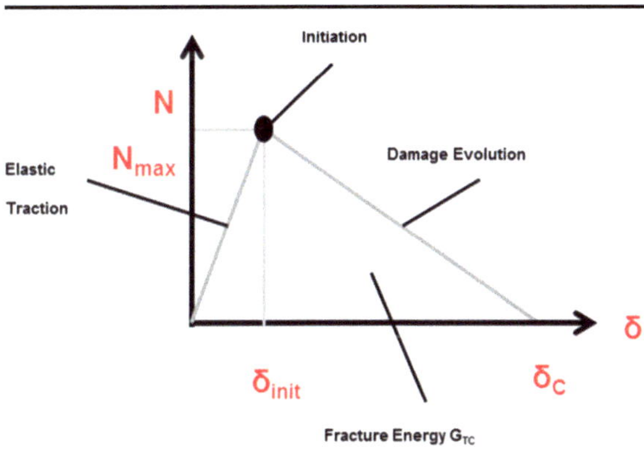

Sec. 16.3.2 can be used for specimen in different sizes, going from the nanoscale to the macroscale to test specimen in a practical size. First the averaged conformation was tested with different EM of the peptide. Comparisons are shown in the next part.

A three point bending specimen that consists of bulk ZnO bonded to bulk peptide was bended. The cohesive elements are found in-between the two materials (compare Fig. 16.5, which also shows the stress distribution in the model), because the authors are mainly interested in the connection between ZnO and peptides. The model is a 2D model and consists out of two element types: 23 elements of quadratic cohesive elements for the cohesive zone and 2053 quadratic plane strain elements for the ZnO and peptide areas.

16.3 Results and Discussion

16.3.1 MD-Simulations

MD simulations were performed to characterize the molecular mechanism of the adsorption of a peptide to the ZnO surface.

16.3.1.1 Adsorbed conformations of the peptide on the ZnO surface

Three independent steered MD simulations were performed to obtain different conformations of peptide 31-6 bound to the ZnO(0001)–O surface. The adsorbed conformations differed in anchor amino acid residues and the backbone structure. Their snapshot structures (Fig. 16.2) and their different shape and volume (Tab. 16.4) indicated that a single peptide could adsorb in different conformations onto the ZnO surface.

The contact area A_{contact} of ZnO and the adsorbed peptide in three conformations was calculated from A_{peptide}, A_{ZnO}, and A_{total} (the solvent accessible surface area of the peptide, the ZnO slab and the ZnO-peptide complex, respectively) by

$$A_{\text{contact}} = ((A_{\text{peptide}} + A_{\text{ZnO}}) - A_{\text{total}}))/2\,. \tag{16.3}$$

A_{peptide}, A_{ZnO}, and A_{total} were determined by averaging over 11 simulation snapshots of the peptide from the first 100 ps of the MD simulation of the adsorbed peptide using the GROMACS tool g_sas.

Tab. 16.4: Contact area of the three peptide conformations.

	Conformation 1	Conformation 2	Conformation 3
Area A_{contact} [nm^2]	1.68	1.47	1.88

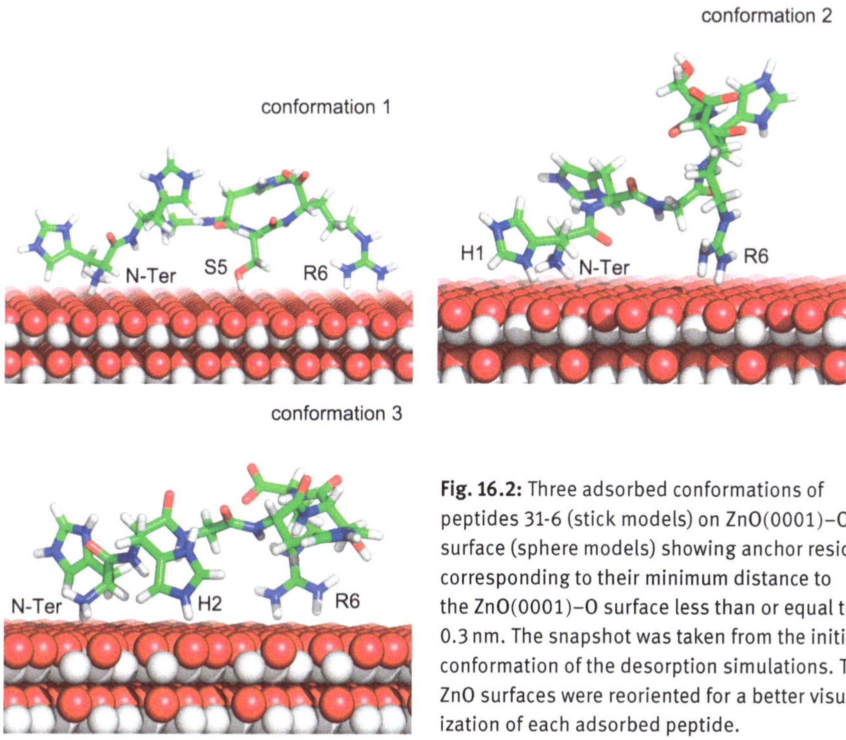

Fig. 16.2: Three adsorbed conformations of peptides 31-6 (stick models) on ZnO(0001)–O surface (sphere models) showing anchor residues corresponding to their minimum distance to the ZnO(0001)–O surface less than or equal to 0.3 nm. The snapshot was taken from the initial conformation of the desorption simulations. The ZnO surfaces were reoriented for a better visualization of each adsorbed peptide.

For the three conformations, the contact areas were 1.68 nm^2, 1.47 nm^2, and 1.88 nm^2. Despite their different conformation, the contact area was similar (1.5 nm^2) and deviated by only 15 %.

16.3.1.2 Adsorption force

Steered MD simulations of the desorption process were performed to determine the adsorption force of peptide 31-6 on the ZnO(0001)–O surface. It has been observed previously that different pulling velocities in steered MD simulations lead to different rupture forces [43–45]. Therefore, four different pulling velocities of 0.25, 0.5, 1.0, and 1.5 nm/ns were performed in 40 different steered MD simulations (Fig. 16.11, Appendix). Although the adsorption force curves obtained from the desorption simulations shared a similar shape, high pulling velocities results in higher peaks of the adsorption force curve. A linear fitting of four average maximum forces revealed an extrapolated rupture force for the pulling velocity approaching zero (Fig. 16.12, Appendix). The detachment of anchor residues (Fig. 16.2) from the surface gave the rupture events that suggest a critical role of these residues in the desorption process of peptide 31-6 from the ZnO(0001)–O surface.

The adsorption force profile also showed three distinct phases for all three peptide conformations (Fig. 16.11, Appendix). In the first phase, the force increased rapidly to a maximum peak, where the rupture event happened at a center of mass distance of 1.2–1.5 nm from the ZnO surface, slightly depending on the adsorbed conformation and the pulling velocity. The force then decreases to lower peaks or plateaus in the second phase, before reducing to a constant value near zero in the third phase. In the first phase, the adsorption force increased until it was large enough to detach the first anchor residue from the ZnO surface. In the second phase, further anchor residues were detached subsequently from the surface. The force, therefore, showed smaller peaks or remained constant in this phase. The second phase ended with the detachment of the last anchor residue. In the third phase, the peptide was not in contact with the ZnO surface. The maximum force needed for the first detachment of anchor residues and the maximum displacement before complete detachment of the peptide varied by only 10–20 % for the three conformations.

16.3.2 Multiscale simulations

16.3.2.1 Three point bending probe simulations

The three point bending test was chosen to analyze the influence of the Elastic moduli of the peptide on the crack opening displacement (COD). In Fig. 16.3 the von-Mises stress distribution in the middle of the simulation is shown.

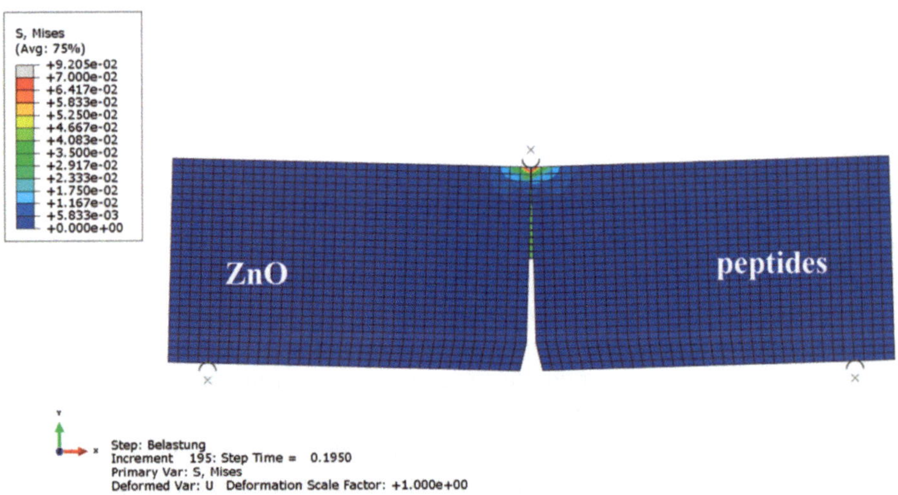

Fig. 16.3: The three point bending test model with colored materials and (von-Mises-) Stress-distribution with half-opened crack (Peptide COD = 2.7 nm). On the right side of the crack are peptides found, on the left side bulk ZnO and green (the bright colored elements in the middle) represents the cohesive elements.

Two different approaches were chosen to analyze the influence of the Elastic modulus of the peptide: "peptide COD" and "full COD". First the horizontal movement of the marked node on the peptide material (compare Fig. 16.4) relative to the baseline is measured and compared in Fig. 16.6 in a load/displacement diagram to analyze the influence of the elastic modulus of the peptide ("peptide COD"). Second the "full COD" is analyzed (see Fig. 16.5) which is defined through the change in distance between the marked node on the peptide material and the marked node on the ZnO material. The influence of the different Elastic moduli of the peptide on the "full COD" is then compared in Fig. 16.7.

The comparison of the results shows that the influence of the Elastic modulus (EM) is linear (see Figs. 16.6 and 16.7). The materials used in the simulations were considered to be fully elastic. With a higher EM, a higher load is required to bend the specimen as far as with a peptide with a lower Elastic modulus. Also, the curves show first an elastic behavior, then the cohesive elements get bended, then the crack occurs (the first cohesive element loses its connection, see marked area in Figs. 16.6 and 16.7) and then the curve experiences an exponential softening. The differences in the EM of the peptide did not result in a big change of the force/COD curve for values of the EM above 6 GPa.

An Elastic Modulus below 6 GPa results in a different material behavior. This is particularly visible in the comparisons of the results of the peptide COD analysis. For example the curve of the peptide with an EM of 0.52 GPa shows the following different result: first the curve shows an elastic behavior until it depicts negative displacements. Here the influence of the ZnO deformation gets visible, which results in the negative displacement of the peptide node. In that case, no crack formation occurs, because the simulation cannot run to the end because of the very high and unrealistic Possion ratio of 0.49 and the very low EM of 0.52 GPa (compare Fig. 16.6).

The peptide COD curve with 1.5 GPa displays a comparable behavior: first the curve shows an elastic behavior until it depicts negative displacements based on the same reasons as mentioned before with the low EM of 0.52 GPa. The energy first gets dissipated during the deformation of the peptides, but then the behavior of the ZnO elements starts to influence the results through their elastic deformation. The cohesive elements open up at a force of 0.08 GPa. But in comparison to the results of the peptide with 0.52 GPa a crack occurs (with forces around 0.9 GPa). With higher EM the curves converge more and more to each other and finally again to zero (the crack is then fully opened).

The full COD curve shows that all values of the Elastic modulus of the peptide result in comparable force/COD curves without any negative displacement. Differences exist also with values of the Elastic modulus below 6 GPa. Here, the maximal forces differ by 0.64 GPa (from 0.52 to 5.69 GPa), whereas the differences of the maximal forces of the different curves above 6 GPa account only for 0.008 GPa.

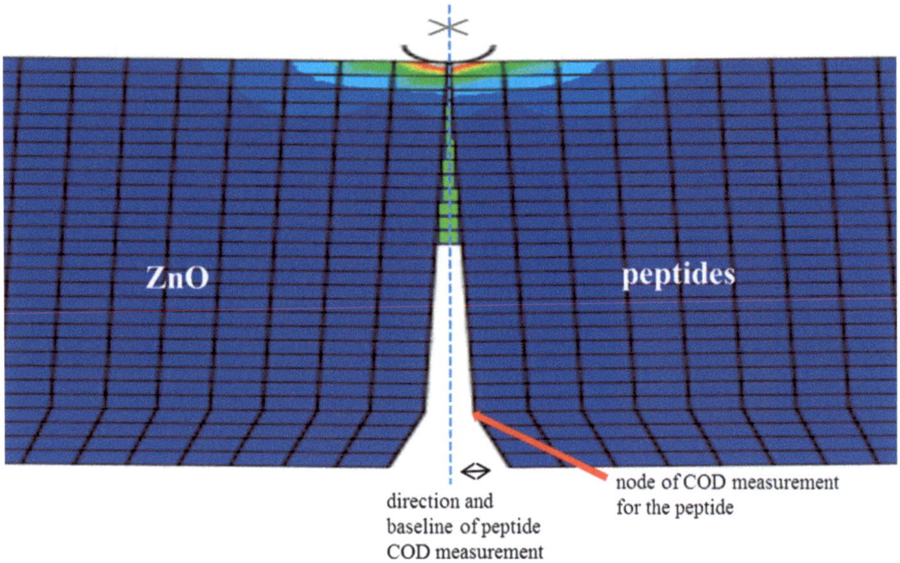

Fig. 16.4: Peptide COD: Description of the measurement method of the COD measurement of the COD of the peptides. The dotted line represents the ground line to which the distance of the marked node is measured.

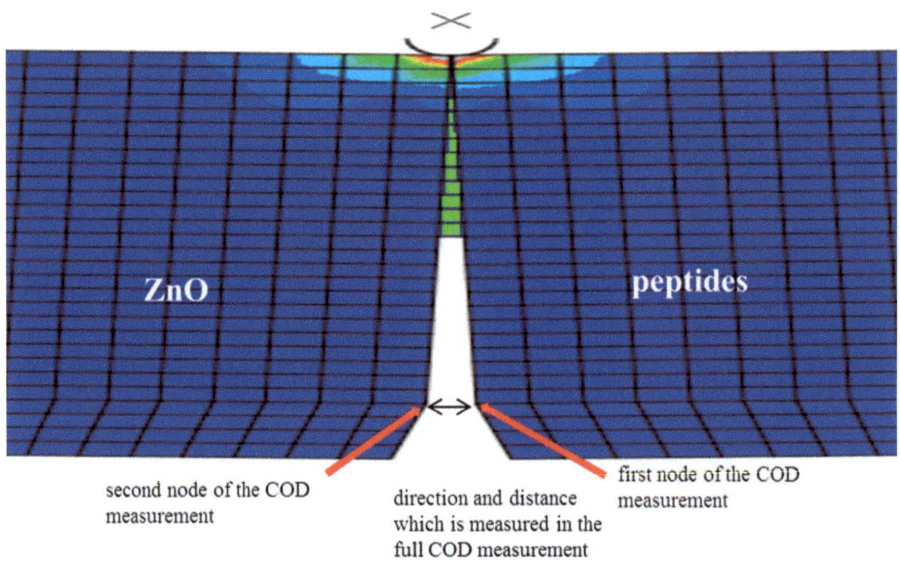

Fig. 16.5: Full COD: Description of the measurement method of the full COD measurement where the distance between the marked nodes of both materials is used in the investigation.

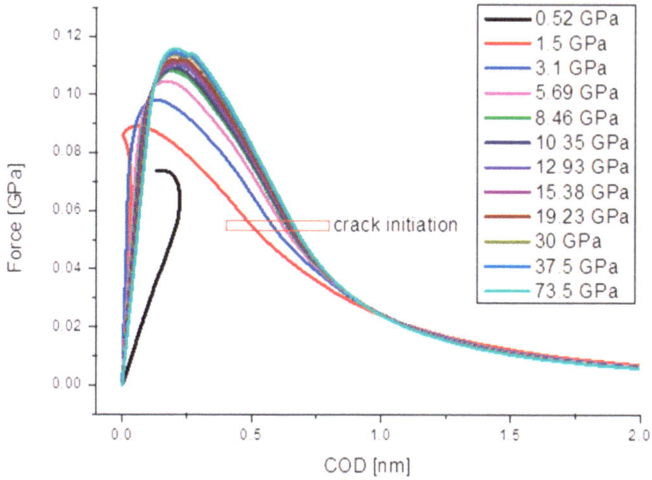

Fig. 16.6: Peptide COD: Influence of the EM of the peptide with averaged conformation on the force/COD behavior.

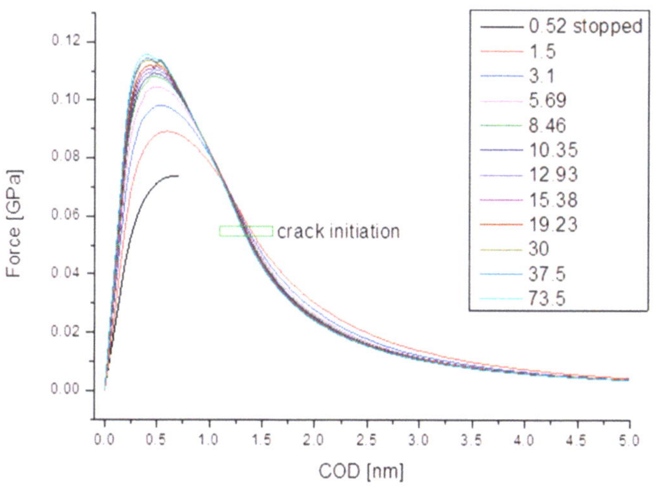

Fig. 16.7: Full COD: Influence of the EM of the peptide with averaged conformation on the force/COD behavior.

16.3.2.2 Distribution of the crack extension

The development of the crack opening can be represented by plotting the force and the crack front (delta a). The Fig. 16.8 represents the results of selected EM of the peptide. The influence on the curves of the EM of the peptide is low but exists. With a lower EM, a higher force is needed to start the crack but to the end of the curve the results are converging.

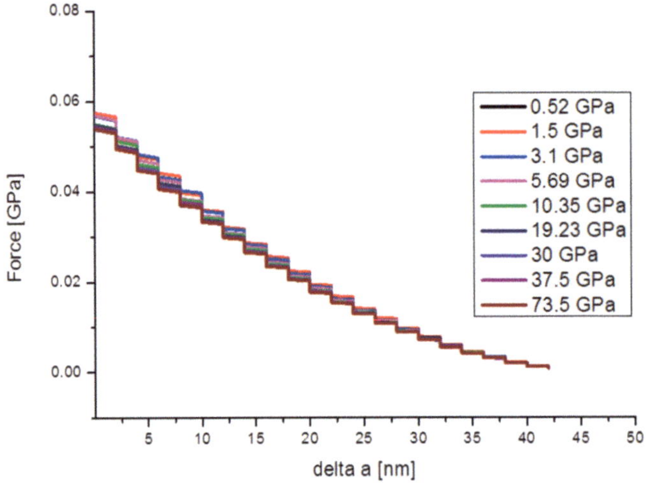

Fig. 16.8: Development of the crack opening of the averaged conformation.

16.3.2.3 Influence of the stiffness of the damage zone

The influence of the peptide Elastic modulus will increase if the stiffness of the damage zone (SDZ) change. With more than one peptide (e.g. a layer of peptides) the Elastic modulus of the SDZ likely will increase. This results in a different situation, the EM of the other components perhaps will influence the simulation differently. A FEM simulation with an EM value of 10 GPa and an EM of the peptides of 15.38 GPa and other examples are presented in the following figures:

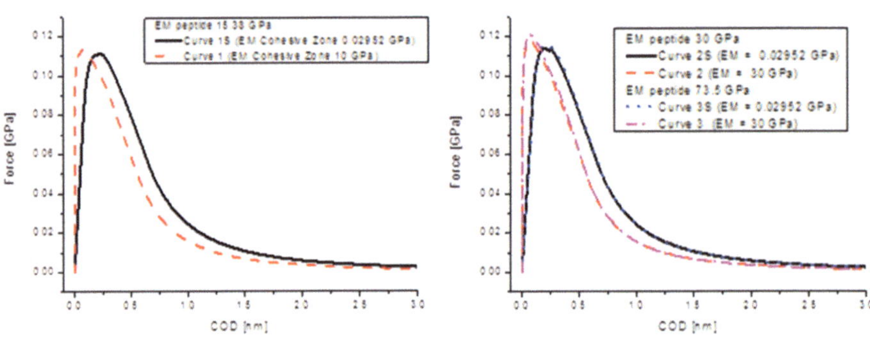

(a) (b)

Fig. 16.9: Peptide COD: (a) Comparison between different values (0.2952 GPa and 10 GPa) for the SDZ for an EM of the peptide of 15.38 GPa. (b) Comparison between different values for the SDZ (30 GPa and 0.2952 GPa) for an EM of the peptide of 30 GPa and 73.5 GPa.

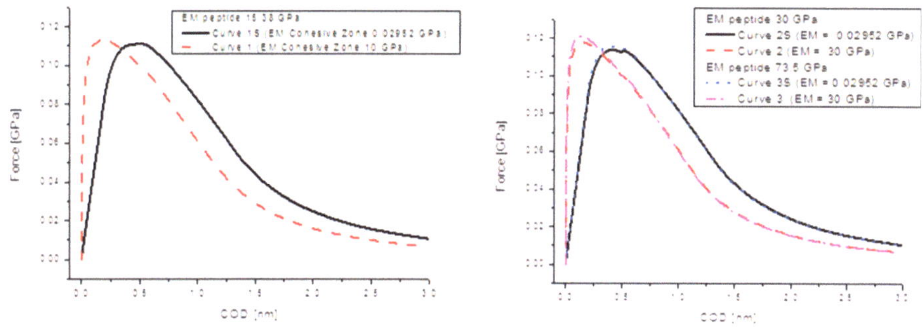

(a) (b)

Fig. 16.10: Full COD: (a) Comparison between different values (0.2952 GPa and 10 GPa) for the SDZ for an EM of the peptide of 15.38 GPa. (b) Comparison between different values for the SDZ (30 GPa and 0.2952 GPa) for an EM of the peptide of 30 GPa and 73.5 GPa.

The standard value for the SDZ (EM value) is 0.2952 GPa for the averaged configuration (a capital S behind the curve name marks the standard cohesive zone value for the EM). The Figs. 16.9 (a) (peptide COD) and 16.10 (a) (full COD) display the peptide conformation with an EM of the protein of 15.38 GPa (Curve 1S) which shows a lower maximum and a flat slope than the same conformation with a EM value of 10 GPa (Curve 1). This accounts also for the comparisons between two peptide conformations with an EM of the peptide of 30 GPa (Curves 2S and 2) and 73.5 GPa (Curves 3S and 3) which are presented in Figs. 16.9 (b) and 16.10 (b). With a higher value for the SDZ, the slope of the curve gets steeper and a higher maximum value of force is reached. E.g. the maximum of the force for the peptide with an EM of the peptide of 73.5 GPa and a EM value of 30 GPa has the maximum of 0.121 GPa (Curve 3) which is about 0.005 GPa higher than the value of Curve 3S with a EM value of 0.116 GPa in the comparison of the Peptide COD. The differences in the maxima of the forces between the different curves (1S and 1, 2S and 2, 3S and 3) account for less than 5 % in the peptide COD. For the full COD comparisons the results are comparable, with a higher value for the SDZ, the slope of the curve gets steeper and a higher maximum value of force is reached.

16.4 Conclusions

Steered MD simulations were applied to model the adsorption force and the binding affinity of a model peptide adsorbed to the ZnO(0001)–O surface. Parameters such as contact area, maximum force, displacement at maximum force and maximum displacement at failure were extracted from the simulation and used as parameters for subsequent FEM simulations. The cohesive element analysis and the traction separation law showed a good consistency. Despite the fact that the force curves generated by the MD simulations were approximated by the FEM simulations and that the peptide was bound in different conformations, the results were robust and deviated by less than 20 % (not shown here). In the future, a more realistic simulation of a whole layer of peptide is promising to account for peptide-peptide interactions.

The three point bending test with the sensitivity analysis of the material with different Elastic moduli revealed that influence of the Elastic modulus of the peptide on the simulation results exists. The use of the averaged results of the MD simulations mirrors the real situation in a bulk peptide material: Even with only one pre-defined crystal orientation of the ZnO material, the peptides will be absorbed in a large number of different conformations. All three different conformations will influence the material behavior differently. With the here presented approach three different conformations and their binding energies averaged and then used as one averaged peptide conformation in the FEM simulations to show the general idea. Ideally one should consider even more configurations, which would involve high computational costs and manpower which was not able to provide in this project.

The results showed a great influence of the used Elastic modulus of the peptide layer.

With lower EM higher forces are needed to introduce a crack as shown in Sec. 16.3.2.2. Also the unrealistic low values of 0.52 GPa and 1.5 GPa as shown in Fig. 16.4 are influencing the simulation so that the behavior of the composites differs greatly from the composites with higher values for the EM of the peptide.

With a thicker layer of peptides, the influence of stiffness of the damage zone, as shown in the last chapter also needs to be considered wisely, while planning and designing a composite like the one presented here.

In future works, the different peptide conformations will be simulated with a crack simulation procedure like XFEM and again for comparison with cohesive elements, to generate materials with the highest crack resistance possible while using ZnO and peptides as components. Therefore, the plasticity information of the material needs to be researched on and the possible range of the values of the Elastic moduli of the peptide will be estimated. Especially for a layer of peptides the stiffness of the damage zone will be interesting, as shown in the last chapter. Also more complicated structures will be analyzed, like a so called brick and mortar structure which can be found in natural nacre and other promising arrangements. This can lead to development of crack resistance ceramics out of ZnO and peptides, which do not exist until now.

Acknowledgment: The authors acknowledge funding by the German Research Foundation (Deutsche Forschungsgemeinschaft, DFG) project number Schm 746/88-1 and PL 145/8-1.

16.A Appendix

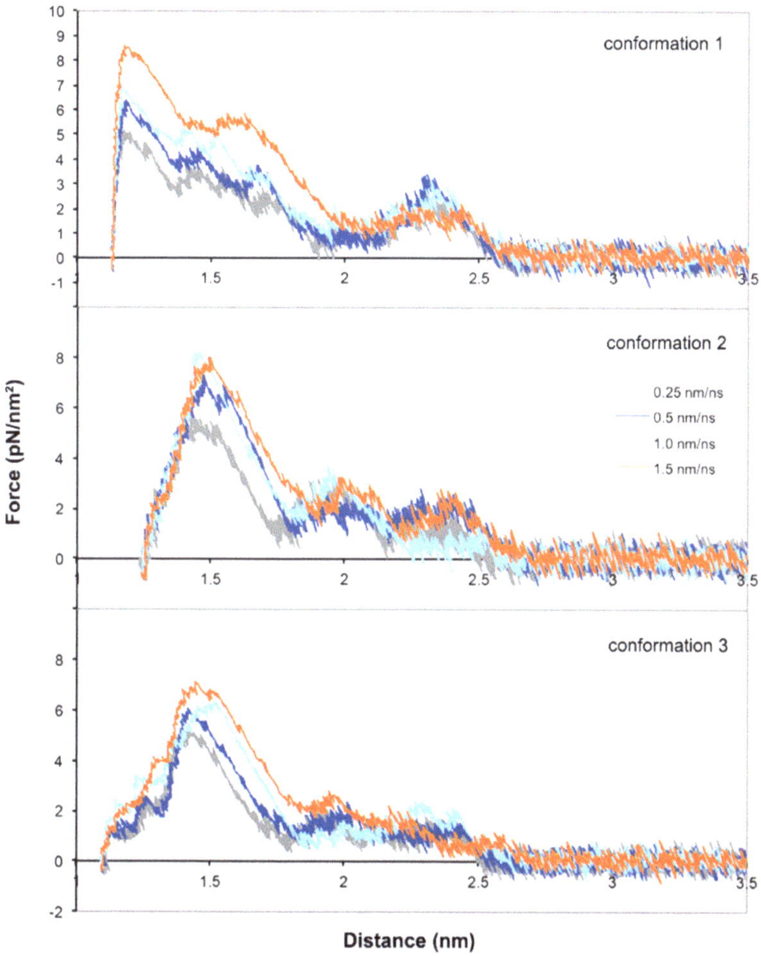

Fig. 16.11: Adsorption force vs COM distance plots of peptide 31-6 on the ZnO(0001)-O surface obtained from the steered MD simulations. Each adsorbed conformation was performed using four different pulling velocities of 0.25, 0.5, 1.0, and 1.5 nm/ns. At each velocity, 10 different simulations were averaged giving a curve. The curves show a same overall shape, but they are different in magnitude of the adsorption force depending on the pulling velocity: a higher pulling velocity leading to a higher adsorption force. Due to the fact that there was a small difference between the forces at 0.25 and 0.5 nm/ns, the other steered MD simulations were conducted at the velocity of 0.5 nm/ns for in order to save the computation time.

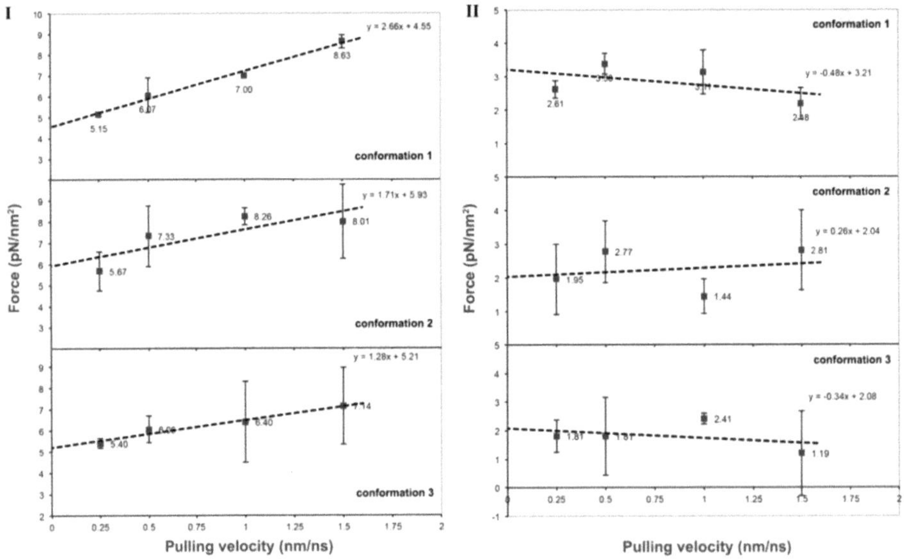

Fig. 16.12: Average rupture force vs pulling velocity occurred at point B (I) and point C (II). The maximum forces for the rupture event have been averaged from 10 steered MD simulation repetitions performed using four different pulling velocities (0.25, 0.5, 1.0, 1.5 nm/ns). The error bars show two standard deviations from the mean values.

References

[1] Molnar D, Mukherjee R, Choudhury A, Mora A, Binkele P, Selzer M, et al. Multiscale simulations on the coarsening of cu-rich precipitates in α-fe using kinetic monte carlo, molecular dynamics and phase-field simulations. Acta Materialia. 2012;60:6961–6971.

[2] Siddiq A, Schmauder S. Interface fracture analyses of a bicrystal niobium/alumina specimen using a cohesive modelling approach. Modelling and Simulation in Materials Science and Engineering. 2006;14:1015–1030.

[3] Karplus M, Petsko G. Molecular dynamics simulations in biology. Nature. 1990;347:631–639.

[4] Karplus M, McCammon J. Molecular dynamics simulations of biomolecules. Nature Structural Biology. 2002;9:646–652.

[5] Zwier M, Chong L. Reaching biological timescales with all-atom molecular dynamics simulations. Current Opinion in Pharmacology. 2010;10:745–752.

[6] Ji B, Gao H. Mechanical properties of nanostructure of biological materials. Journal of the Mechanics and Physics of Solids. 2004;52:1963–1990.

[7] Sikdar D, Pradhan S, Katti D, Katti K, Mohanty B. Altered phase model for polymer clay nanocomposites. Langmuir?: the ACS Journal of Surfaces and Colloids. 2008;24:5599–5607.

[8] Özgür U, Alivov Y, Liu C, a Teke, a Reshchikov M, Dog?an S, et al. A comprehensive review of zno materials and devices. Journal of Applied Physics. 2005;98:041301.

[9] Wang Z. Zinc oxide nanostructures: growth, properties and applications. Journal of Physics: Condensed Matter. 2004;16:829–858.

[10] Fang X, Bando Y, Gautam U, Ye C, Golberg D. Inorganic semiconductor nanostructures and their field-emission applications. Journal of Materials Chemistry. 2008;18:509–522.

[11] Jensen R, Ryswyk HV, She C, Szarko J, Chen L, Hupp J. Dye-sensitized solar cells: sensitizer-dependent injection into zno nanotube electrodes. Langmuir: The ACS Journal of Surfaces and Colloids. 2010;26:1401–1404.
[12] Gerstel P, Hoffmann R, Lipowsky P, Jeurgens L, Bill J, Aldinger F. Mineralization from aqueous solutions of zinc salts directed by amino acids and peptides. Chemistry of Materials. 2006;18: 179–186.
[13] Umetsu M, Mizuta M, Tsumoto K, Ohara S, Takami S, Watanabe H, et al. Bioassisted room-temperature immobilization and mineralization of zinc oxide – the structural ordering of zno nanoparticles into a flower-type morphology. Advanced Materials. 2005;17:2571–2575.
[14] Okochi M, Sugita T, Furusawa S, Umetsu M, Adschiri T, Honda H. Peptide array-based characterization and design of zno-high affinity peptides. Biotechnology and Bioengineering. 2010; 106:845–851.
[15] Togashi T, Yokoo N, Umetsu M, Ohara S, Naka T, Takami S, et al. Material-binding peptide application – zno crystal structure control by means of a zno-binding peptide. Journal of Bioscience and Bioengineering. 2011;111:140–145.
[16] Baier J, Naumburg T, Blumenstein N, Jeurgens L, Welzel U, Do TA, et al. Bio-inspired mineralization of zinc oxide in presence of zno-binding peptides. Biointerface Research in Applied Chemistry. 2012;2:380–391.
[17] Rothenstein D, Claasen B, Omiecienski B, PLammel, Bill J. Isolation of zno-binding 12-mer peptides and determination of their binding epitopes by nmr spectroscopy. Journal of the American Chemical Society. 2012;134:12547–12556.
[18] Tomczak M, Gupta M, Drummy L, Rozenzhak S, Naik R. Morphological control and assembly of zinc oxide using a biotemplate. Acta Biomaterialia. 2009;5:876–882.
[19] Zhang Y, Ram M, Stefanakos E, Goswami D. Synthesis, characterization, and applications of zno nanowires. Journal of Nanomaterials. 2012;2012:1–22.
[20] Bhushan B, Li X. Nanomechanical characterisation of solid surfaces and thin films. International Materials Reviews. 2003;48:40.
[21] Bradby J, Kucheyev S, Williams J, Jagadish C, Swain MV, Munroe P, et al. Contact-induced defect propagation in zno. Applied Physics Letters. 2002;80:4537.
[22] Sarikaya M. An introduction to biomimetics: a structural viewpoint. Microscopy Research and Technique. 1994;27:360–375.
[23] Waraich S, Hering B, Ž Burghard, Bill J, Behrens P, Menzel H. Fabrication and characterization of biocompatible nacre-like structures from α-zirconium hydrogen phosphate hydrate and chitosan. Journal of Colloid and Interface Science. 2012;367:74–82.
[24] Barthelat F, Zhu D. A novel biomimetic material duplicating the structure and mechanics of natural nacre. Journal of Materials Research. 2011;26:1203–1215.
[25] Kjaergaard K, Sørensen J, Schembri M, Klemm P. Sequestration of zinc oxide by fimbrial designer chelators. Applied and Environmental Microbiology. 2000;66:10–14.
[26] Kjaergaard K, Schembri M, Klemm P. Novel zn(2+)-chelating peptides selected from a fimbria-displayed random peptide library. Applied and Environmental Microbiology. 2001;67:5467–5473.
[27] Thai C, Dai H, Sastry M. Identification and characterization of cu2o and zno binding polypeptides by escherichia coli cell surface display: toward an understanding of metal oxide binding. Biotechnology and Bioengineering. 2004;87:129–137.
[28] Spoel DVD, Lindahl E, Hess B, Groenhof G, Mark A, Berendsen H. Gromacs: fast, flexible, and free. Journal of Computational Chemistry. 2005;26:1701–1718.
[29] Weiner S, Kollman P, Case D, Singh U, Ghio C, Alagona G, et al. A new force field for molecular mechanical simulation of nucleic acids and proteins. Journal of the American Chemical Society. 1984;106:765–784.

[30] Duan Y, Wu C, Chowdhury S, Lee M, Xiong G, Zhang W, et al. A point-charge force field for molecular mechanics simulations of proteins based on condensed-phase quantum mechanical calculations. Journal of Computational Chemistry. 2003;24:1999–2012.

[31] Xu YN, Ching W. Electronic, optical, and structural properties of some wurtzite crystals. Physical Review B. 1993;48:4335–4351.

[32] Raymand D, van Duin A, Baudin M, Hermansson K. A reactive force field (reaxff) for zinc oxide. Surface Science. 2008;602:1020–1031.

[33] Kornherr A, French S, Sokol A. Interaction of adsorbed organosilanes with polar zinc oxide surfaces: a molecular dynamics study comparing two models for the metal oxide surface. Chemical Physical Letters. 2004;393:107–111.

[34] SIMULIA. Abaqus analysis user's manual. Tech. rep., Dassault Systèmes Simulia Corp., Providence, RI, USA; 2011.

[35] Martin L, Dadon D, Rosen M. Evaluation of ultrasonically determined elasticity porosity relations in zinc oxide. Journal of the American Ceramic Society. 2005;79:1281–1289.

[36] Lipowsky P, Ž Burghard, Jeurgens L, Bill J, Aldinger F. Laminates of zinc oxide and poly(amino acid) layers with enhanced mechanical performance. Nanotechnology. 2007;18:345707–345712.

[37] Sadd M. Elasticity: theory, applications, and numerics. Elsevier Butterworth-Heinemann; 2009.

[38] Pérez-Rigueiro J, Viney C. Mechanical properties of silkworm silk in liquid media. Polymer. 2000;41:8433–8439.

[39] Brydson J. Plastics materials. 7th ed. Butterworth-Heinemann; 1999.

[40] Cowie J. Polymers: Chemistry and Physics of Modern Materials. 2nd ed. CRC Press Inc.; 1991.

[41] Liu B, Zhang L, Gao H. Poisson ratio can play a crucial role in mechanical properties of biocomposites. Mechanics of Materials. 2006;38:1128–1142.

[42] Mintchev O, Schmauder S, Weber U. Finite element simulation of crack initiation and growth based on a critical traction criterion? – some examples –. In: Proceedings of an International Workshop on Nonlocal Modelling of Failure of Materials. pp. 73–82.

[43] Krüger J, Fels G. Potential of mean force of ion permeation through alphs6 nachr ion channel. In: Proceedings of IWPLS. 4, pp. 1–7.

[44] Lemkul J, Bevan D. Assessing the stability of alzheimer's amyloid protofibrils using molecular dynamics. The Journal of Physical Chemistry B. 2010;114:1652–1660.

[45] Hung SW, Hsiao PY, Chieng CC. Dynamic information for cardiotoxin protein desorption from a methyl-terminated self-assembled monolayer using steered molecular dynamics simulation. The Journal of Chemical Physics. 2011;134:194705.

Index

Bei Fragen zur Produktsicherheit wenden Sie sich bitte an:
If you have any questions regarding product safety,
please contact:

Walter de Gruyter GmbH
Genthiner Straße 13
10785 Berlin
productsafety@degruyterbrill.com